Wild Philippines

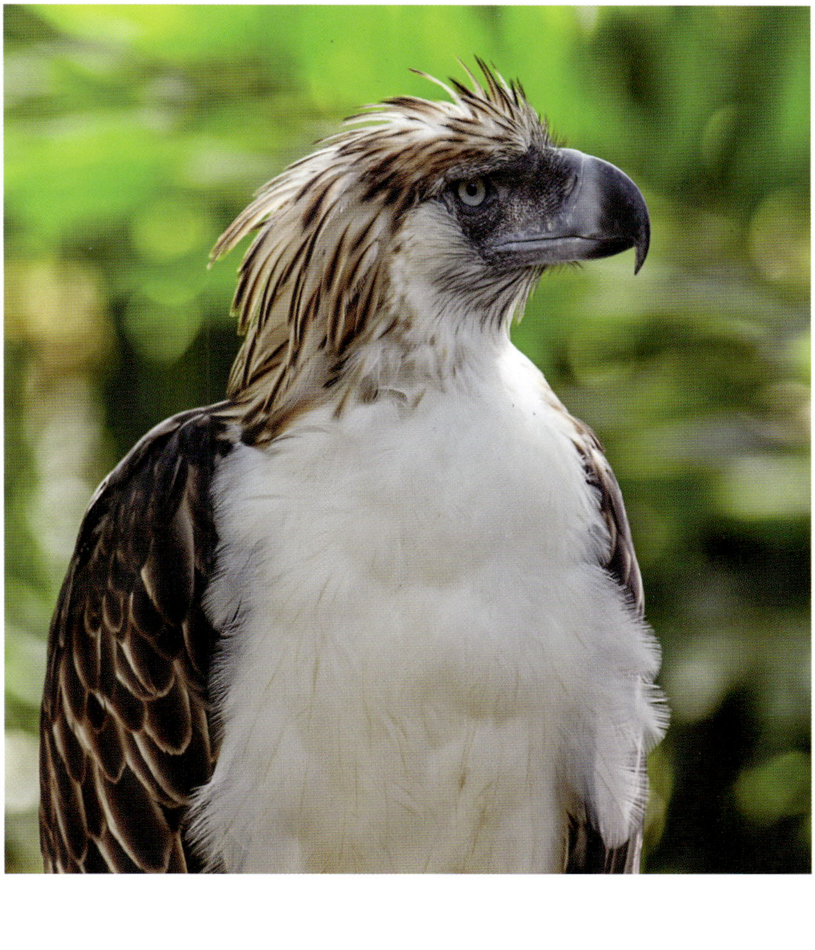

This book is dedicated to the memory of William Oliver, whose passion and commitment to the protection of Philippine wildlife has inspired an entire new generation of Filipino conservationists.

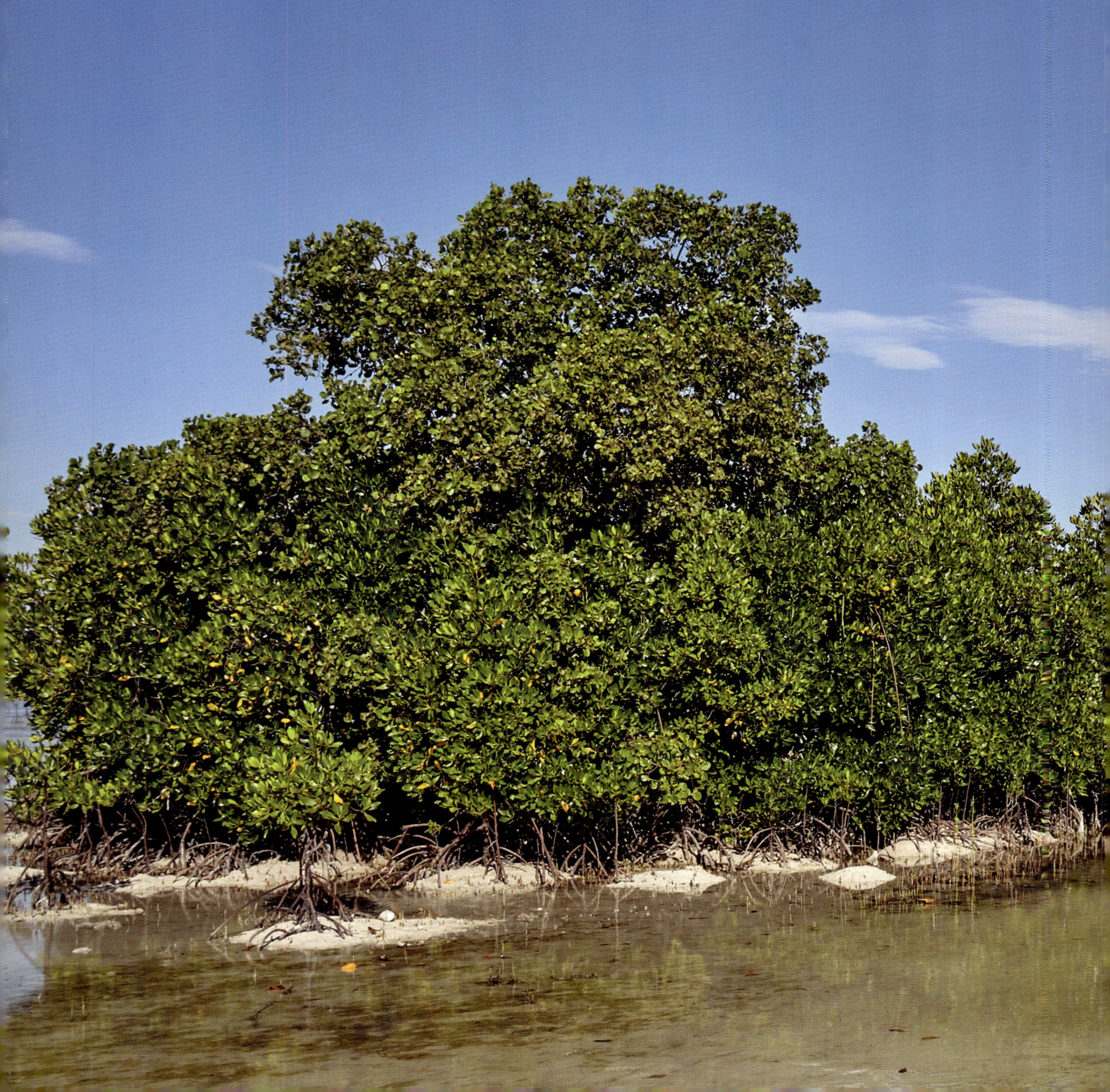

Wild Philippines

The landscapes, habitats and wildlife of the Philippine islands

NIGEL HICKS

JOHN BEAUFOY PUBLISHING

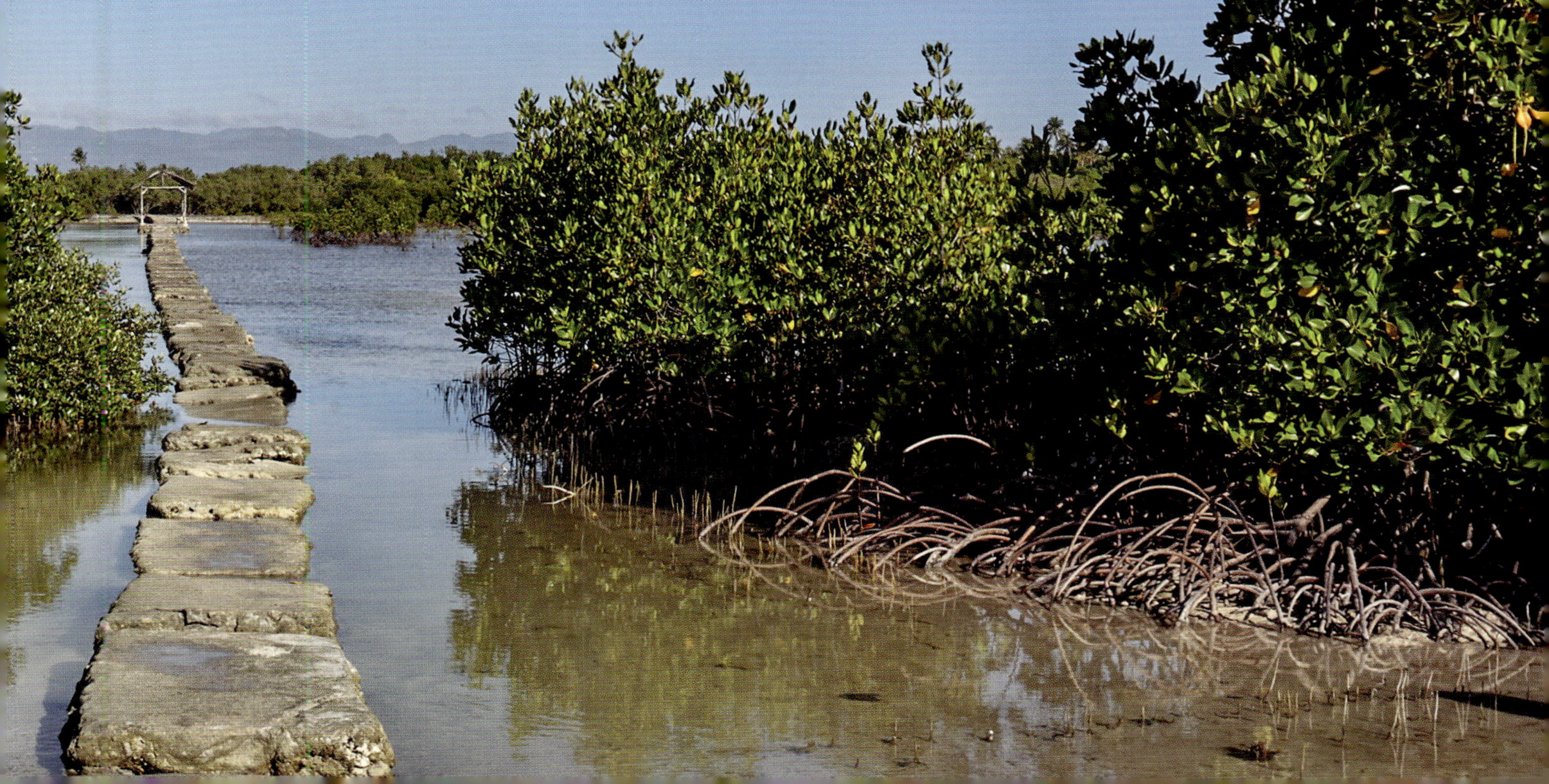

Acknowledgements

The author would like to give a huge thank you to the following people and organizations for their assistance with this project:

Professor Juan Carlos Gonzalez, Director of the Natural History Museum, University of the Philippines Los Baños and Professor, Institute of Biological Sciences, University of the Philippines Los Baños.

Offices of the Philippine Department of Tourism London; Cordillera Administrative Region; Region VIII (Samar and Leyte); Region XI (southern Mindanao); Region XIII (north-east Mindanao/Caraga).

Philippine provincial tourism offices Negros Occidental Tourism Division; Samar Provincial Tourism.

Hotels and resorts Microtel by Wyndham, Manila; Hotel Jen, Manila; Asia Divers and El Galleon Beach Resort, Puerto Galera, Mindoro; Plantation Bay, Mactan Island, Cebu; Quo Vadis Dive Resort, Moalboal, Cebu; Bohol Beach Club, Panglao Island, Bohol; Loboc River Resort, Bohol; Lagen Island Resort, El Nido Resorts Company, El Nido, Palawan; Stunning Republic Beach Resort, El Nido, Palawan; Microtel by Wyndham, Puerto Princesa, Palawan.

Environmental organizations Natural History Museum, University of the Philippines Los Baños; El Nido Resorts Environmental Team; Mabuwaya Foundation; Katala Foundation; Wild Bird Club of the Philippines.

Main picture: *A dusk view of Panlabuhan Lake, surrounded by swamp forest, in Agusan Marsh Wildlife Sanctuary, Mindanao.*

Insets (opposite from top: *Philippine Tarsier (*Carlito syrichta*), a tiny primate found only in the forests of the south. The white flower of* Sonneratia alba*, a common mangrove tree. Collared Kingfisher (*Halcyon chloris*), a common bird of the coast, especially mangroves. Shoal of Yellow-finned Snapper (*Lutjanus xanthopinnis*), seen off the coast of Miniloc Island, El Nido, Palawan.*

Half-title page: *The Philippine Eagle (*Pithecophaga jefferyi*), is both one of the world's largest eagles and one of its most endangered. It is very much the icon of the Philippine conservation movement.*

Title page: *Mangroves growing on the sand flats of Olango Island Wildlife Sanctuary, a hugely important habitat for migratory birds, on Olango Island, Cebu.*

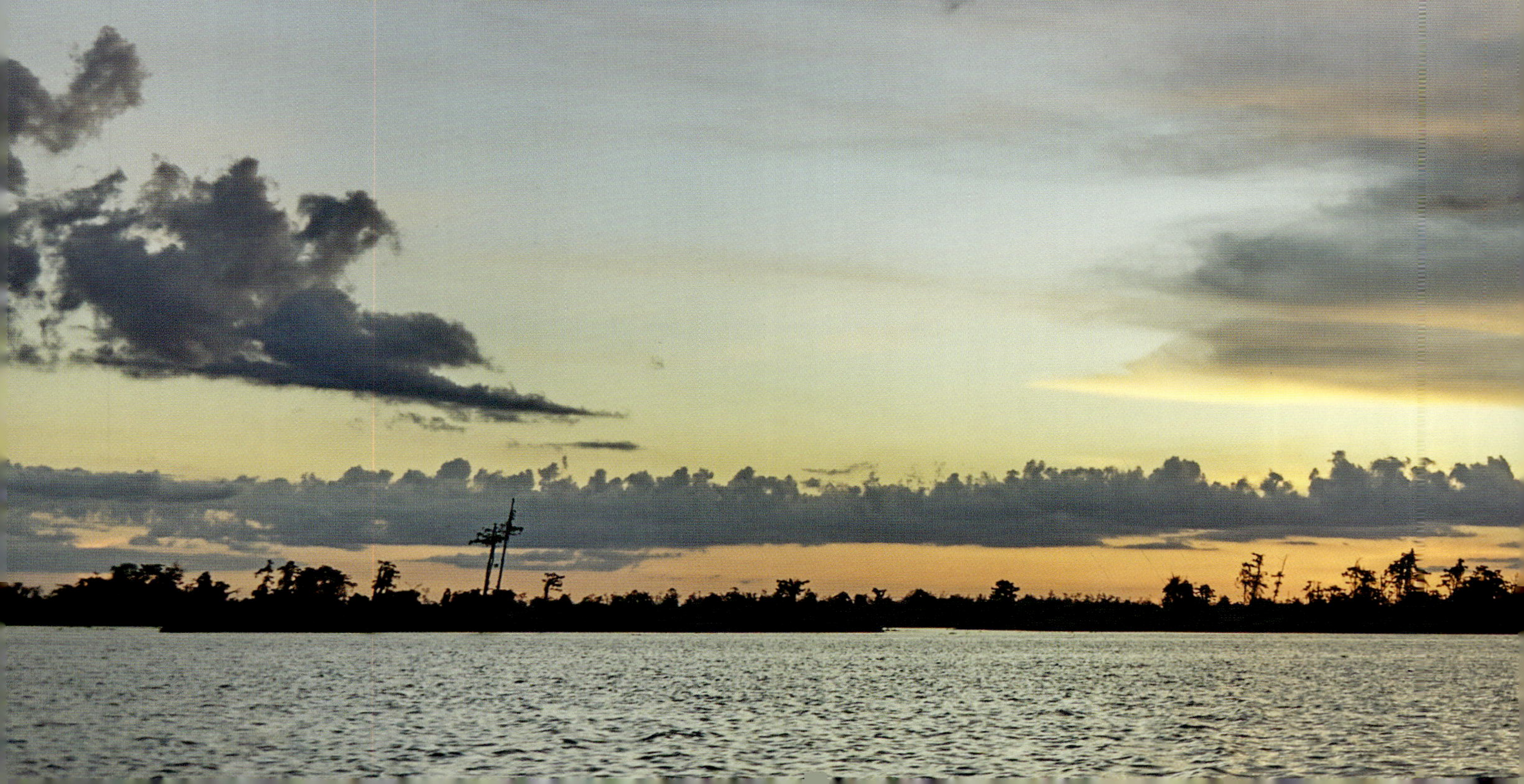

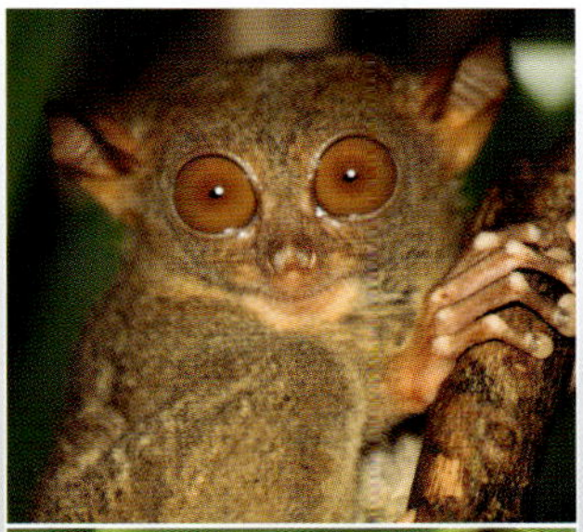
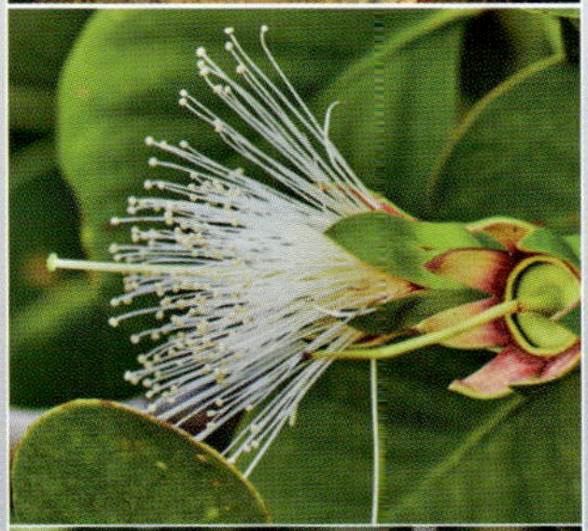
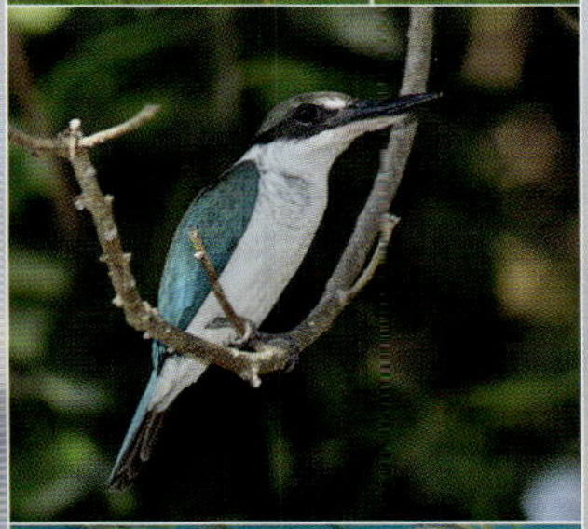
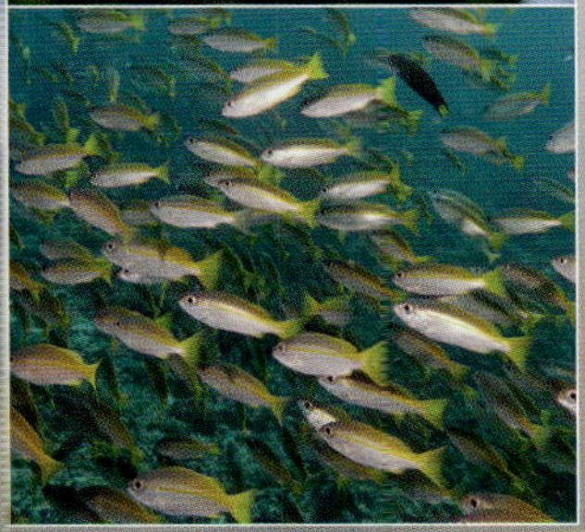

Contents

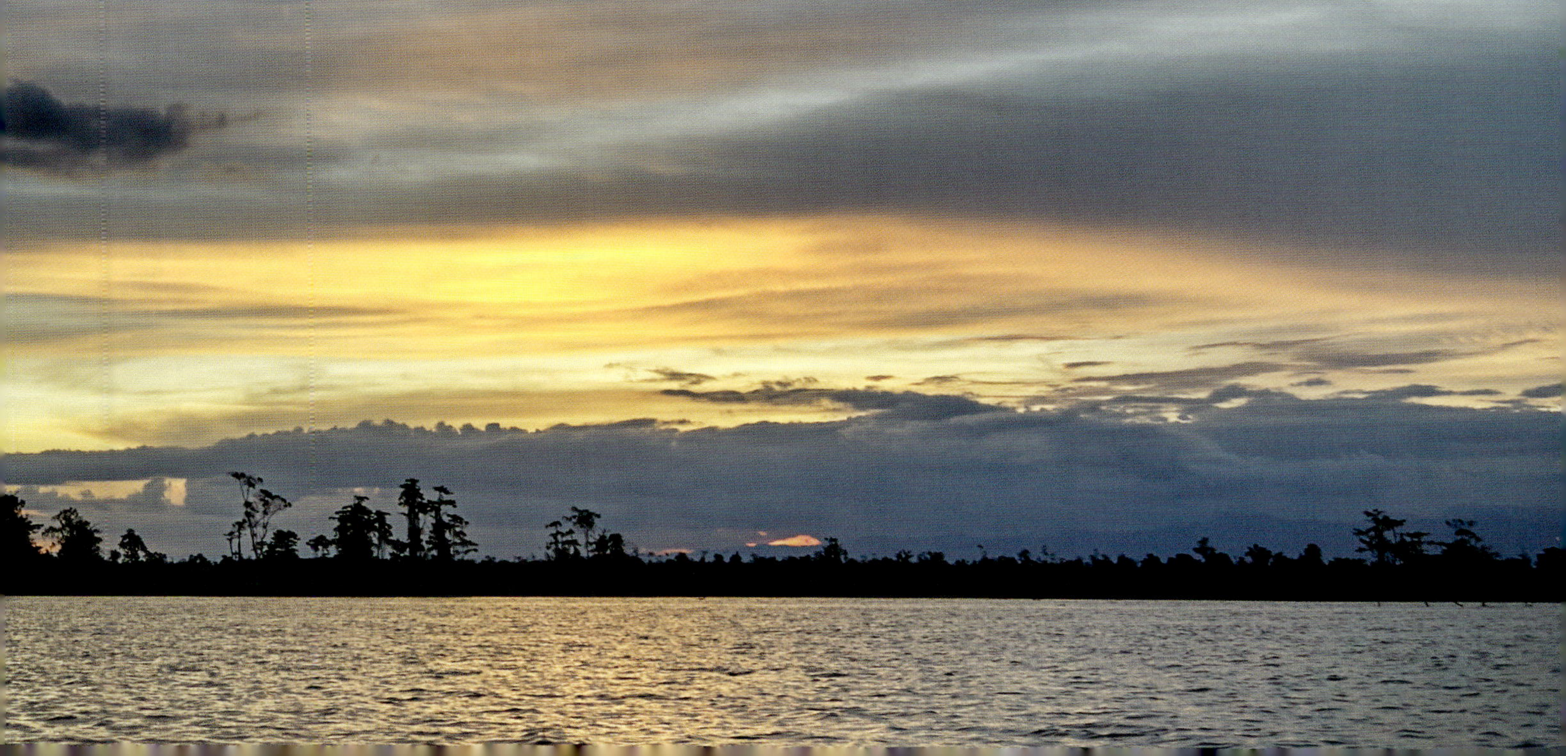

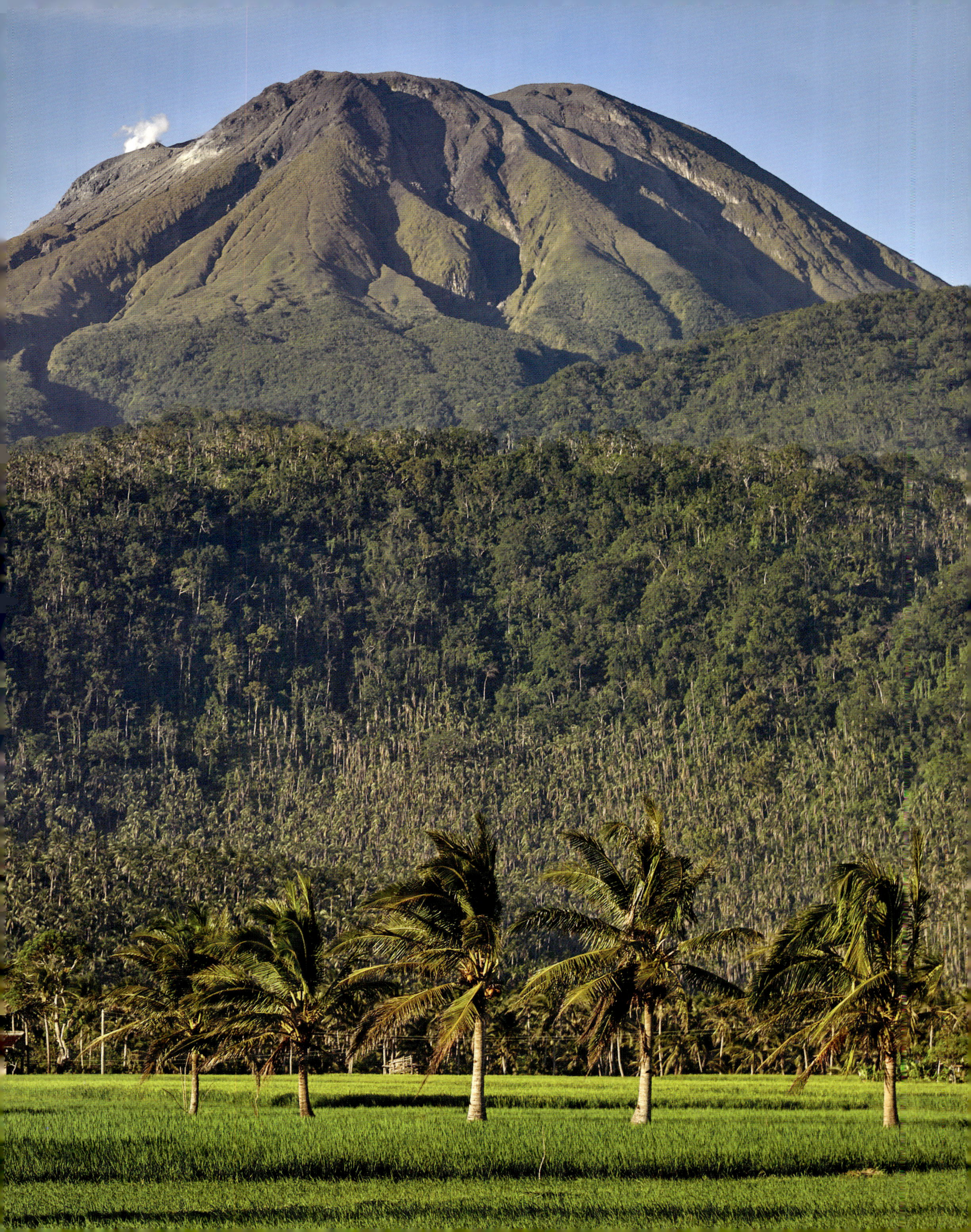

Foreword

Wild Philippines contains an amazing collection of images, supported by informative text, documenting the majestic wild flora and fauna found in the Philippine archipelago. It is an ideal rendition of artistic prowess and the extensive travel experience of seasoned photographer and globetrotting writer Nigel Hicks, delivering a superb representation of the country's natural heritage and landscapes. The book features wildlife commonly seen around the islands, both in its natural habitats and around human habitation, including species easily seen around ecotourism locations. It features noteworthy rare or threatened wildlife, a diverse representation of native residents, migratory and introduced species, and more importantly species known to be endemic to the Philippines.

The exceptionally rich biological diversity of the Philippines is magnified by its global importance due to the high degree of endemism across a wide range of organisms, from microbes to mammals, found within a wide diversity of habitat types, from fringing coral reefs, to mossy cloud forests. The country is globally known to be the last refuge for some of the world's endangered organisms. It is a biodiversity hotspot due to the continued threats on its finite but unique natural ecosystems on scattered small islands, which harbour many of the threatened endemics, inherently limited in population and having restricted ranges.

Despite the ever-increasing and expanding human population and the associated pressures of development, the Philippines' red list of threatened organisms includes a long array of species. However, only a handful can be regarded as nearly extinct or extinct in the wild – so there is still hope. Sustainable conservation efforts can help to reverse and improve their beleaguered status through protection and education, spreading awareness (as does this book) of this rich tropical oasis between the West Philippine Sea and the Pacific Ocean. Indeed, *Wild Philippines* is a picturesque narrative of the natural history of the Philippines, including many of the unique iconic species, such as the Philippine Eagle, Philippine Tarsier, Philippine Crocodile, Red-vented Cockatoo, Northern Luzon Cloud Rat, Golden-crowned Flying Fox, Tamaraw, Gray's Water Monitor, Jade Vine, Waling-waling orchid and magnificent Kapa-kapa.

Wild Philippines also showcases the equally rich ecosystem diversity, featuring a myriad of spectacular natural seascapes and landscapes across this charmed archipelago of more than 7,600 islands. The book pays homage to the many acclaimed natural wonders and critical wildlife habitats around the country, such as World Heritage Sites, Key Biodiversity Areas, ASEAN Heritage Parks, Marine Protected Areas, Ramsar Sites and other invaluable protected areas and wildlife sanctuaries. Very much akin to the mission and aims of natural history museums, *Wild Philippines* documents an important snapshot in time for the state of the Philippine environment, and showcases the resilience of its wildlife amid the increasing pressures from habitat conversion, pollution, human encroachment and impacts of climate change.

The wildlife of the Philippines is definitely beautiful and interesting, and provides crucial linkages in maintaining the world's biological diversity. It is worthy of conservation and admiration for generations to come.

Juan Carlos T. Gonzalez, DPhil
Director, UPLB Museum of Natural History and Professor, Institute of Biological Sciences, University of the Philippines Los Banos

Opposite: *The smouldering summit of Mt Bulusan, one of the Philippines' most active volcanoes, and at the heart of a forested national park.*

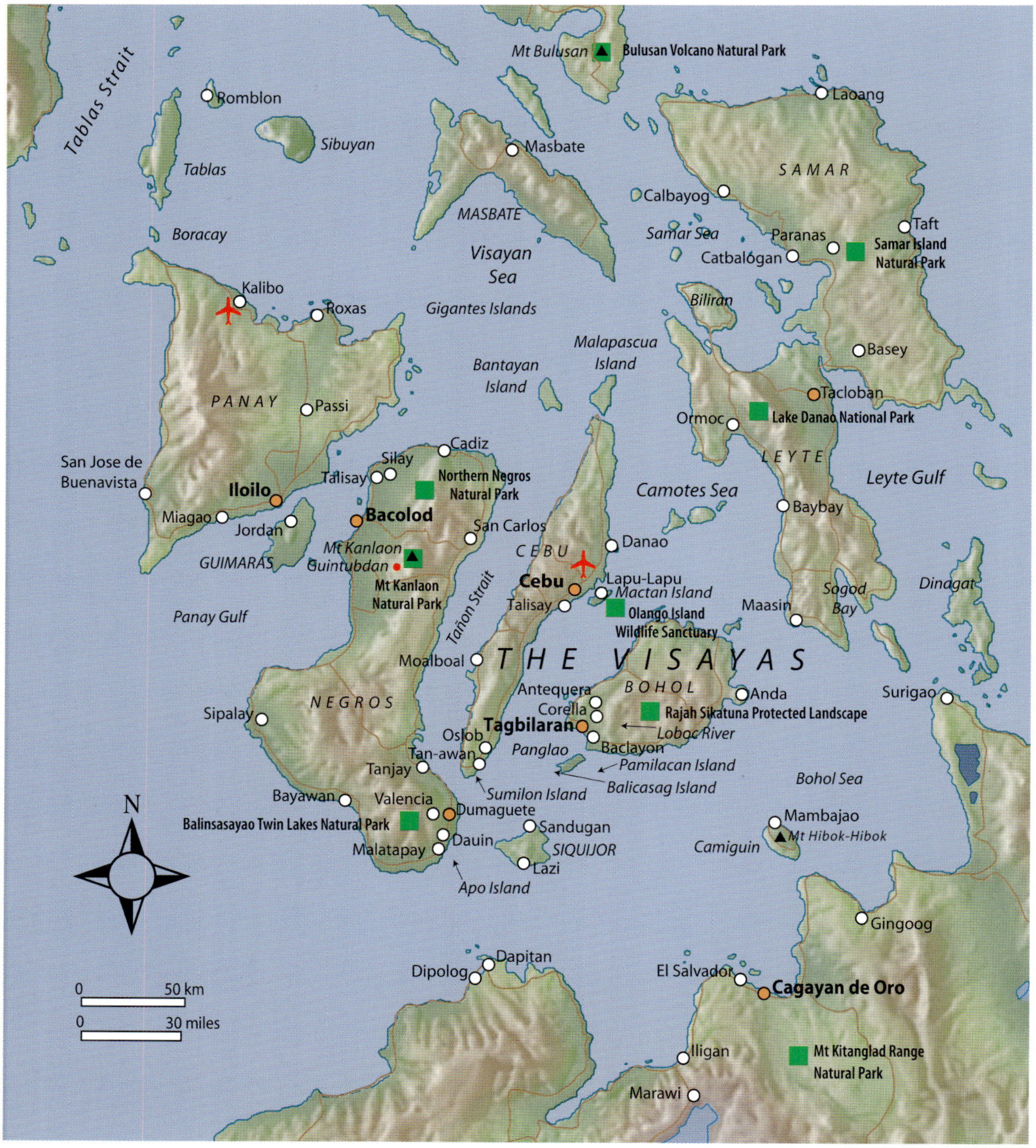
Tablas Strait
Romblon
Sibuyan
Tablas
Boracay
Mt Bulusan
Bulusan Volcano Natural Park
Masbate
MASBATE
Visayan Sea
Gigantes Islands
Laoang
SAMAR
Calbayog
Samar Sea
Paranas
Catbalogan
Taft
Samar Island Natural Park
Biliran
Basey
Kalibo
Roxas
PANAY
Passi
Bantayan Island
Malapascua Island
Tacloban
Ormoc
Lake Danao National Park
LEYTE
Leyte Gulf
San Jose de Buenavista
Iloilo
Miagao
Jordan
GUIMARAS
Panay Gulf
Cadiz
Silay
Talisay
Northern Negros Natural Park
Bacolod
San Carlos
Mt Kanlaon
Guintubdan
Mt Kanlaon Natural Park
Tañon Strait
Camotes Sea
Baybay
Danao
CEBU
Cebu
Lapu-Lapu
Mactan Island
Talisay
Olango Island Wildlife Sanctuary
Maasin
Sogod Bay
Dinagat
THE VISAYAS
Moalboal
Antequera
BOHOL
Anda
Corella
Rajah Sikatuna Protected Landscape
Tagbilaran
Loboc River
Baclayon
Surigao
Sipalay
NEGROS
Oslob
Tan-awan
Panglao
Pamilacan Island
Balicasag Island
Sumilon Island
Tanjay
Bayawan
Valencia
Dumaguete
Balinsasayao Twin Lakes Natural Park
Malatapay
Dauin
Sandugan
SIQUIJOR
Lazi
Apo Island
Bohol Sea
Mambajao
Mt Hibok-Hibok
Camiguin
Gingoog
N
0
50 km
0
30 miles
Dapitan
Dipolog
El Salvador
Cagayan de Oro
Iligan
Mt Kitanglad Range Natural Park
Marawi

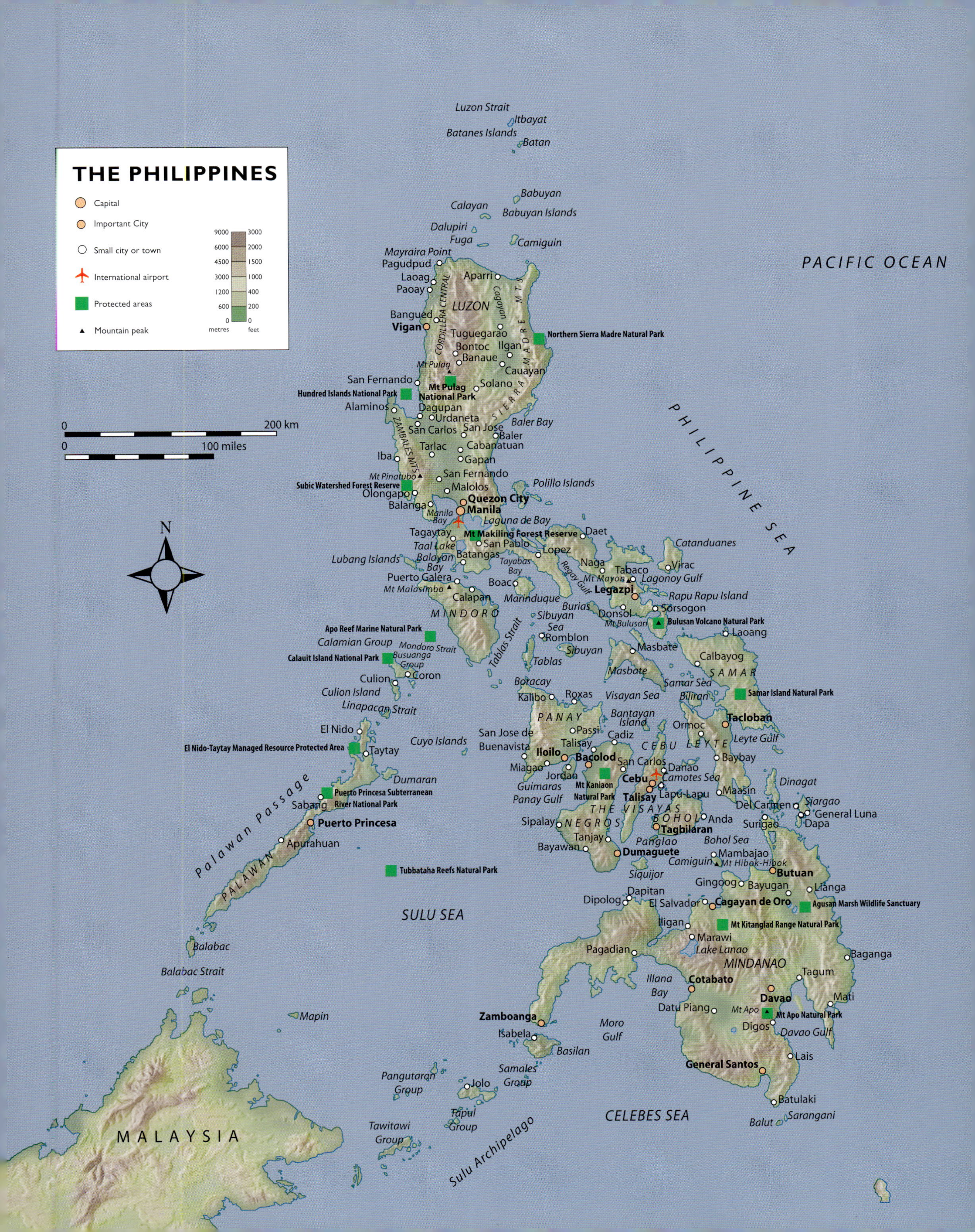

THE PHILIPPINES
Capital
Important City
Small city or town
International airport
Protected areas
Mountain peak
9000
6000
4500
3000
1200
600
0
metres
3000
2000
1500
1000
400
200
0
feet
0
200 km
0
100 miles
N
PACIFIC OCEAN
PHILIPPINE SEA
SULU SEA
CELEBES SEA
MALAYSIA
Luzon Strait
Itbayat
Batanes Islands
Batan
Babuyan
Calayan
Babuyan Islands
Dalupiri
Fuga
Camiguin
Mayraira Point
Pagudpud
Laoag
Paoay
Aparri
LUZON
Bangued
Vigan
CORDILLERA CENTRAL
Cagayan
SIERRA MADRE MTS
Tuguegarao
Bontoc
Ilgan
Banaue
Mt Pulag
Cauayan
Northern Sierra Madre Natural Park
San Fernando
Mt Pulag National Park
Solano
Hundred Islands National Park
Alaminos
Dagupan
Urdaneta
San Carlos
San Jose
Baler Bay
Baler
ZAMBALES MTS
Tarlac
Cabanatuan
Iba
Gapan
Mt Pinatubo
San Fernando
Subic Watershed Forest Reserve
Malolos
Polillo Islands
Olongapo
Balanga
Quezon City
Manila
Manila Bay
Laguna de Bay
Tagaytay
Mt Makiling Forest Reserve
Daet
Taal Lake
San Pablo
Catanduanes
Lubang Islands
Balayan Bay
Batangas
Lopez
Tayabas Bay
Naga
Virac
Tabaco
Ragay Gulf
Puerto Galera
Mt Mayon
Lagonoy Gulf
Mt Malasimbo
Boac
Legazpi
Calapan
Marinduque
Rapu Rapu Island
MINDORO
Burias
Sorsogon
Sibuyan Sea
Donsol
Mt Bulusan
Bulusan Volcano Natural Park
Apo Reef Marine Natural Park
Tablas Strait
Romblon
Laoang
Calamian Group
Mondoro Strait
Busuanga Group
Sibuyan
Masbate
Calauit Island National Park
Tablas
Calbayog
Coron
Masbate
Samar Sea
SAMAR
Culion
Boracay
Culion Island
Kalibo
Roxas
Visayan Sea
Biliran
Samar Island Natural Park
Linapacan Strait
PANAY
Bantayan Island
Tacloban
El Nido
Passi
Ormoc
San Jose de Buenavista
CEBU
LEYTE
Leyte Gulf
Cuyo Islands
Cadiz
El Nido-Taytay Managed Resource Protected Area
Taytay
Talisay
Iloilo
Bacolod
San Carlos
Danao
Baybay
Miagao
Jordan
Camotes Sea
Dinagat
Dumaran
Guimaras
Cebu
Puerto Princesa Subterranean River National Park
Mt Kanlaon Natural Park
Talisay
Lapu-Lapu
Maasin
Panay Gulf
Siargao
Sabang
THE VISAYAS
Del Carmen
General Luna
Puerto Princesa
Sipalay
NEGROS
BOHOL
Anda
Dapa
Palawan Passage
Tagbilaran
Surigao
Apurahuan
Tanjay
Panglao
Bohol Sea
Bayawan
Dumaguete
Mambajao
Camiguin
Mt Hibok-Hibok
PALAWAN
Tubbataha Reefs Natural Park
Siquijor
Butuan
Gingoog
Bayugan
Dapitan
Lianga
Dipolog
El Salvador
Cagayan de Oro
Agusan Marsh Wildlife Sanctuary
Iligan
Mt Kitanglad Range Natural Park
Marawi
Balabac
Pagadian
Lake Lanao
Baganga
MINDANAO
Balabac Strait
Tagum
Illana Bay
Cotabato
Mati
Davao
Datu Piang
Mt Apo
Mt Apo Natural Park
Mapin
Zamboanga
Moro Gulf
Digos
Davao Gulf
Isabela
Basilan
Lais
Samales Group
General Santos
Pangutaran Group
Jolo
Batulaki
Tapul Group
Sarangani
Tawitawi Group
Balut
Sulu Archipelago

Introduction

The spectacular jumble of Southeast Asian islands that we call the Philippines collectively makes up one of the most remarkable places on Earth. Having risen up from the sea bed through a mixture of tectonic force, volcanic eruption and coral reef 'construction', today we see an island environment that is a hugely varied mix of tropical landscapes and habitats, ranging from submarine coral reefs, through vibrantly green countryside, to dense forests and high, rugged mountains topped by smouldering volcanoes.

If this were not enough, having risen up from the sea bed and developed with very little direct physical contact with any of the world's main land masses over a prolonged period, the Philippine wildlife – both plant and animal – has developed down its own evolutionary path. This has resulted in not only a stupendously varied fauna and flora, but also an incredibly high rate of *endemism*; that is, species that occur nowhere else on Earth. These two factors together have led the Philippines – rather belatedly – to be recognized as one of the world's greatest megadiverse regions. So rich is the country's wildlife diversity that, when its relatively small size is taken into consideration, it surpasses all

Below: Mangrove forests are a vital habitat, protecting both the coast and the young of a huge number of marine species.

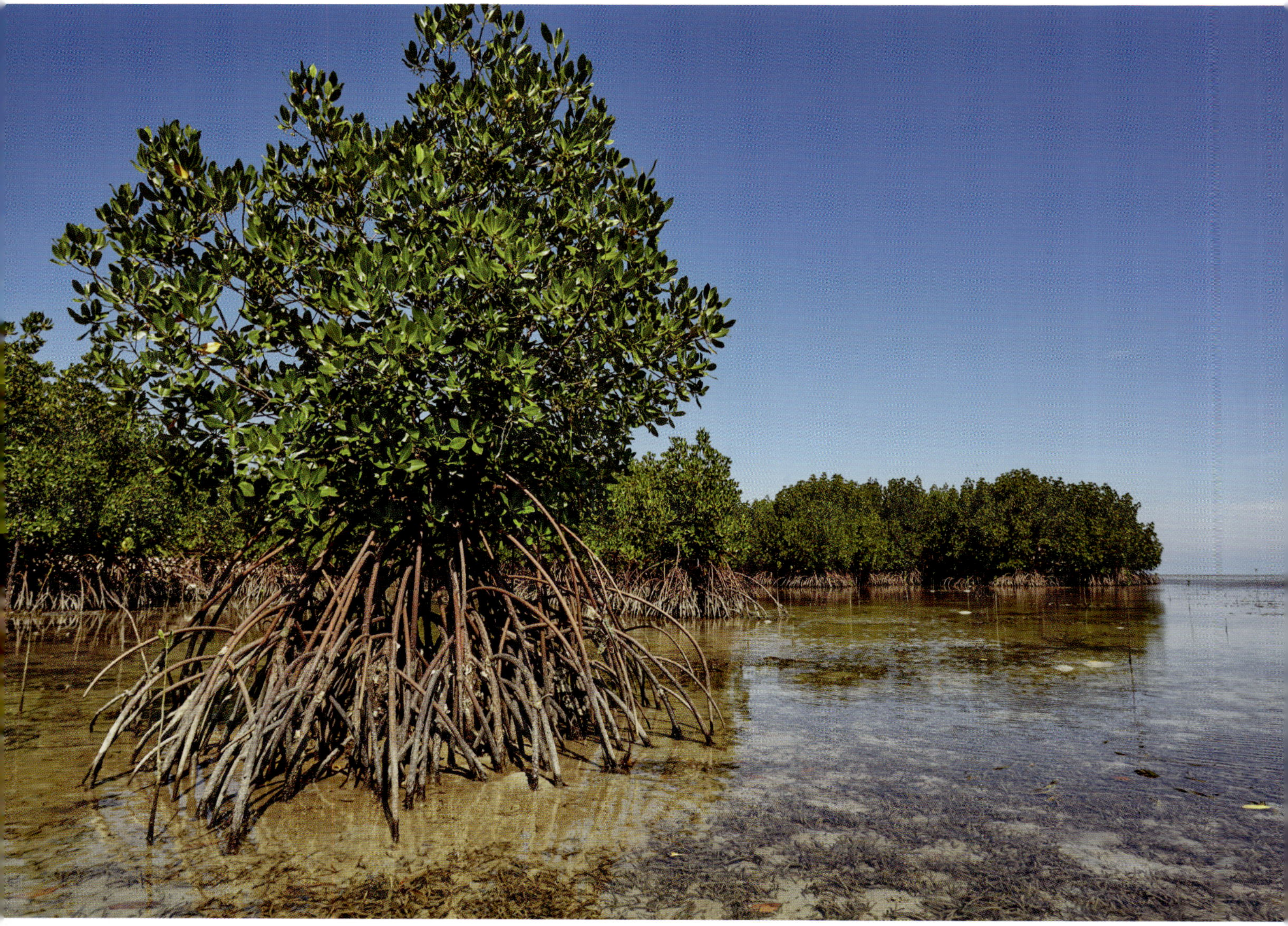

other, much more famous centres of biodiversity such as the Amazon, the Congo basin and Madagascar.

Unfortunately, not all is well with this treasure house of life, with pressure from the Philippines' large and still growing population causing severe environmental devastation over the past 100 years, bringing the natural environment and its precious wildlife to the brink of collapse. However, starting largely in the final decade of the twentieth century and growing rapidly since then, an increasingly active and strong environmental movement has begun to have success in stemming the tide of destruction, giving some hope for the future survival of this beautiful country's natural environment.

This book aims to highlight just what is at stake, showcasing the wide range of the Philippines' natural environments, from the coral reefs to the volcano summits, along with some examples of the islands' incredible unique plant and animal wildlife, plus a number of the most important and representative protected areas. Many of the most critical environmental threats, both historical and current, are included, along with an introduction to some of the conservation projects working to improve the situation.

It has to be added that much of the science surrounding our understanding of the Philippines' natural environment is still far from complete. Not only is knowledge about many of the habitats and their interactions still patchy, but new species are being discovered at quite an astonishing rate. So much remains uncertain, and no one and no publication yet has a full picture, including this book. Nevertheless, every piece of information in this book has been researched in depth, and confirmed with as much certainty as is possible through multiple sources, ranging from scientific journals, to a host of books and personal communication with scientists working in the field. It is beyond the scope of a book such as this one to give a complete list of all its sources, but the Resources section (see page 204) provides a useful selection.

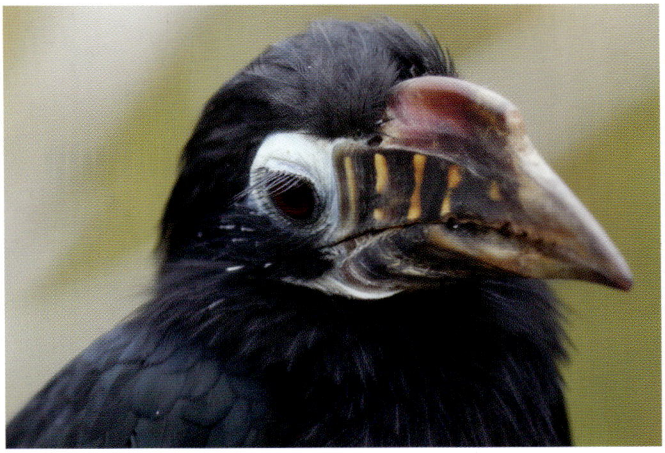

Above: *A female Visayan Hornbill, a highly threatened species that is unique to the Philippine islands of Panay and Negros.*

ACRONYMS USED IN THIS BOOK

ASEAN Association of Southeast Asian Nations
BFAR Bureau of Fisheries and Aquatic Resources
BMB Biodiversity Management Bureau
CBD Convention on Biological Diversity
CH Critical Habitat
CPPAP Conservation of Priority Protected Areas Program
CTI Coral Triangle Initative
DENR Department of Environment and Natural Resources
EDC Energy Development Corporation
ENTMRPA El Nido-Taytay Managed Resource Protected Area
IBA Important Bird Area
ICCA Indigenous Community Conservation Area
IUCN International Union for the Conservation of Nature
KBA Key Biodiversity Area
LCA Local Conservation Area
MAB Man and the Biosphere
MPA Marine Protected Area
NFEFI Negros Forests and Ecological Foundation, Inc.
NewCapp New Conservation Areas in the Philippines Project
NGO Non-government organization
NIPAP National Integrated Protected Areas Program
NIPAS Philippines' National Integrated Protected Areas System
NNNP Northern Negros Natural Park
NSMNP Northern Sierra Madre Natural Park
PAWB Protected Areas and Wildlife Bureau
PEFI Philippine Eagle Foundation
PHIVOLCS Philippine Institute of Volcanology and Seismology
PTFI Philippine Tarsier Foundation
PWRCC Palawan Wildlife Rescue and Conservation Center
TSPS Tañon Strait Protected Seascape
UNESCO United Nations Educational, Scientific and Cultural Organization
WWF World Wide Fund for Nature

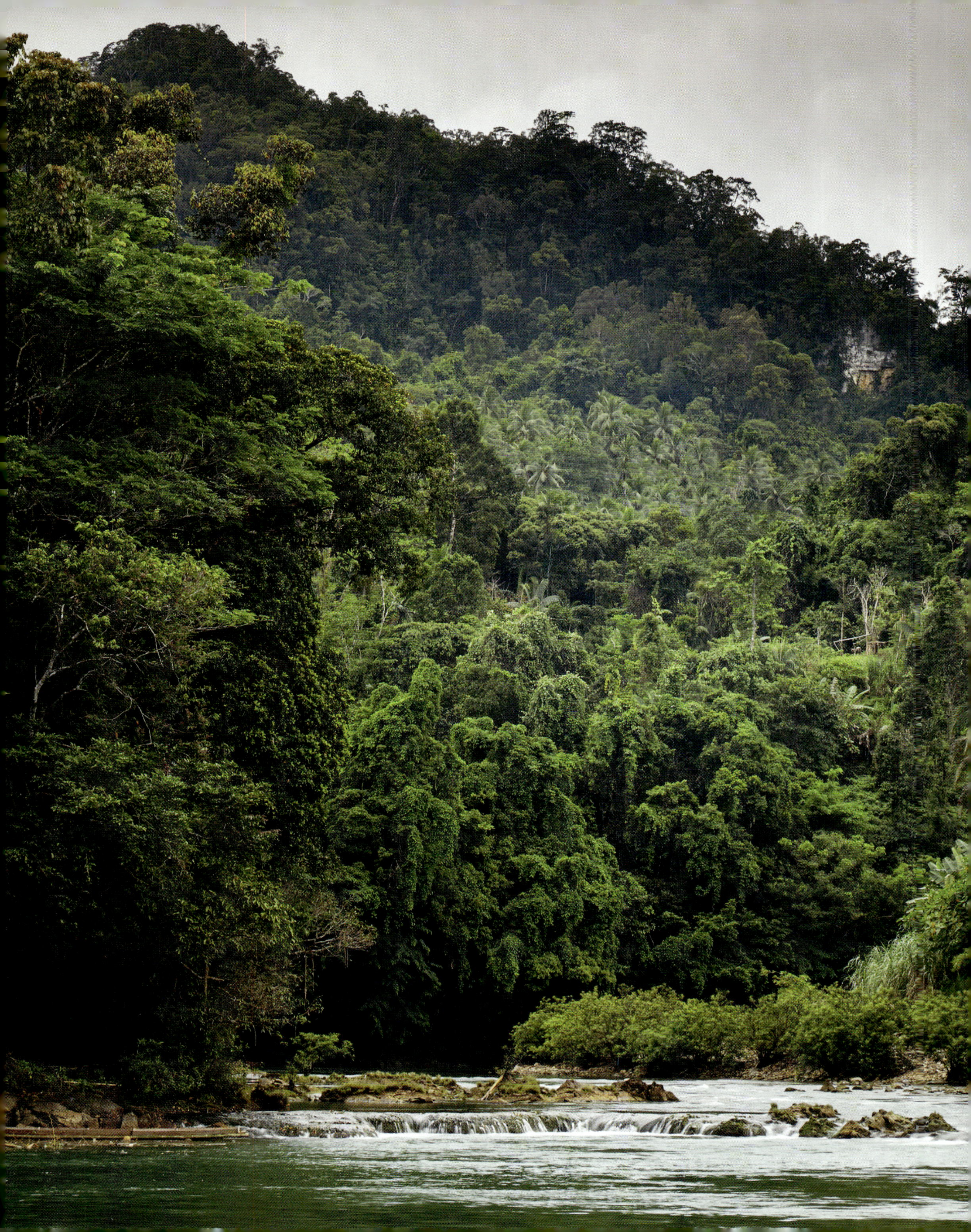

1
The Philippines
A treasure house of life

The Philippines is the quintessential island nation, a fragmented mass of tropical islands sitting at the southeastern edge of Asia. The country faces the huge expanse of the Pacific Ocean to the east, and is separated from the main Asian land mass by the South China Sea to the west. The 7,000-plus islands of this country stretch for more than 1,200 km (745 miles) from north to south without a single land border, its nearest neighbours Taiwan some distance to the north, Malaysian Borneo to the south-west, and Indonesian Sulawesi to the south.

Despite the huge number of islands, more than 90 per cent of the country's 300,000 km^2 (115,800 mi^2) land area is made up by just 10 islands. The largest is Luzon, the 'anchor' for the northern half of the country and the site of Manila, the national capital, followed by Mindanao, in the far south. Between the two sits a mass of islands that make up the central Visayas region, while sandwiched between the Visayas and Luzon is the rugged island of Mindoro. South-west of Mindanao stretch the Jolo and Sulu Islands, an island chain that reaches almost all the way to Borneo. The westernmost part of the country consists of Palawan, a long, pencil-thin island province that runs north-east to south-west, stretching much of the way from Mindoro down to Borneo.

***Opposite:** A rugged forested landscape along the Ulot River, in the heart of Samar Island Natural Park, one of the Philippines' largest remaining areas of lowland rainforest, a home to a large number of endemic plant and animal species.*

Tectonics and volcanoes

Not surprisingly, with so many widely flung islands and island groups there is quite a diversity of landscapes and habitats, though all of them are tropical in nature. Much of the Philippines, with the exception of Palawan and Mindoro, have risen up from the sea bed, starting – so it is believed – about 20 million years ago. The driving force behind this rise was the Australasian tectonic plate, pushing the far smaller Philippine plate northwestwards, then crushing it against the vast Eurasian plate to the west. This upwards and westwards pressure continues to this day, and the islands' mountain ranges are sites of frequent earthquakes and regular volcanic eruptions. The entire country sits on the western arc of the Pacific Ring of Fire, its unstable mountain ranges pockmarked with an estimated 200 volcanoes, 22 of which are currently classed as active.

The result on many of the larger islands is a landscape dominated by high, rugged mountain ranges, many of the peaks quiescent or active volcanoes, their slopes sweeping down to coastal plains that have been created by lava or ash fallout from repeated eruptions down the millennia. The country's highest mountain, Mt Apo (2,954 m/9,689 ft) in southern Mindanao, is one such volcano, still active though with no recorded eruptions in the historic period. The Philippines' most famous volcano, however, must be Mt Mayon, its majestic, almost textbook-perfect conical shape sweeping up from the coast to its single, almost perfectly pointed summit, towering over the city of Legazpi in the south of Luzon. Beautiful it certainly is, but also rather ominous, for it

Below: *The textbook-perfect conical shape of Mt Mayon, the Philippines' most active volcano, towers over the city of Legazpi, in southern Luzon.*

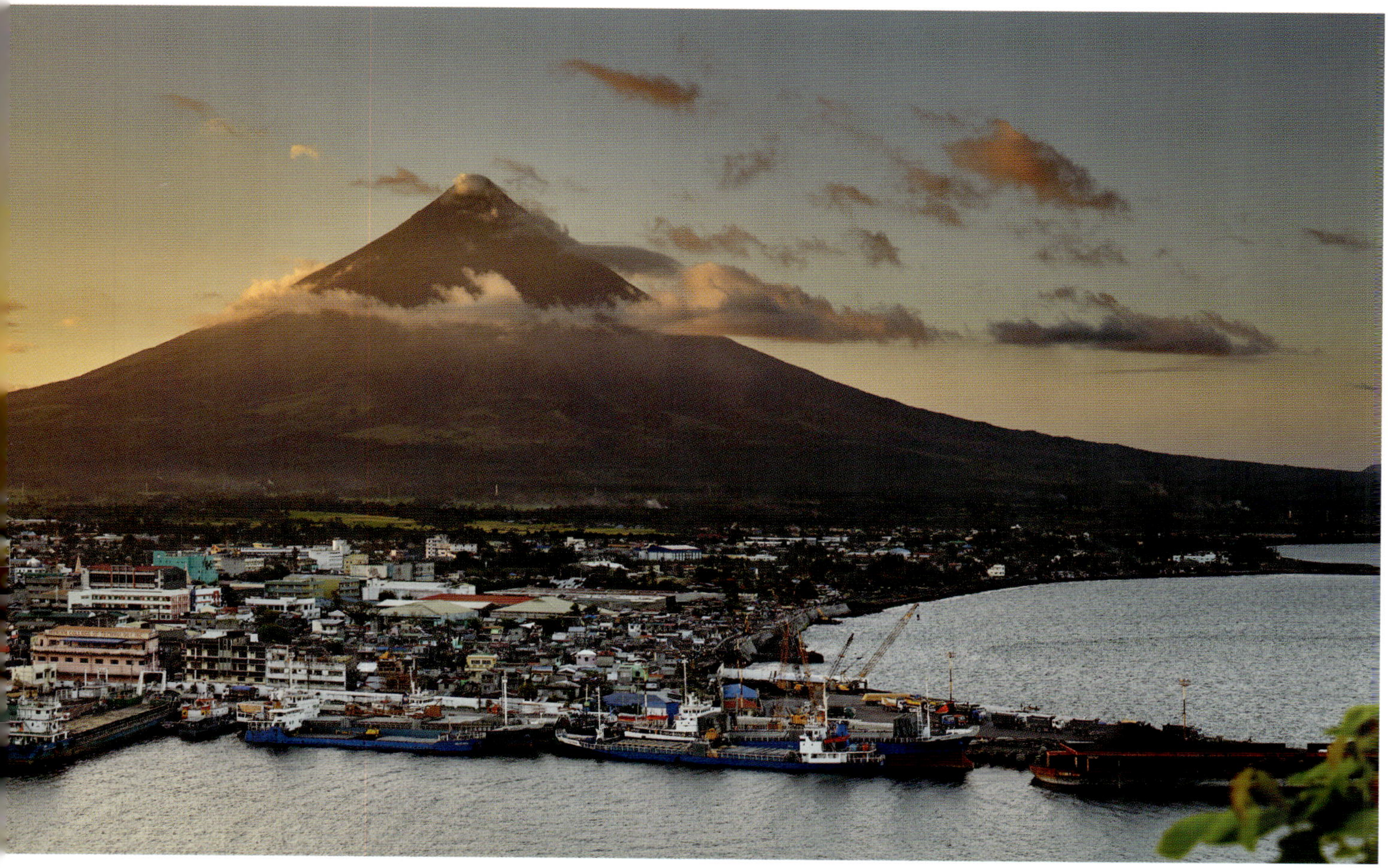

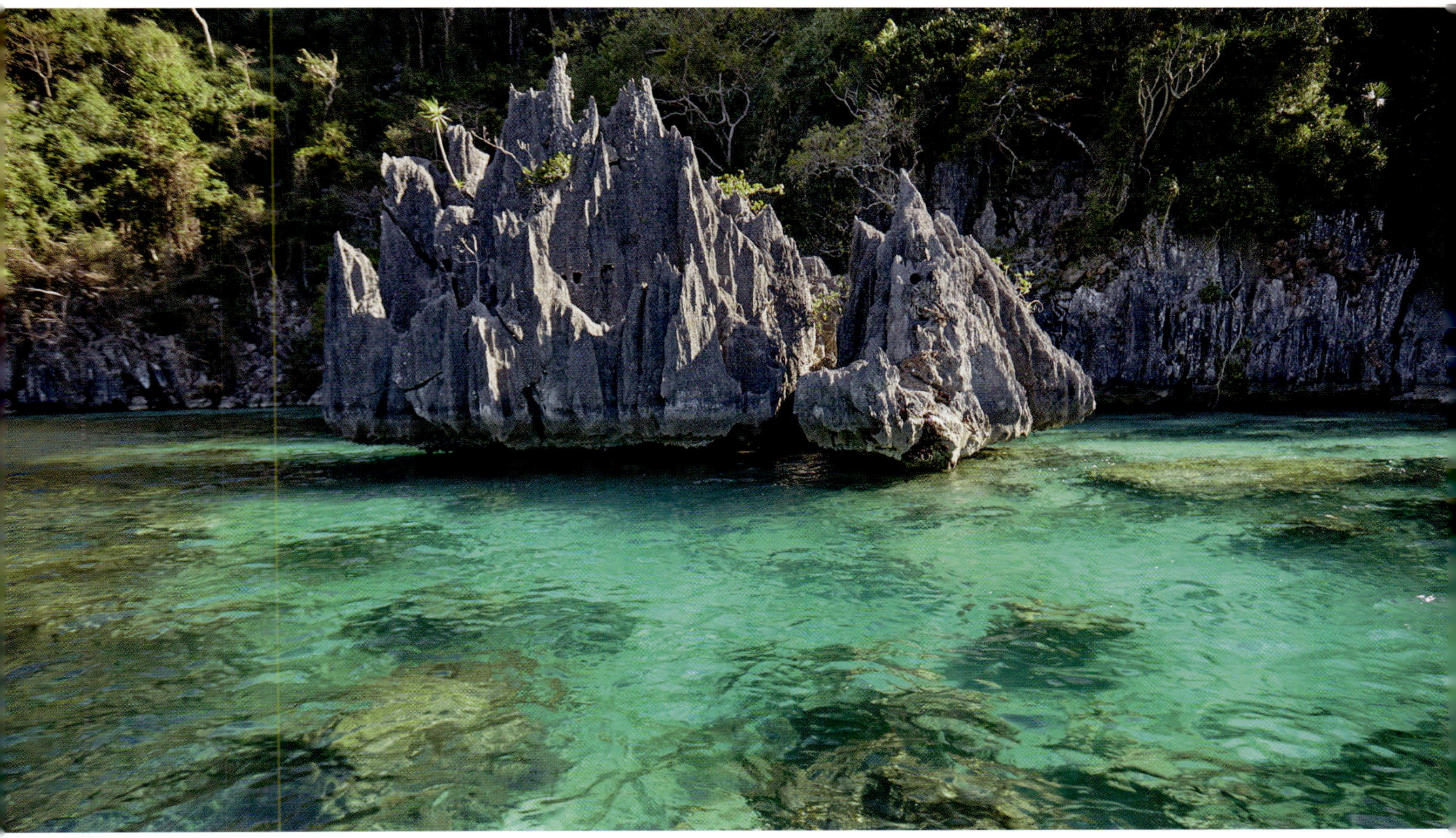

Above: Jagged karst limestone rocks and cliffs, all that remain of coral reefs built millions of years ago, form the landscape of Coron Island, in the north of Palawan.

constantly smoulders, and is continually under the watchful scrutiny of the scientists at the Philippines Institute of Volcanology and Seismology (PHIVOLCS). And no wonder, for it is the country's most active volcano, with no less than 50 recorded eruptions to its name, the most recent in 2018.

Island-building with corals

Though most of the Philippine landscape has been created by the combination of tectonic forces and volcanic eruptions, a third significant island-building force has been at work: repeated cycles of coral reef formation, growth and death. The calcareous 'skeletons' the coral animals create when building their reefs remain long after the animals themselves have died, leaving solid structures on which successive generations of corals can build ever-larger reefs, as long as water depth allows. If water levels rise then the corals continue building upwards, remaining just below the sea's surface. If sea levels drop, or tectonic forces push the reef upwards, the reef becomes exposed, the corals die and an island is created. Exactly these processes have happened time and again across millions of years, with the result that many of the Philippines' smaller islands, as well as a number of quite large ones, such as Siquijor, Guimaras and parts of Palawan and Samar, were created in this way.

Such coral islands do not rise up to create the huge mountain ranges seen on the biggest islands – that is left to plate tectonics and volcanoes – but they can rise to several hundred metres above sea level. It can come as some surprise to be able to find seashells on even the highest parts of Siquijor, for example. They can also create some very dramatic scenery, typified by sheer cliffs pockmarked with caves, and razor-sharp crags and ridges devoid of a single piece of flat ground.

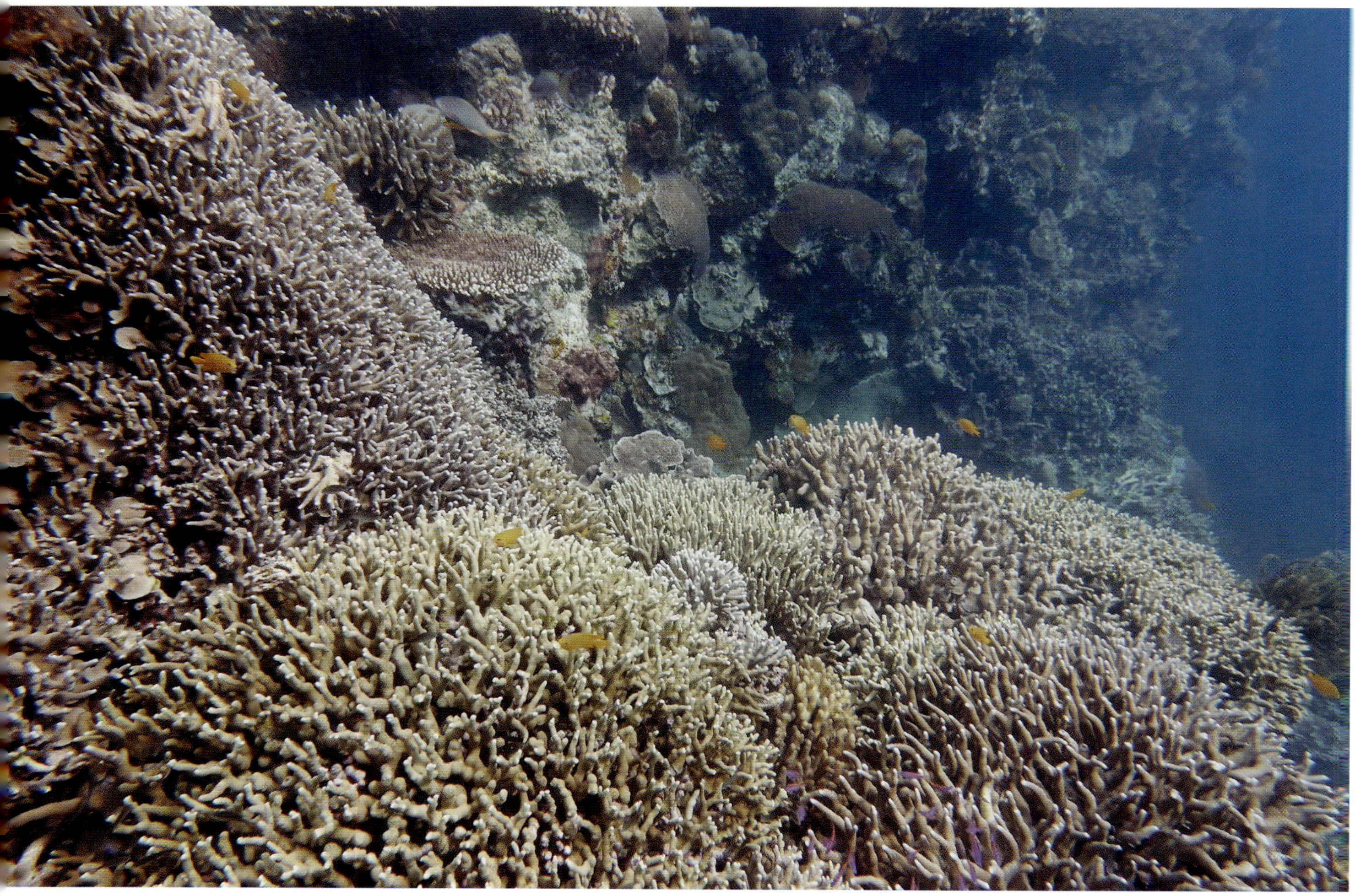

Above: A mass of hard corals, mostly Acropora *species, crowd onto a sheer submarine wall. It is corals such as these that are responsible for building the huge reef structures.*

Below: Much of the landscape of northern Palawan is built from karst limestone, the remains of ancient coral reefs.

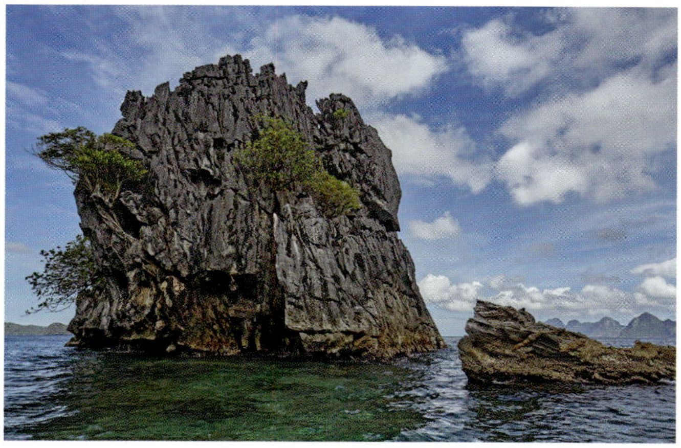

These are the karst limestone landscapes commonly seen across Southeast Asia, and in the Philippines the hallmark of such places as Coron and El Nido, both in Palawan.

The Philippines' marine environment

Being an island nation, the sea of course has a huge role to play in the country's natural environment, and not just in the island-building activities of its corals. Indeed, the country's 300,000 km^2 (115,800 mi^2) land area is dwarfed by the nearly 2 million square kilometres of sea that the Philippine government claims as territorial waters, emphasizing the massive importance placed on the country's marine environment.

These seas vary from calm, sheltered waters shallow enough to walk through, to some of the world's deepest oceanic regions. The tectonic forces that have created the Philippines have also gouged out some phenomenally deep ocean trenches, the biggest of them all the Philippine Trench, which parallels much of the country's Pacific coast, reaching its deepest point in the Galathea Depth, a staggering 10,540 m (34,571 ft) below sea level, barely 120 km (74 miles) off the coast of Mindanao. This is, of course, quite exceptional, but even among the crowded islands the seas can be remarkably deep in places. The Sulu Trench, for example, which runs north–south through the Sulu Sea, separating Palawan from the main body of the Philippines, drops down to several thousand metres deep. Even the Tañon Strait, the narrow channel running between the major Visayan islands of Negros and Cebu, reaches a depth of more than 500 m (about 1,650 ft). The presence of such deep waters separating island groups has had a dramatic effect on the country's terrestrial wildlife, as discussed later.

The country's 7,000-plus islands have a combined coastline length of about 35,000 km (21,700 miles), significant stretches of it – though not all by a long way – protected from the worst that the sea can throw at it by fringing coral reefs, estimated to total about 25,000 km² (9,600 mi²). While for the most part the deeper seas are rather poor in nutrients, and hence tend to have rather limited life, the coral reefs are the exact opposite; they are the real powerhouse of the tropical marine environment, teeming with a huge array of animal life of all descriptions.

The coral reef environment

Although corals are animals, most are able to survive only in areas where light levels are comparatively high, namely relatively shallow waters less than about 50 m (164 ft) deep, and where water clarity is good. This is because for many of them their nutrition depends on a symbiotic relationship with a class of algae called zooxanthellae that live within the coral's calcareous structures and which are able to capture the sun's energy through photosynthesis. If the light levels are too low, or the water is too cloudy, the zooxanthellae cannot photosynthesize, and as a consequence the corals cannot grow.

Fortunately for the Philippines, conditions around many parts of its coast are ideal, leading to rich coral

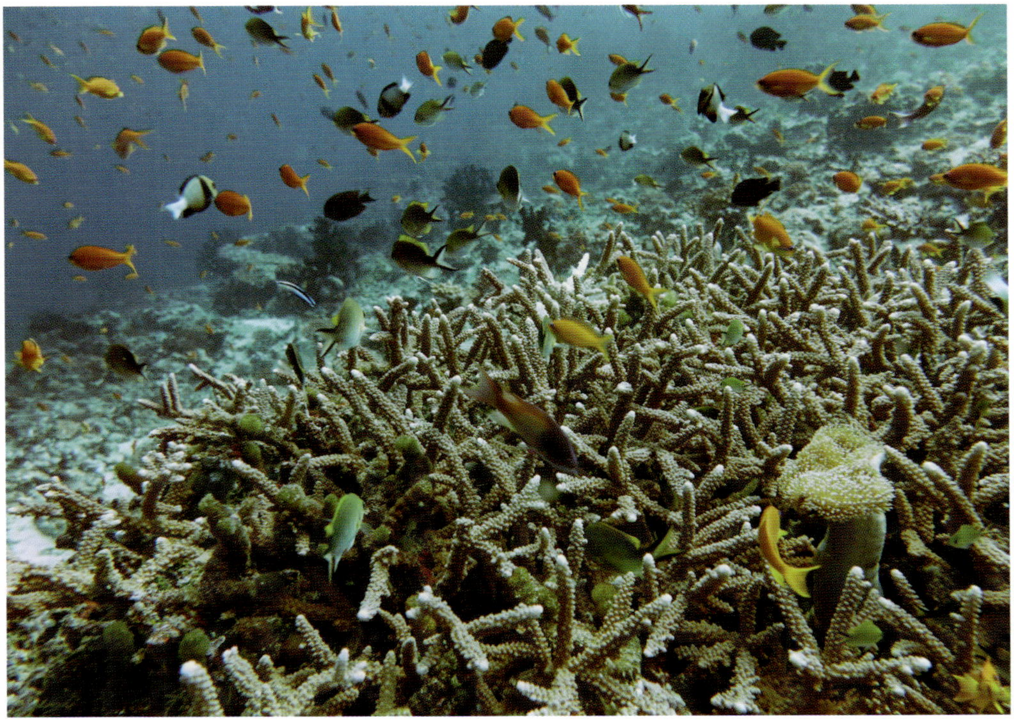

***Right:** The Philippines hosts a staggering 25,000 km² of coral reefs around its coasts. Much of it is not in as pristine a state as this reef, this image showing reef fish clustering around a stand of* Acropora *species staghorn hard corals.*

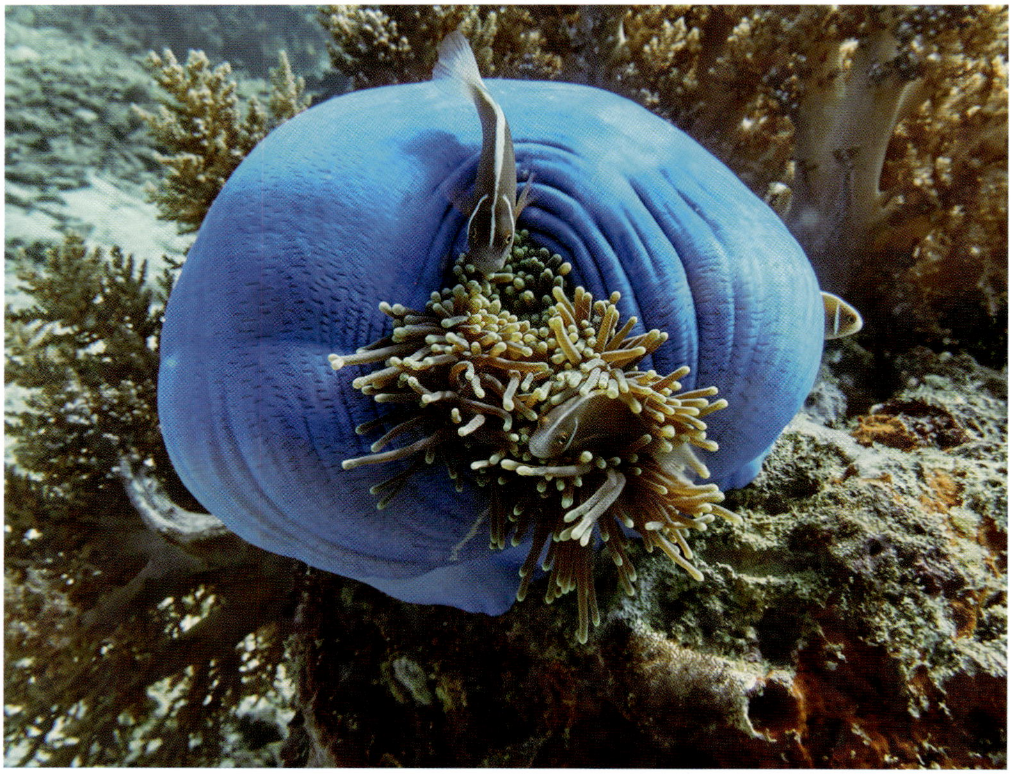

Left: *Sea anemones (in this case* Heteractis magnifica*) are a common sight on Philippine reefs, usually with resident clownfish (also called anemonefish), here Pink Anemonefish (*Amphiprion perideraion*), in attendance, which will defend their homes vigorously against all potential attackers, including divers.*

Below: *A very common sight on Philippine reefs are crinoids, or feather stars, a hugely diverse group of starfish that have multiple delicate arms.*

growth and extensive reefs. The reefs take on different forms depending on local conditions, and can be found growing on sheer walls that plunge directly from shallow waters straight down into great depths, on moderate sandy and/or rocky slopes, and on flat or gently sloping sandy areas, usually in the shallowest and most sheltered areas close to shore. This latter area is often protected from strong wave action by the 'reef crest', a large mound of corals that forms at the boundary between a gently sloping, sandy area on the landward side, and a steep or sheer, rocky drop-off into deeper waters on the seaward side.

Corals growing in the sheltered, shallow areas often have quite delicate structures, while those on the reef crest are usually made of sterner stuff designed to resist the power of the waves. Those in deeper waters, especially on the steep drop-offs below a depth where wave action is significant, again may consist of delicate forms, including beautiful fans designed to catch food as it floats by on sometimes quite strong currents.

Most coral reefs in the Philippines lie off sandy or rocky shores, or are sometimes associated with particularly stable and usually extensive mangrove forests, areas where sediment coming off the land is restricted, ensuring high water clarity. It is rare to see coral reefs close to river estuaries, areas where large amounts of sediment are discharged into the sea, making it too cloudy for reefs to grow.

The Philippines and the Coral Triangle

The coral reefs of Southeast Asia in general, but particularly around the coasts of the Philippines, Indonesia, Malaysia, East Timor, Papua New Guinea

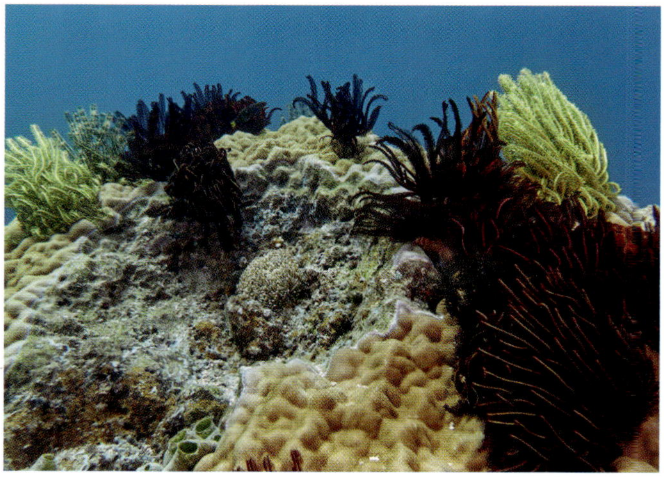

and the Solomon Islands – a region whose reefs are now called the Coral Triangle – are enormously rich in wildlife biodiversity. So rich are they that the Coral Triangle area is now believed to have been the original focus for coral evolution, and to this day it still is the main central region for distribution of corals across most of the world's tropical seas, and certainly the Indo-Pacific region.

Geographically, the Philippines lies towards the northern side of the Coral Triangle, but in terms of biodiversity it is right at its heart. Indeed, it has been suggested that the waters around the Visayas may have the greatest biodiversity of any region within the Coral Triangle. Whether that is true or not, the Philippines' coral reefs are certainly teeming with life, home to nearly 600 coral species (80 per cent of the world's total) and more than 3,000 fish species, along with a huge array of sponges, crustaceans, echinoderms and many other animals that make the coral reefs their homes. This topic is covered in more detail in Chapter 3.

Moving towards the land

Apart from the coral reefs, several other coastal marine environments are hugely important, particularly the seagrass beds and mangrove forests. Seagrasses, the only flowering plants able to live in salt water, colonize shallow, sandy areas, usually just below the low-tide line. Mangroves, on the other hand, grow right along the shore, generally between the high- and low-tide lines. Sometimes forming quite extensive forests, and in the Philippines consisting of about 40 species, they are a highly specialized group of trees able to tolerate salt water for a number of hours per day. In the Philippines mangroves cover about 310,000 ha (765,700 acres), and are frequently concentrated in river estuaries and very shallow, sheltered bays, often behind a protective coral reef.

Both mangroves and seagrasses are hugely important in stablizing the shore and building new land. They also act as protective nursery grounds for a wide range of economically important fish and shellfish species.

The terrestrial environment and its vegetation

Left alone and without any human interference, the Philippine islands would be almost wholly covered with tropical rainforests of one sort or another. There are actually a number of different types of rainforest, quite apart from the already mentioned mangroves.

***Below:** Lowland evergreen rainforest is characterized by a predominance of huge, straight-trunked trees belonging to the dipterocarp family.*

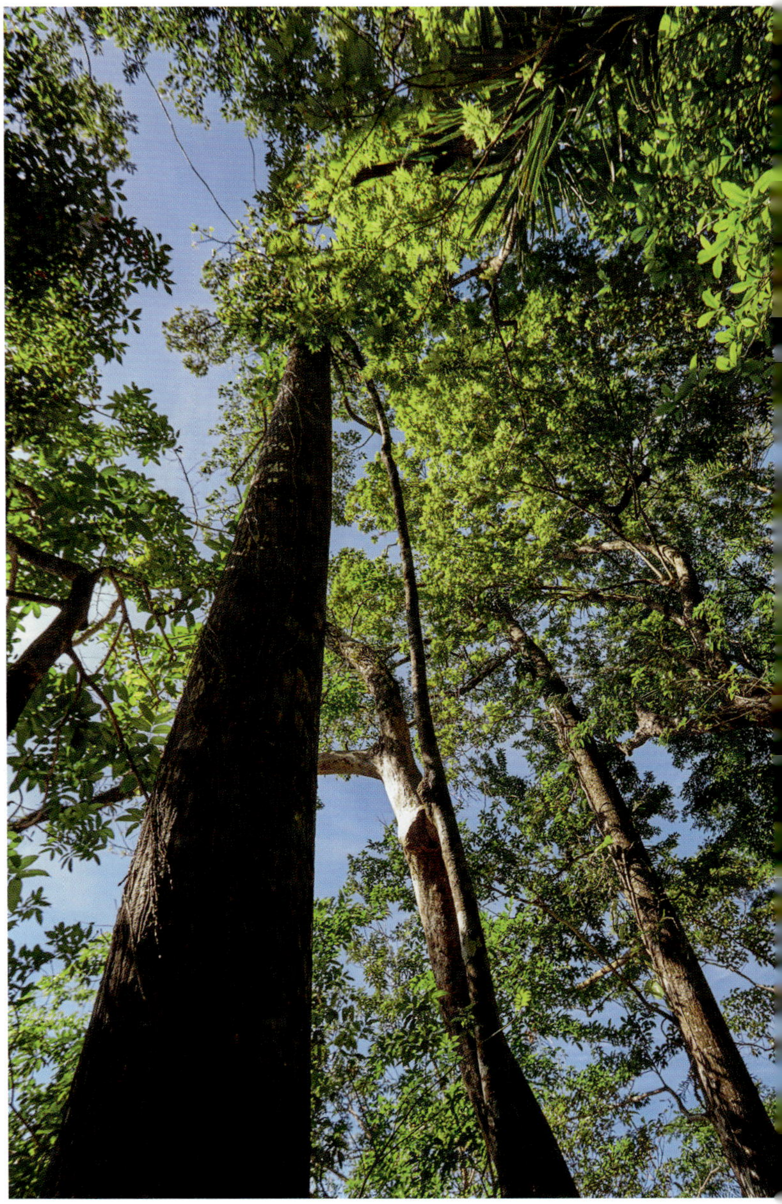

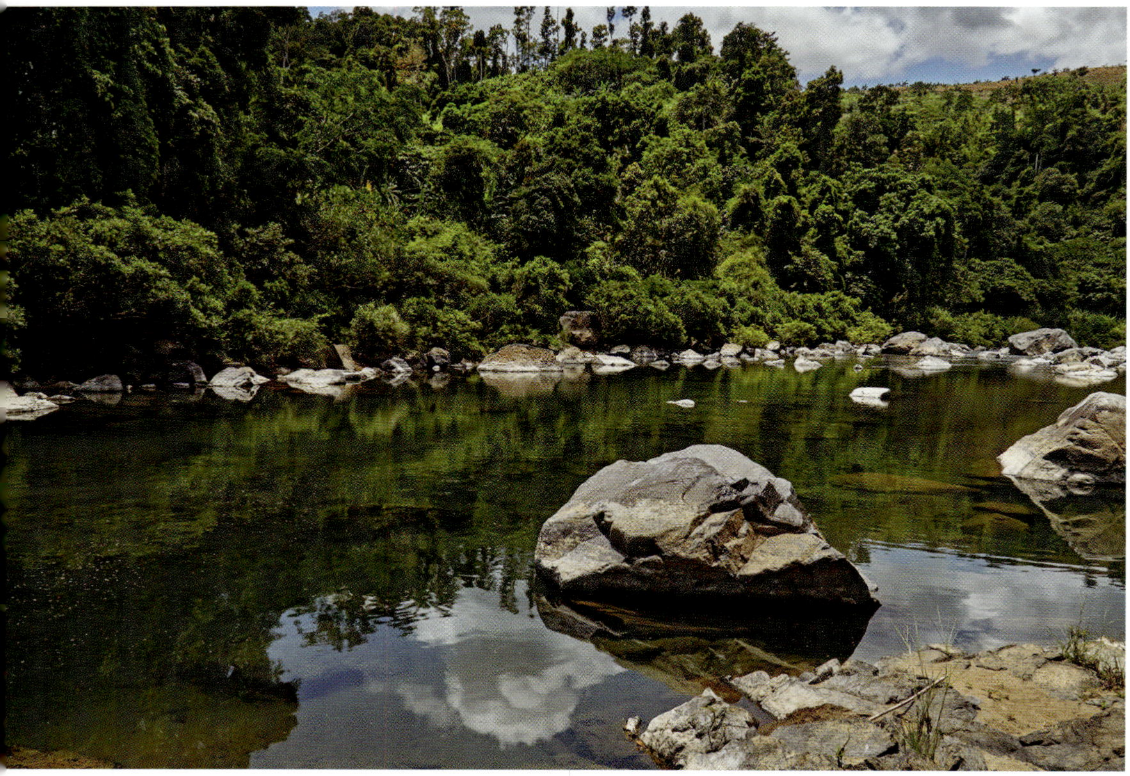

Left: *Riverside areas, as here on the western fringe of the Northern Sierra Madre Natural Park, are often the best places to get a good view of both the forest and its wildlife.*

Below: *Limestone rainforest is characterized by having trees that are smaller than is normal for lowland forest, growing in rather thin soils overlying highly porous limestone. As here, in Rajah Sikatuna Protected Landscape, the landscape is often pockmarked with sinkholes and caves.*

Moving away from the high-tide line, you first come to beach forest, a narrow band of trees able to withstand salt and a life growing in sandy ground, often found along the landward edges of beaches and mangroves.

Inland from here you come to the classic lowland rainforest, usually characterized by a variety of huge trees, the dipterocarps, so named for the structure of their seeds. This is the type of forest most normally associated with the lowlands, dependent on high humidity and rainfall, and a reasonably deep (though not necessarily fertile) soil.

With other conditions come different types of lowland rainforest, including seasonal forest in which the trees may lose their leaves during a prolonged dry season, limestone forest consisting of trees specialized to survive in the poor soils above karst limestone, and swamp forest containing trees able to survive in boggy or flooded areas. While examples of limestone forest can easily be seen across many parts of the country, the best-known example of extensive swamp forest occurs at Agusan Marsh, in Mindanao, though there are other, smaller sites.

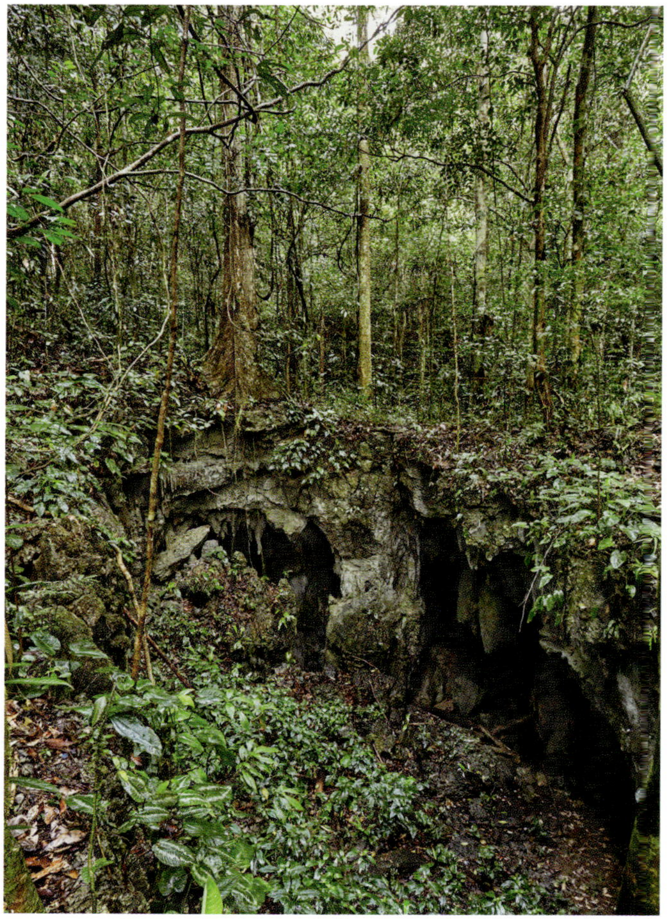

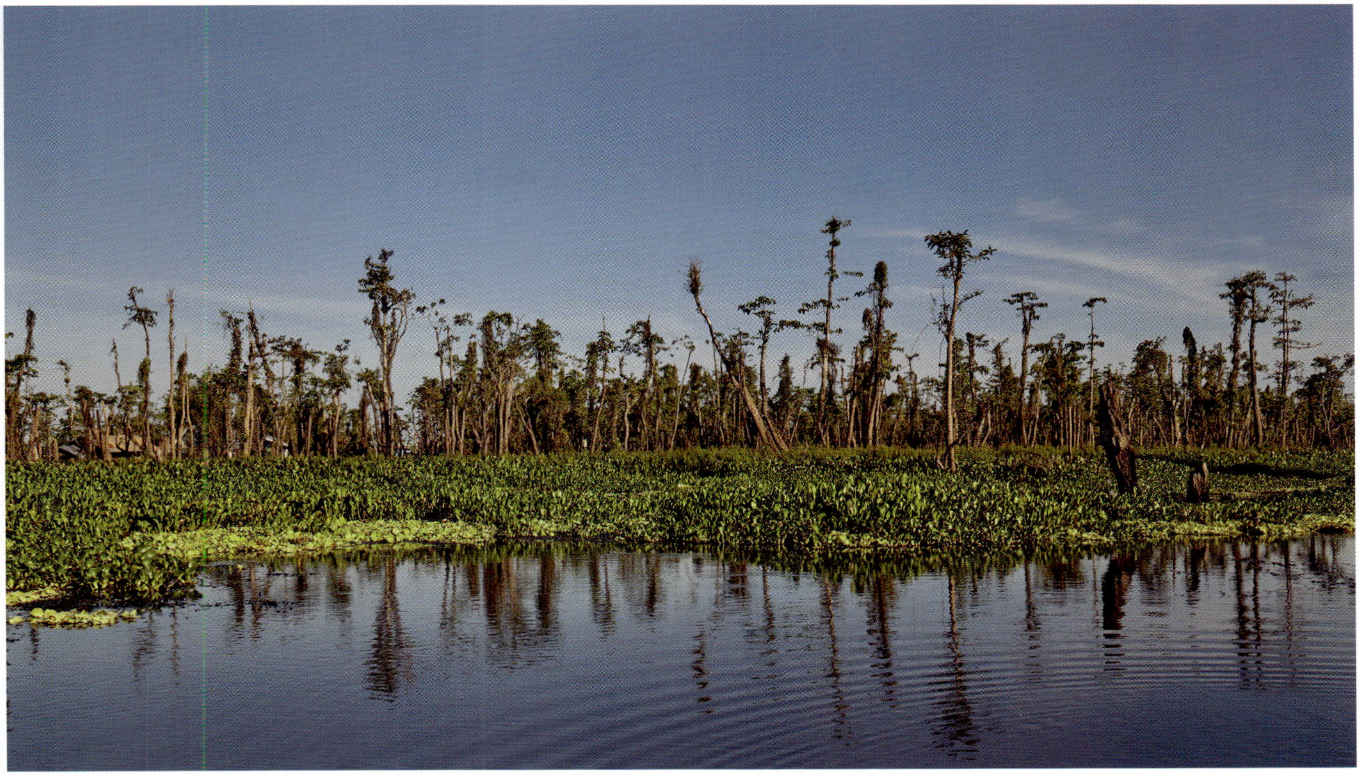

***Above:** Swamp forest is relatively uncommon in the Philippines, but by far the largest area is Agusan Marsh, in eastern Mindanao, a place of lakes and dense forest growing out of the swamp floor.*

Apart from the trees, a wide variety of palms is common in these rainforests, including rattans, an incredibly thorny climbing palm, as well as a host of different types of shrub, liana and vine, and herbaceous plants.

On climbing into the mountains, things start to change above an altitude of about 1,000 m (3,300 ft). As temperatures drop and rainfall increases, so the trees become smaller and the dipterocarps slowly disappear, replaced by increasing numbers of tropical oak and laurel species. The palms also disappear, replaced by tree ferns and many more vines. Mosses start to appear, along with huge numbers of ferns and orchids, mostly growing on the tree trunks and branches as epiphytes, using the trees simply as a convenient perch, but having no actual interaction with them. This is the montane forest, so typical of high-altitude rainforests across much of the Philippines. At the highest altitudes, in what is known as upper montane or mossy forest, from around 1,500 m (4,900 ft) or more, annual rainfall typically rises to a staggering 10,000–12,000 mm (395–474 in), up from a usual 2,000 mm (78 in) in the lowland forests. With temperatures typically down at no more that 12–18°C (54–64°F), and often a lot lower, both growth and decompositon are slow. The trees grow to no more than a few metres high and are often gnarled and twisted; thick wads of moss grow around the branches and hang downwards, constantly dripping wet. In some parts of the Philippines shrubs more associated with the temperate world, such as rhododendrons, appear. The ground is often covered with a thick layer of half-rotted vegetation, sometimes more than 1 m (3 ft) thick. This is a fog-enshrouded, soggy primeval place, very different from the lowland forests far below.

Finally, around many of the highest summits, at an altitude above about 2,500 m (8,225 ft), the montane forest may give way to open grassland, including on a few mountains dwarf bamboo, growing up to about 40 cm (16 in). Probably the mountain summit best known for this is Mt Pulag, Luzon's highest peak, and at 2,922 m (9,584 ft) the third highest in the country.

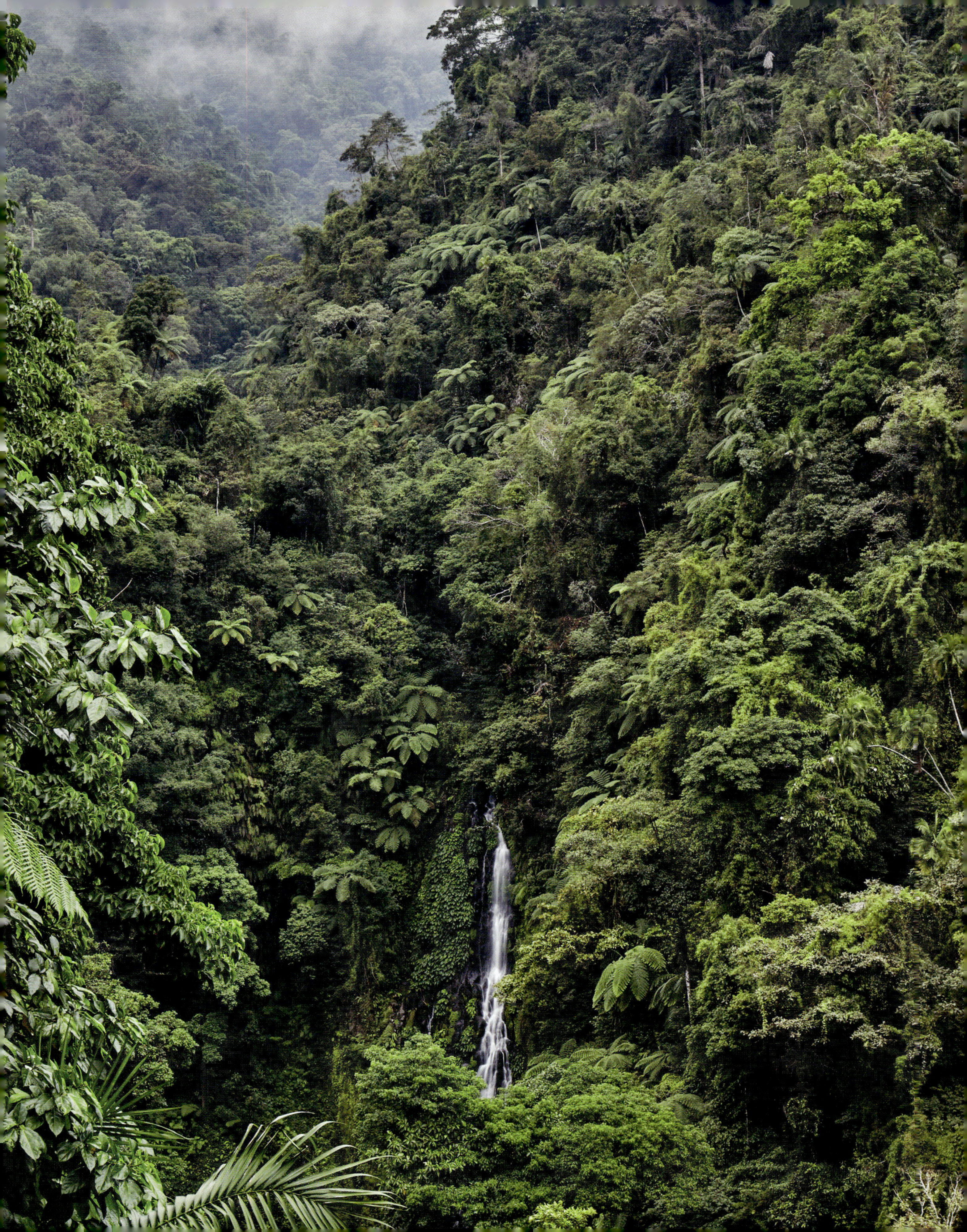

Opposite: *Montane rainforest, typified by large numbers of tree ferns, clings to impossibly steep slopes on Mt Kanlaon, on the island of Negros.*

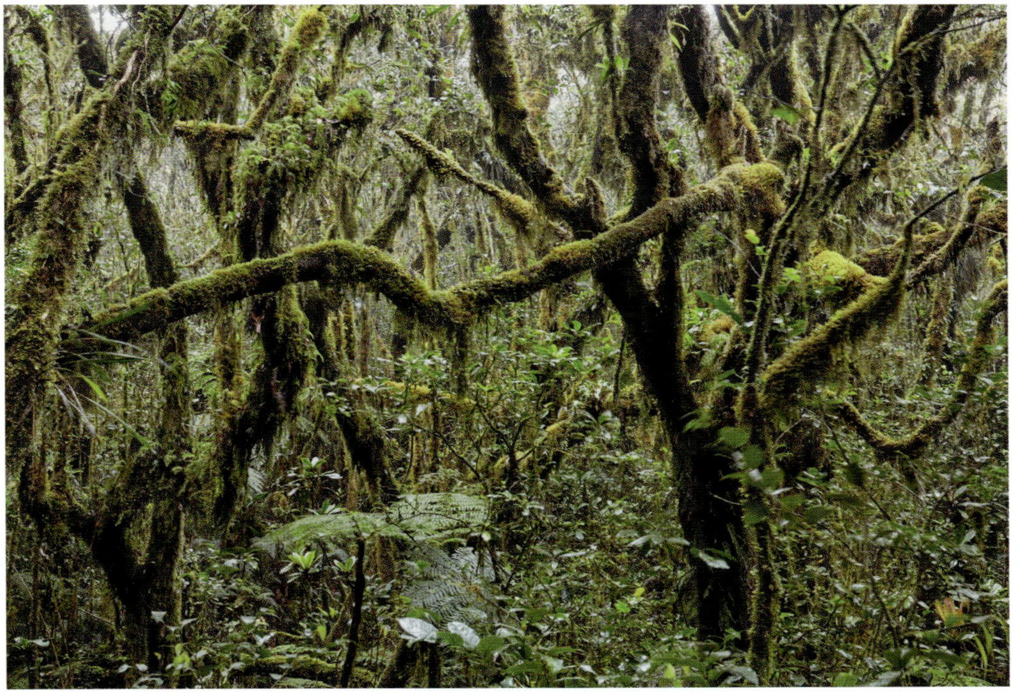

Right: *A classic scene in upper montane, or mossy, rainforest on Mt Pulag, on Luzon, with stunted and gnarled trees festooned with hanging mosses, along with epiphytic ferns and orchids.*

In certain mountain areas, namely in western Mindoro and many valleys and mountain slopes of the Cordillera Central and Zambales mountains of northern Luzon, the usual montane rainforest has been replaced by pine forests. These occur partly within localized rain shadows where rainfall is not high enough to support montane rainforest, but also where repeated man-made fires have destroyed the latter.

A staggering wildlife diversity

With the Philippines' natural environment so dominated by tropical rainforest habitats of one kind or another, it is not surprising that the terrestrial wildlife – both plant and animal – is largely adapted to this forest environment. Furthermore, being an archipelago most of whose islands have had little direct land link with any major land mass, the fauna and flora have evolved down their own paths quite separate from the world at large. Those instances where there has been external input must have been via species that were able to either cross large stretches of open water, or make use of the few land bridges that did open up between islands at various times during the periodic ice ages in the far past, when sea levels dropped significantly.

The result is not only that the Philippine plant and animal wildlife is adapted to forest habitats, but also that it is hugely diverse both as a result of the warm tropical climate and due to variations from one island to another. Moreover, it is also quite distinct from plant and animal wildlife found anywhere else in either mainland Asia or Australasia. In other words, the Philippines has a very high rate of species uniqueness, or endemism. Taken together, all this makes the country one of the world's greatest megadiverse regions.

Unfortunately, this crucial fact was overlooked around the world until comparatively recently, perhaps in part because the Philippines lacks any of the famous large animals – particularly mammals – that are such a popular part of wildlife across much of Asia. Being an isolated archipelago, none of the large, non-flying species was able to reach the Philippines, and typically for an island environment most of the mammals that are found here are rather small and inconspicuous.

The one area of the Philippines that does have a wildlife that overlaps with that of mainland Southeast Asia is Palawan. Sitting in the far west of the country, at various points in the past it did have a land bridge to

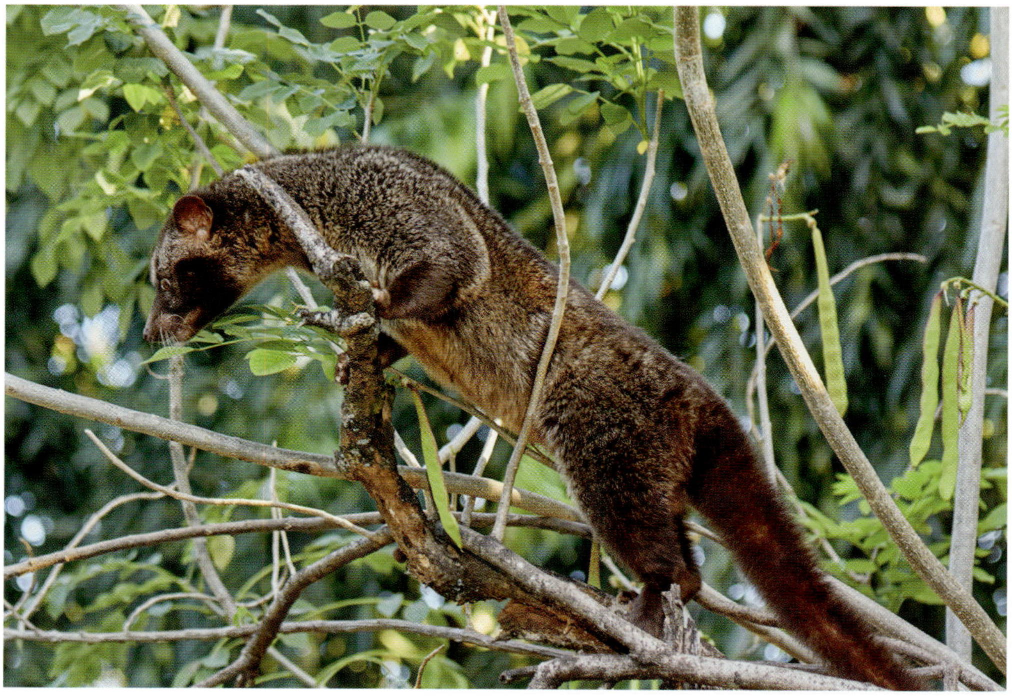

Left: *The Philippine Palm Civet is one of the Philippines' larger mammals, originally introduced to these islands, but long since evolved into a separate species. It is widespread across the country.*

Below: *Fruit bats are among the most common of the Philippines' mammalian wildlife, and among the most important in forest ecology. This image shows a juvenile Island Flying Fox, a species found across Southeast Asia.*

Borneo, which in turn was linked to the mainland, allowing some wildlife to filter across to Palawan. In the rest of the Philippines, a comparatively small number of mammals occur that are identical to those found in mainland Southeast Asia, including Long-tailed Macaques, Malay and Palm Civets, and Leopard Cats, but it is believed that these were brought to the islands by people at some point in the distant past.

A great many Philippine plant and animal species, however, are likely to have been derived from progenitors that, though they made it here from elsewhere originally, have long since evolved into species related to but quite distinct from those found anywhere else.

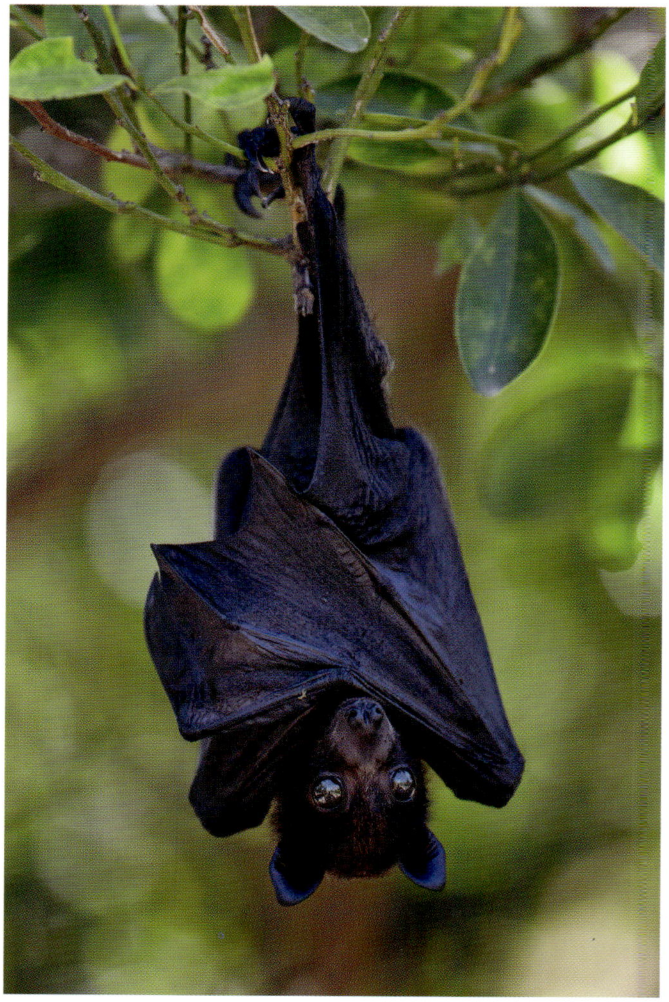

Endemism across the Philippines

A quick glance at the species counts for the Philippines is all that is needed to drive home the uniquness of its fauna and flora. Many bird species can, of course, fly across large stretches of water, creating the possibility that endemism might be comparatively low among this group. Despite this, 190 of the 600-plus species found in the Philippines are endemic (or unique) to the country. Many of these are relatively small forest birds, though the endemic group also includes larger birds such as numerous hornbills and the mighty Philippine Eagle (*Pithecophaga jefferyi*), the world's second largest eagle, as well as several other raptors.

Among the nearly 200 species of land mammal, roughly 60 per cent are endemic. Only one of these is relatively large, though even this is a dwarf species by

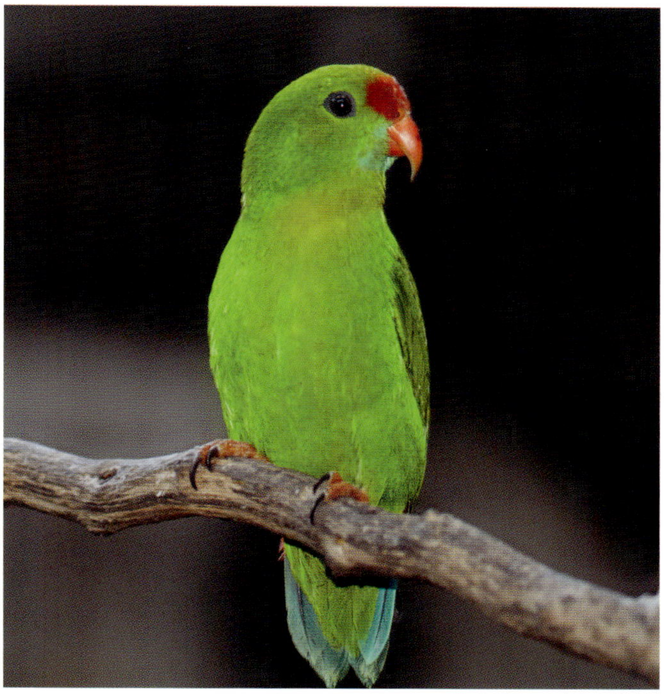

Right: The Colasisi is endemic to the Philippines, and is the country's most common member of the parrot family, spread across the country, except Palawan.

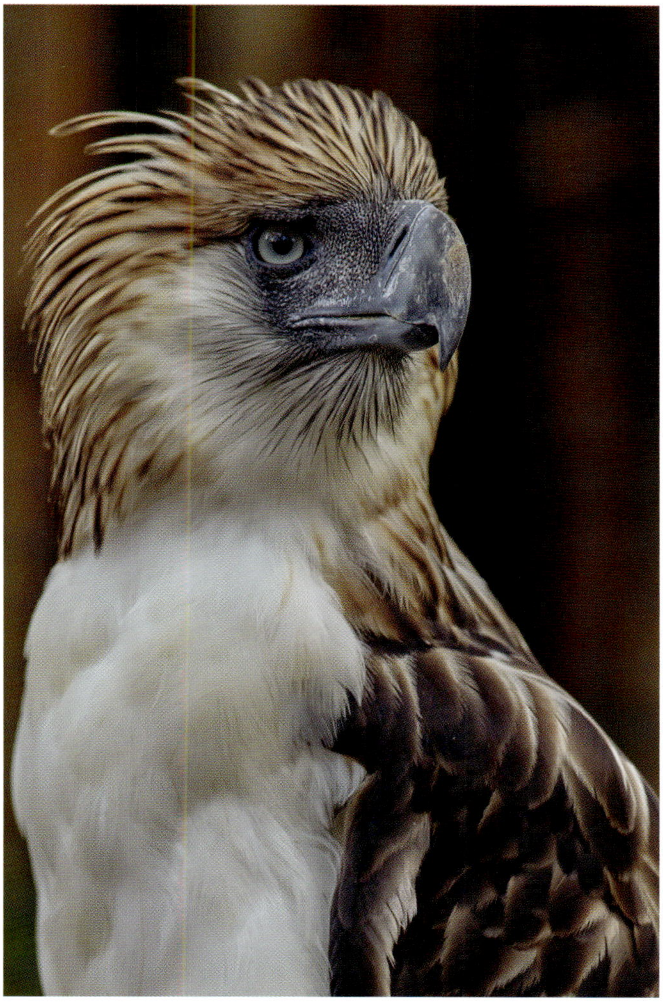

Below: The Philippine Eagle is also endemic to the country. It is seriously endangered, with just a few hundred birds left in the wild.

international standards. This is the Tamaraw (*Bubalus mindorensis*), a dwarf water buffalo endemic not only to the Philippines but actually to just Mindoro. The majority of Philippine endemic mammals are either bats or rodents, the former ranging from tiny insectivorous forms to large fruit-eating species. The rodents are quite varied and remarkable animals specialized to, and often endemic to, different forest areas. What almost all the bats and rodents have in common is a need for good-quality rainforests.

Among the reptiles and amphibians, it is estimated that about 70 per cent of the roughly 350 species are endemic, though it is still hard to be sure as new species are still being discovered quite regularly. This is even more the case for insects, whose vast species count – and endemism with it – is growing almost daily.

Among the thousands of plant species, from trees to mosses, endemism rates are high, running at about 70 per cent, and even higher for some plant families. Many are, nevertheless, related to species found in mainland Asia, among which there is a geographic pattern. A large number of plant species found in the north of the Philippines have relatives in Taiwan and China, while those in the south often have links to species in Borneo, Sulawesi and other islands of

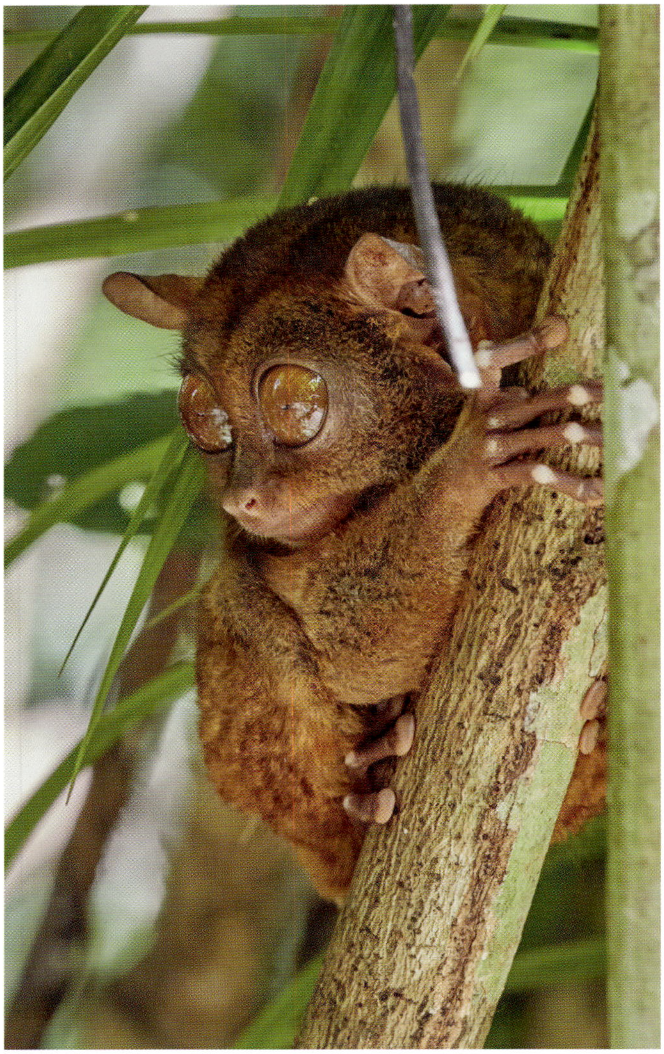

Above: The tiny Philippine Tarsier is the world's smallest primate, and is found only in the south of the Philippines.

eastern Indonesia. Something similar can be found among some animals. The Philippine Tarsier (*Carlito syrichta*), the world's smallest primate at just a few centimetres long, for example, and found only in the south of the country – specifically Mindanao, Bohol, Samar and Leyte – is related to tarsier species found in Borneo and Sulawesi.

Endemism and variation among island groups

Why is it that not only is Philippine wildlife distinct from that of mainland Asia, but also that there are differences from one island group to another within the Philippines? The difference between Palawan and the rest of the Philippines, and the tarsier's restriction to the south of the country, have already been mentioned. But there are many other examples, such as the bleeding-heart pigeons, of which there are five species each restricted to different parts of the country. Similarly, the tarictic hornbills, also with five species, are again restricted to different specific regions.

The explanation is found in changing sea levels and the depth of those inter-island water channels described earlier. During the last ice age, which ended about 10,000 years ago during the Pleistocene Epoch, sea levels were 120 m (394 ft) lower than they are today. When a line marking the 120-m depth point is drawn around the Philippine islands, the country becomes aggregated into six distinct islands, which have been found to correspond exactly with the restricted distributions of the Philippines' localized endemic species. So, during the last ice age animals were able to move around their own large island, but were unable to cross the deep-water channels that still existed and separated each of the major Philippine island regions, resulting in the rise of separate species on each of the islands. Once the ice had melted and the sea level had risen, these larger islands became fragmented into smaller ones, but the record of their unified past has remained in their endemic species distributions.

The six island groups recognized today, called Major Faunal or Biogeographic Regions, are:

- Greater Luzon (Luzon and many of the off-islands, such as Catanduanes, Marinduque and Polillo).
- Mindoro (just Mindoro and a few minor islands).
- The Western Visayas (also called Greater Negros-Panay, and consisting of Negros, Panay, Guimaras, Masbate and Cebu).
- Greater Mindanao (Mindanao, Basilan, Bohol, Samar and Leyte).
- Greater Sulu (the Sulu and Jolo islands).
- Palawan (including the Calamian group and Balabac).

In addition, a number of smaller islands that have remained separate from any of the main island groups from long before the last ice age fall into their own 'mini' faunal regions. Examples include Sibuyan (in the Visayas), Camiguin del Sur (off the north coast of Mindanao), and the Batanes and Babuyan island groups, both off the northern tip of Luzon.

Once this division and grouping of the islands is understood, the distribution of the Philippines' endemic species (or at least its animals) makes complete sense, and underpins all studies into the evolution and distribution of the Philippines' fauna. It is not yet clear how closely the country's plants conform to these divisions.

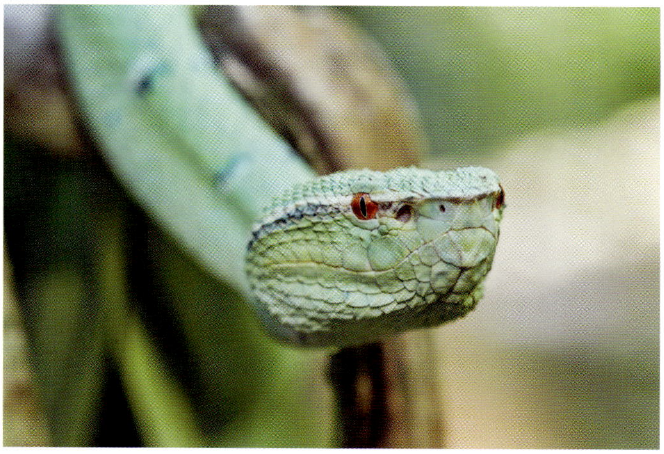

***Above:** The Northern Philippine Temple Pit Viper is a Philippine endemic, found in forests around many parts of the country.*

The Wallace Line and Wallacea

Within the Philippines there is one more important biogeographical division, one that has huge global importance. It is the Wallace Line, named after the nineteenth-century British biologist Alfred Russel Wallace. His studies into the wildlife of eastern Indonesia and New Guinea generated much of the data that allowed Charles Darwin to propose the Theory of Evolution.

In the course of his work, Wallace found that a biological transition zone exists between the wildlife of mainland Asia and that of Australasia, a zone that has become known as Wallacea. The Wallace Line was originally drawn by Wallace himself, with Asian, or Oriental, wildlife to its west and Wallacea to the east. He drew the line running to the south of the Philippines, putting the country firmly in the Oriental faunal camp. However, this was subsequently modified by T. H. Huxley, who drew the line that is mostly used today, running north–south between Palawan and the main body of the Philippines, leaving Palawan in the Oriental region, but moving the main body of the country, often called oceanic Philippines, into Wallacea.

***Below:** The Luzon Forest Frog (*Platymantis luzonensis*) is endemic to the Philippines, and is restricted to forests on Luzon.*

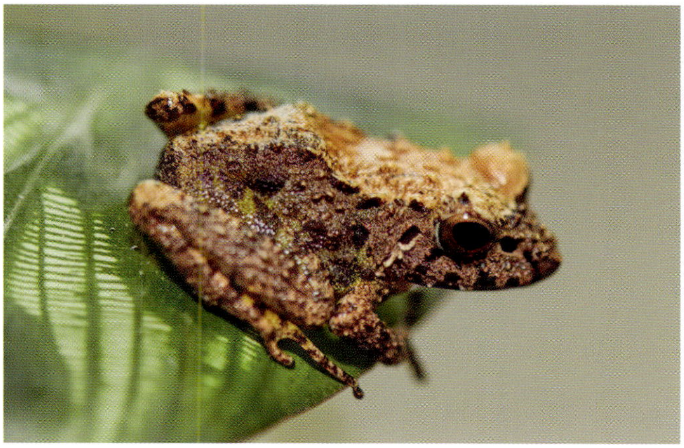

Although the Wallace Line marks the Philippines as part of Wallacea, due to its rather unique wildlife, many scientists still consider the 'true' Wallacean transition zone to be restricted largely to the islands of eastern Indonesia. While the wildlife of much of the Philippines is clearly quite distinct from that of both mainland Asia and Australasia, it is admittedly difficult to see it as transitional between the two. Some Australasian-derived species can be found in the Philippines, but the two quintessentially Australasian groups, namely eucalypt trees and marsupial mammals, failed to make the crossing. The nearest they managed to come was the Talaud Islands to the north-east of Sulawesi, barely 200 km (120 miles) from Mindanao's south coast. So, although the Philippines is formally part of Wallacea, it is clearly rather different from the 'true' Wallacea of eastern Indonesia.

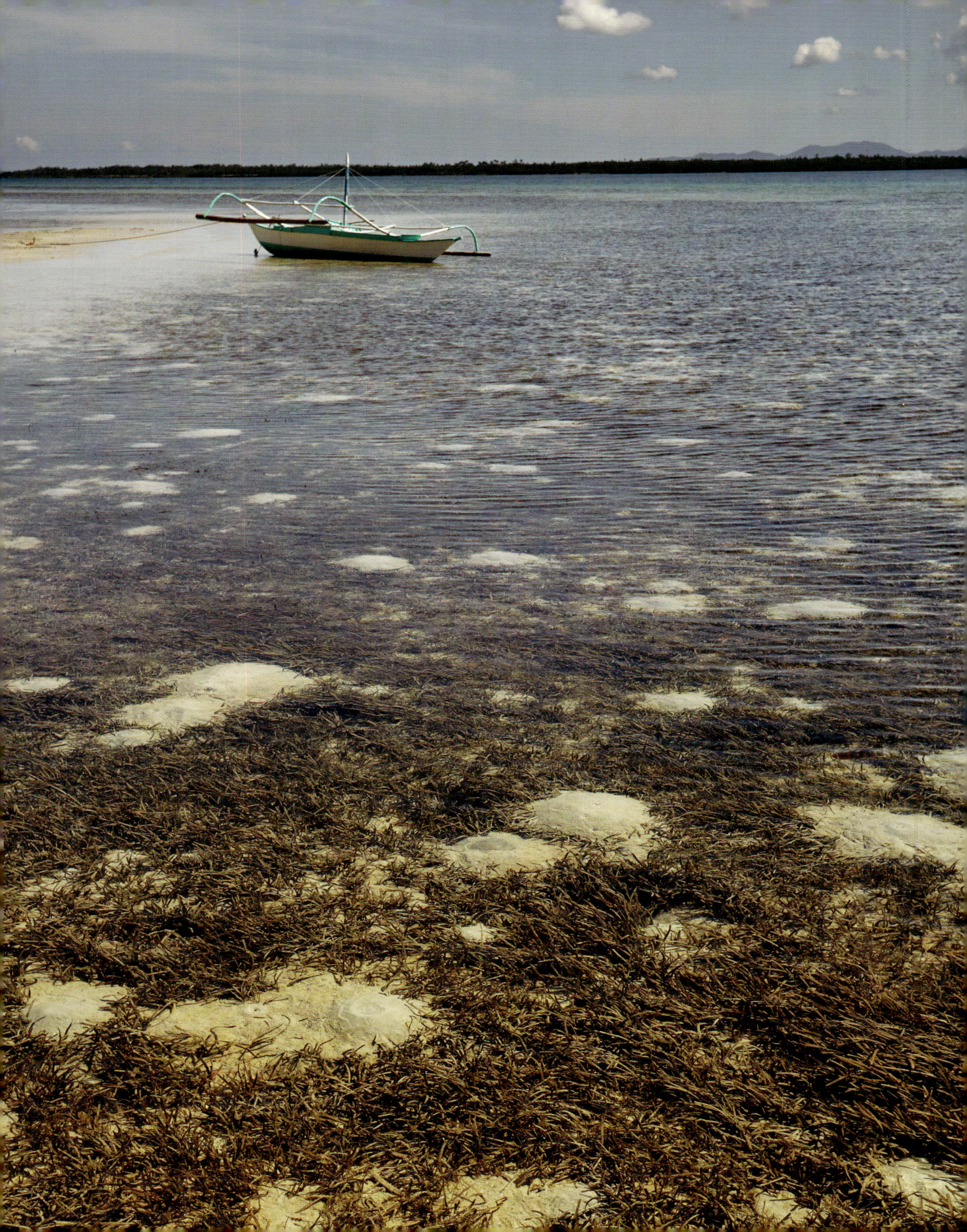

2

Protecting the Treasure House

Philippine conservation past and present

It is an unfortunate fact that the Philippines' natural environment and its wildlife are far from being in a pristine condition. With a human population of more than 100 million and still growing, it is hardly surprising that the country's natural resources are under enormous strain.

Over-harvesting of resources from the wild, conversion of forest land to agriculture and housing, industrialization, pollution, and the struggles of the large numbers of rural and coastal poor simply to eke a survival from the land and sea, have all taken their toll. At times it seems as though the entire biological system is not only badly out of balance but actually teetering on the brink of collapse. If this were to happen, it would be a total disaster, not only for the survival of untold numbers of the Philippines' utterly unique plant and animal wildlife species, but also for the millions of Filipinos who depend on the natural world for their livelihood.

Since the early 1990s there have been growing efforts to launch a widespread and effective conservation programme to save what is left. This is being promoted as both part of an international programme to safeguard as much of the world's biodiversity as possible (the Convention on Biological Diversity), and efforts to develop a sustainable economy. To date, the results have been mixed, but they give cause for hope, and there are good reasons for feeling that as we move into the third decade of the twenty-first century things are nowhere near as bad as they would have been without any conservation programme.

Opposite: *Beds of seagrass, the only flowering plant able to live in seawater, are a hugely important marine habitat, stabilizing sandy or muddy shallows close to the shore, and creating a home to the young of a great range of animal species. Seen here in Honda Bay, Palawan.*

Measuring the bad news

Almost all the headline numbers for the Philippines' natural environment make for depressing reading. Studies conducted since 2000 all show a similar kind of pattern, though the exact numbers may vary slightly. Around the coast, 70 per cent of the coral reefs have been assessed to be in a poor condition, with only 5 per cent deemed 'excellent', as defined by the health of the corals, the amount of live coral cover and the numbers and diversity of fish and other marine life. The general trend still seems to be downwards, though there are some hopeful signs, as described later.

On land, where the vast majority of the country's endemic species live, things are even worse. Natural forest cover has shrunk from an estimated 70 per cent of the total land area at the start of the twentieth century, to just 20–24 per cent (depending on what source you use) at the start of the twenty-first century. Less than half of this is thought to be 'old-growth' or primary forest, meaning forest that has never been cut and is therefore in as near pristine a state as is possible. The remainder is secondary forest, areas that have been logged and are now recovering. There are also some areas of forest plantations.

Despite a ban on logging introduced in the 1990s, loss of natural forest cover continues, at roughly 2 per cent loss per year. This is the result mainly of continued logging in areas licensed by the government before the logging ban came into effect, illegal felling of trees elsewhere and the rather general clearance of forest for agriculture.

In a country in which most of the terrestrial wildlife is dependent on healthy forest this is very bad news. It is bad enough for animals and plants that can be found in other parts of Asia, but it is truly serious for endemic species, and potentially catastrophic for those that are limited to just one or two islands, or even areas within a single island. A number of species are believed to have already become extinct, and many more are teetering on the brink.

The situation is even worse for plants than it is for animals. More than 500 plant species are threatened with extinction, mostly dipterocarp trees, orchids and palms, as opposed to a little over 200 animals. The most seriously afflicted animal group is the birds, with nearly 130 threatened species (out of a total of just over 600), though this may be a reflection of the fact that the Philippines' birds are the most well-studied of the country's animal wildlife.

For most threatened animal species, loss of healthy forest habitat is the primary reason for their demise. Just a few examples include the Visayan Tarictic Hornbill (*Penelopides panini*), whose forest ranges across the Western Visayas have largely disappeared; the Philippine Crocodile (*Crocodylus mindorensis*), once widespread in waterways across the country, but now restricted to a few very small areas, particularly in the Sierra Madre mountains of northeastern Luzon; the Tamaraw (*Bubalus mindorensis*), the already mentioned dwarf buffalo that hangs on in a few pockets of Mindoro; and the mighty Philippine Eagle (*Pithecophaga jefferyi*), the world's second largest eagle which survives only in good-quality rainforest in Mindanao, Samar, Leyte and northern Luzon's mountains.

What constitutes good-quality forest varies for different species. While many seem to be quite happy with reasonable secondary forest, others require primary rainforest that has never been logged.

Loss of forest habitat

When it comes to loss of habitat, it is the lowland forests that have suffered the most. These areas are, of course, the most accessible and had the largest trees (the dipterocarps), which were therefore of greatest interest to the timber industry.

The removal of the best trees, and the construction of logging roads deep into the forests, then enabled land-hungry farmers to move in to finish the job, clearing the remaining forest and converting the land to farming. It is a sequence of events that over the years has been repeated time and again, to the point now that most surviving forest areas are either very remote or are on steep mountain slopes high above the

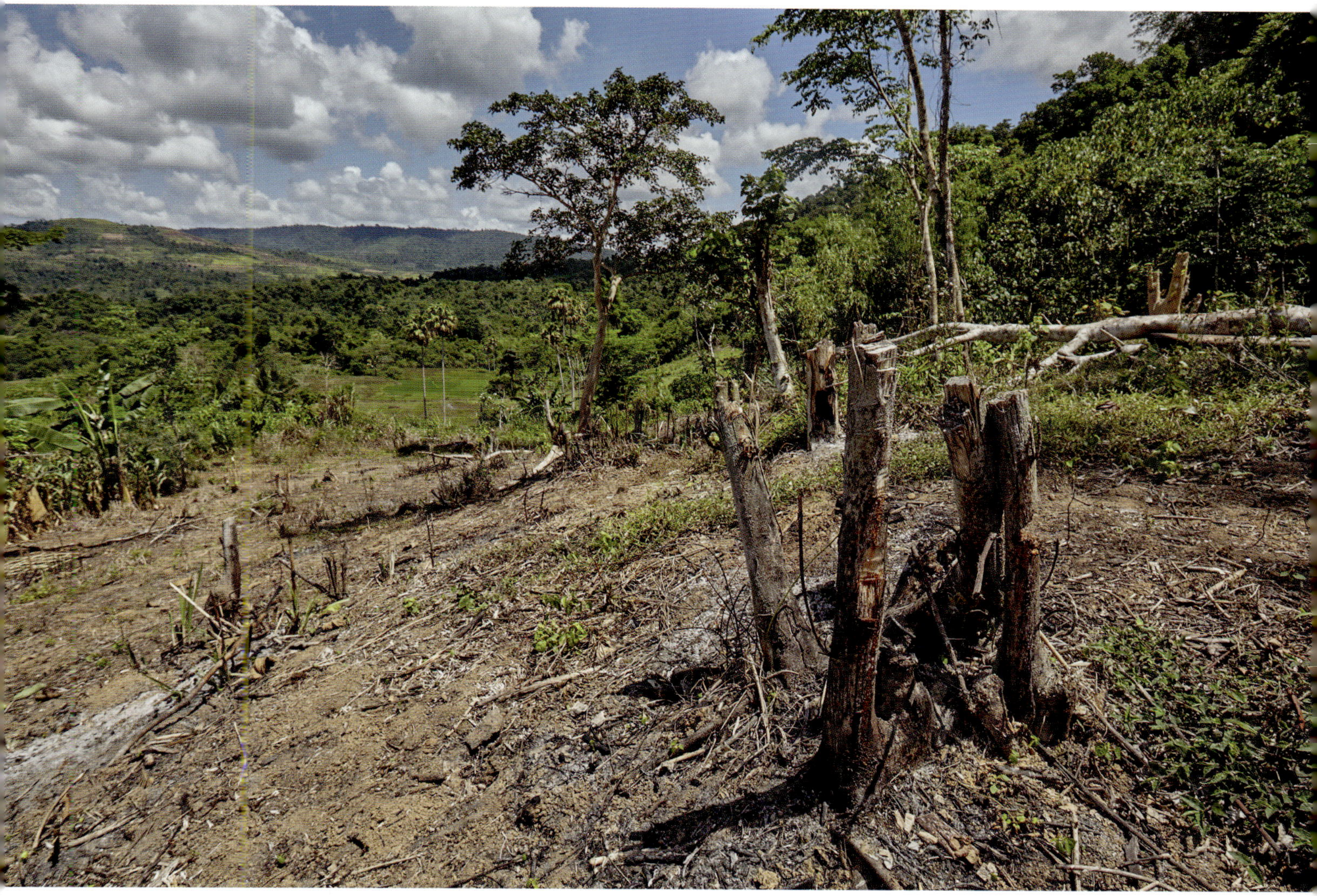

Above: Illegal forest clearance for agriculture is still rampant even within protected areas. Seen here in the Northern Sierra Madre Natural Park, in north-east Luzon.

Below: The illegal felling of trees for timber remains a big issue inside protected areas. Here a valuable lauan (Shorea species) tree has been cut down in the heart of Samar Island Natural Park.

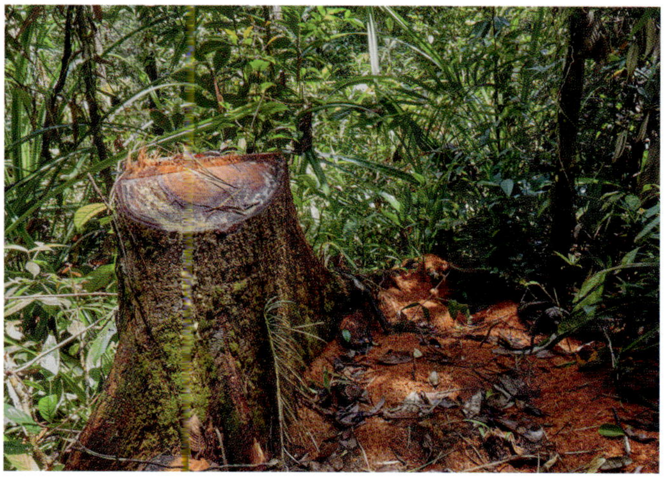

more fertile lowlands. It is difficult to blame the farmers. Most are poor, their numbers are continually growing and they are denied access to the best lowland farmland – the areas with the gentlest slopes, easiest access and best soils – as much of this land is dedicated to plantation crops (such as sugar cane) grown largely for export and owned by a small proportion of the wealthiest class.

Even though commercial logging has stopped in most parts of the country, the loss of forest to agriculture continues, moving increasingly into the mountains. With the rural population still growing and

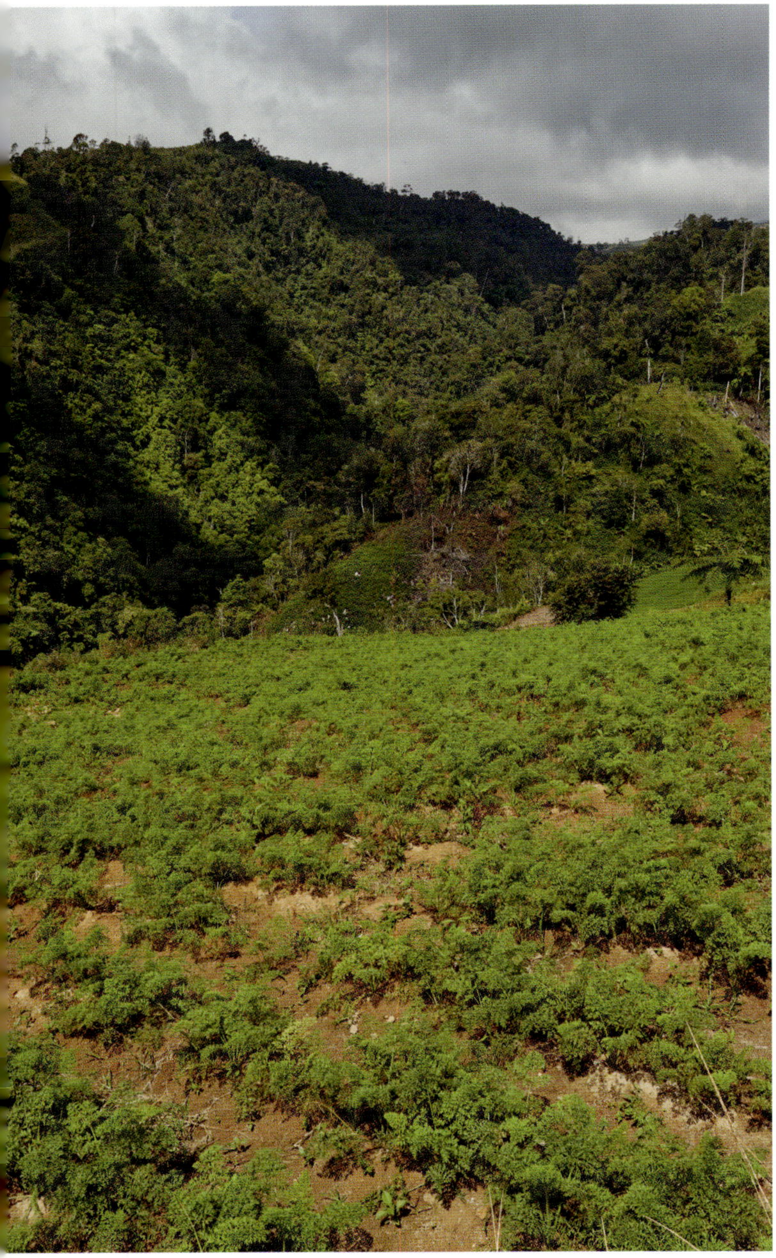

Above: *Farmland regularly encroaches well into many protected areas, seen here inside Mt Apo Natural Park, in Mindanao.*

fertility of farmland already in use declining, there is inevitably pressure to move ever higher up the mountains, cutting further into the remaining forest and attempting to farm steeper, less stable soils, exposed to higher amounts of rainfall and lower temperatures. This inevitably means ever greater inroads into the high-altitude montane forests, and in recent years these have been eaten away at an ever-greater speed. With forest lost from such steep, rain-soaked slopes, the inevitable result is an increase in landslides, especially during the typhoon season. Since the late 1980s there has been a growing incidence of major landslides (with considerable loss of life), and almost all of them have been firmly attributed to deforestation of nearby hills.

Decline of the marine environment

It has been estimated that more than 60 per cent of the Philippines' coastal population depends on the coral reefs, which supply about 70 per cent of the available fish. As is the case with the forests, the reefs' decline is a serious issue not just for biodiversity reasons but also for the economy of one of the poorest sectors of Philippine society.

Again as is the case with the forests, the causes of the problems are multiple and complex, and not all the causes are within the Philippines' control. Those that are the results of local activity include over-harvesting of the fish, sometimes with the use of destructive fishing methods, directly damaging reef structure and reducing both biodiversity and the sustainability of fish supplies. The destructive methods usually consist of explosives and cyanide, the former – produced using fertilizers – designed to make fish float to the surface, but at the same time, not surprisingly, causing massive damage to the reef, which takes years to recover from each blast. Cyanide – usually squirted by a diver into a reef's crevasses – is designed to stun fish that can then be removed by hand and used mainly in the aquarium trade. Unfortunately, while the targeted fish are usually not killed, surrounding corals often are. While both destructive methods are illegal, and have been for quite a number of years, they still continue in some parts of the Philippines, though on what scale is not entirely certain.

The widespread over-harvesting of fish, whether by legal or illegal means, has caused a big drop in biodiversity and fish availability across the majority of reefs. Not only does this make it harder for fish populations to sustain themselves, to generate large individuals that form an essential part of a sustainable fishery, but it can also upset the delicate balance that exists between the fish and corals. Removal of some fish species, such as the Humphead Wrasse (*Cheilinus undulatus*), which feeds on the Crown of Thorns Starfish (*Acanthaster planci*), a voracious eater of corals, can result in proliferation of the latter, and then widespread damage to a reef.

Siltation and loss of water clarity are also major issues that can rapidly cause widespread loss of a reef's health. Their main causes are a combination of pollution from coastal development, whether industrial, housing or tourism infrastructure, and soil erosion as a result of inland deforestation. Both causes are generally on the increase across the Philippines.

Away from the corals, the seagrass beds have also been in decline, the current estimated area over the country of a little more than 27,000 km^2 (10,400 mi^2) representing a 30–40 per cent decline on what was believed to be present 50 years ago. If true, this means a massive loss of potential nursery grounds for juvenile fish and other marine life. The main causes, again, are probably siltation and pollution, not only making photosynsthesis difficult but also actually smothering the plants. Many seagrass beds have also been lost through land reclamation across shallow areas of water, as well as conversion to deeper fish ponds for aquaculture.

Similarly, mangroves have suffered significantly, with their area decreasing from an estimated 450,000 ha

Below: *The collection of shellfish in the shoreline shallows is common among many poor coastal populations.*

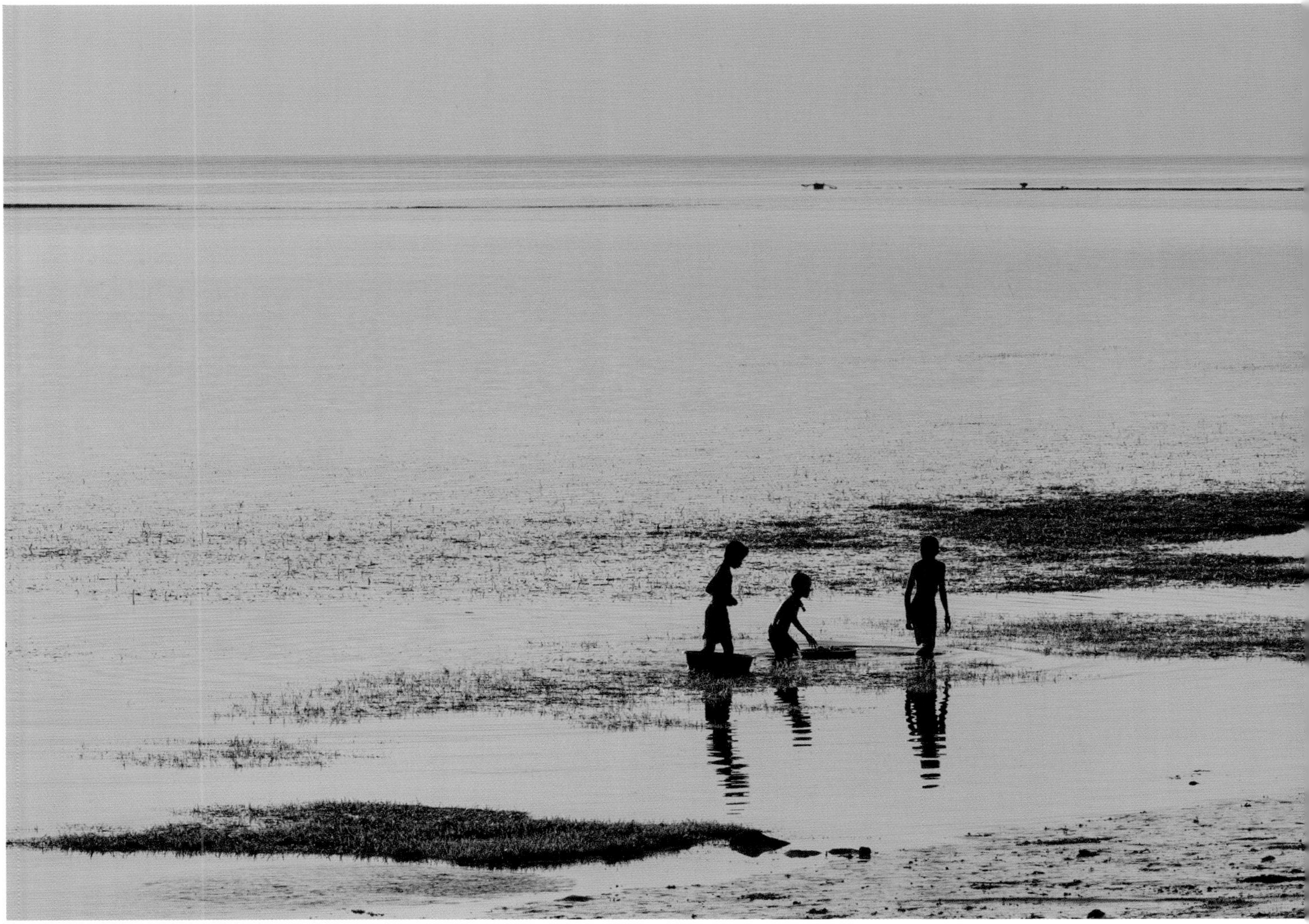

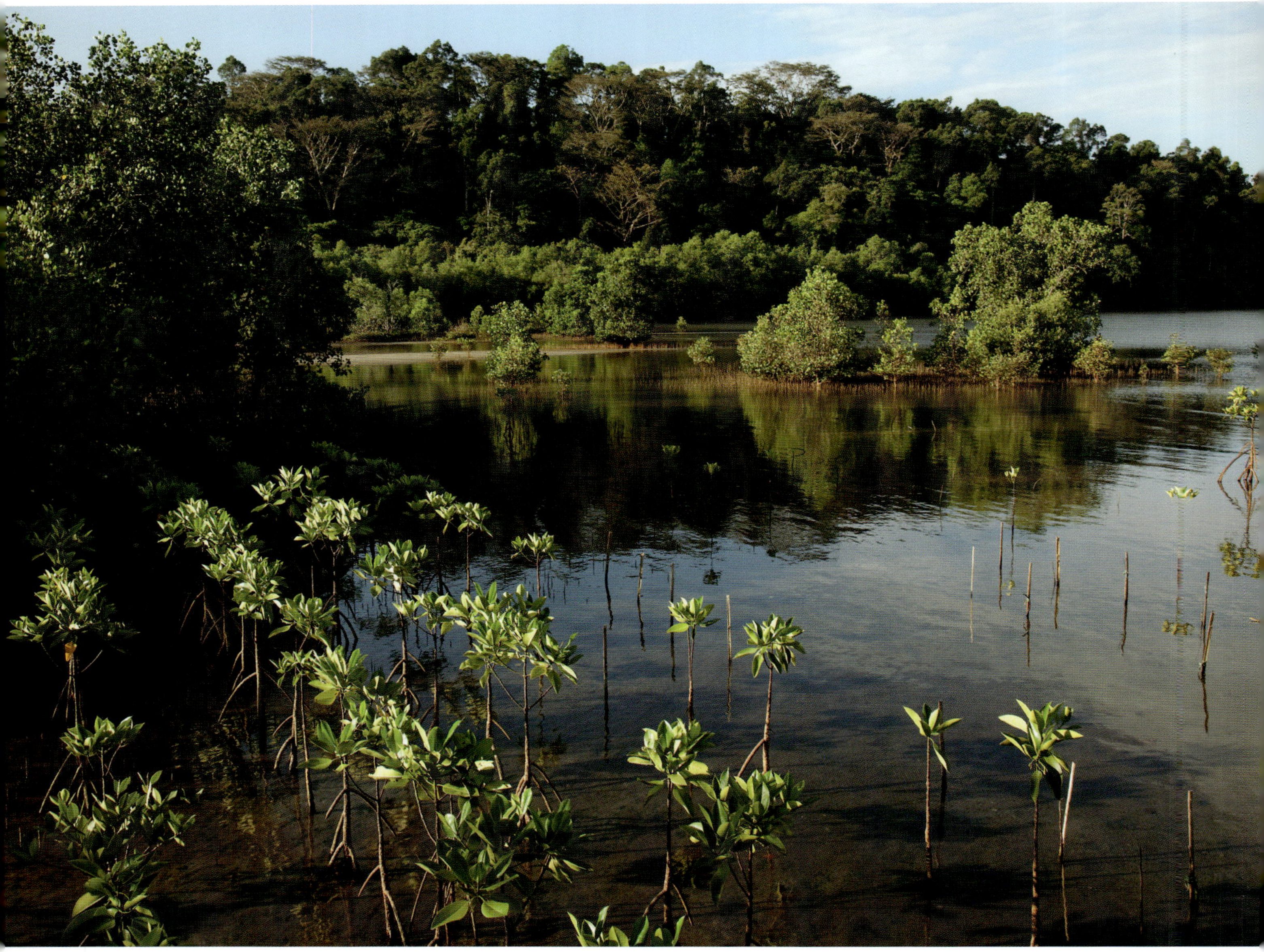

(1,111,500 acres) early in the twentieth century to about 310,000 ha (765,700 acres) today (see also page 59). Hugely important both as a nursery for juvenile fish and for protecting the coast from erosion by the sea, their decline could have an enormous impact on fisheries and the coastal environment. Their decline is due to a combination of cutting for production of both building materials and fuel (particularly charcoal) for largely impoverished coastal villages, and industrial-scale clearances for conversion to fish and prawn farms.

However, the mangroves do represent a bright light in the marine environment, in that over recent years there have been growing efforts to recreate lost

Above: *Widely decimated in the latter years of the twentieth century, mangrove forests are now undergoing extensive restoration.*

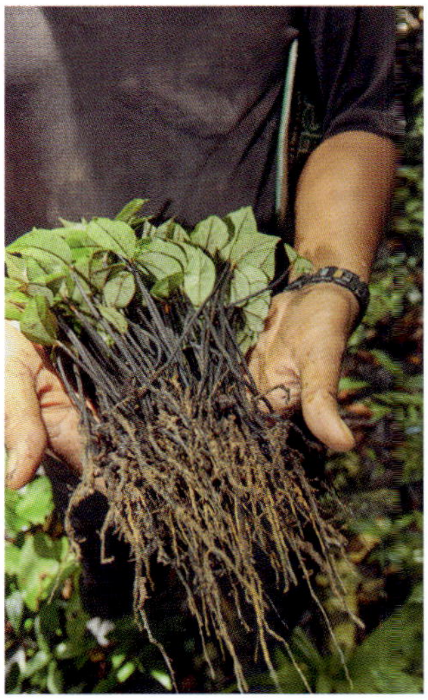

Right: *Increasingly, forest restoration programmes are making use of tree nurseries that cultivate 'wildlings', tree seedlings gathered from the forest, as seen here in Samar Island Natural Park.*

mangrove forests through both planting and management of natural regeneration. As a result, it is not at all uncommon today to see mangroves in places that in the 1990s had been cleared or badly damaged. This is reflected in the current estimated coverage, which represents a significant increase on pessimistic estimates of future coverage made in the 1990s.

Causes of the marine decline that are beyond the Philippines' control are principally of importance to the coral reefs, and are the result of climate change, following from global warming. The principal issue is rising sea temperature, something that can lead to coral bleaching, which is covered in more detail in Chapter 3.

Protecting what is left

The Philippines first developed a network of national parks in the 1930s, while under American colonial rule, making it the first country in Asia to have any kind of conservation programme. However, during the Japanese occupation of the Second World War and the following first few decades of independence, little heed was paid to the parks' existence, leading to logging across many supposedly protected areas. This, combined with the spread of farming, reached a crescendo of destruction in the 1970s and '80s, during the rule of the dictator Ferdinand Marcos, when corruption and self-enrichment caused logging on a massive scale, the income from the timber going to a relatively small number of individuals, not to the country as a whole. As a result, many protected areas became simply paper parks, with boundaries that existed on paper but with no real surviving presence on the ground.

Following the 1986 ousting of Marcos by the People Power Revolution, a huge amount of work went into building a new constitution and a whole new civil society. At the same time, by the late 1980s international scientists were starting to realize just how critical the Philippines is to global biodiversity, leading to growing efforts to rejuvenate the remnants of the country's protected areas. The Philippines became recognized as both a megadiverse region and a biodiversity hotspot, the former denoting its huge species count and diversity, the latter its vast number of seriously threatened endemic species.

In 1992, virtually on the eve of the famous Rio Earth Summit, the Philippines' National Integrated Protected Areas System (NIPAS) law came into force, establishing a whole new conservation and protected areas programme. It has remained the basis of the Philippines' protected areas system ever since, coupled with the country's 'membership' of the Convention on Biological Diversity (which spun off from the Rio Earth Summit), obliging nations to work towards understanding and protecting their most important species and wild areas.

The mid-1990s saw the launch of two large internationally funded and managed conservation programmes: the World Bank-backed Conservation of Priority Protected Areas Program (CPPAP), and the European Union-backed National Integrated Protected Areas Program (NIPAP). The two projects, which ran until the early 2000s, both created a number of new protected areas, and incorporated several of the most important and largest surviving ones from the

Below: *With signage proclaiming protected area boundaries, farmers have little excuse for claiming they do not know where the boundaries are.*

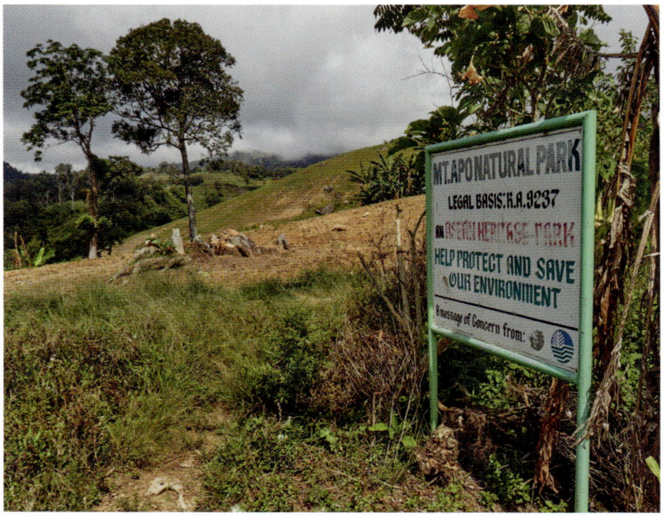

previous system. The result was 18 internationally supported and well-funded large protected areas, aimed at conserving the greatest range of species possible, located throughout five of the country's major faunal regions and incorporating most of its different habitat types.

While these projects were largely successful, once they came to an end in the early 2000s, the protected areas involved suffered the same problems as all those inherited from the pre-NIPAS system; namely lack of funding and expertise, an issue that continues to dog most protected areas to this day.

Today, there are roughly 240 protected areas covering 5.4 million hectares (13.3 million acres) spread across the country officially incorporated under the NIPAS banner, a mix of new areas and many inherited from the old system. Most are terrestrial, but a number – covering over a million hectares of the total – are marine. These are all under the overall management of the Department of Environment and Natural Resources (DENR), and in particular its Biodiversity Management Bureau (BMB), originally called the Protected Areas and Wildlife Bureau (PAWB).

Within the umbrella title of 'protected area', the system has created several different area types, including strict nature reserves, natural parks, wildlife sanctuaries, protected landscapes and seascapes, and resource reserves. Interestingly, there is no specific 'national park' designation, though this is still used in the title of some protected areas (such as Mt Pulag National Park) inherited from the pre-1992 protected areas system.

Hopefully, the old issue of paper parks will slowly pass into history as management, enforcement and cooperation with local people slowly improve. That said, many of the protected areas still include large areas that were cleared in the past, and still suffer regular encroachment and steady, inexorable 'nibbling away' at the edges of what remains. One classic example is the Northern Negros Natural Park, whose apparently impressive size, at over 88,000 ha (217,360 acres), actually includes cleared farmland that amounts to over half of the park's area.

Contradictions and complications in today's protected areas

From the early 2000s onwards, a number of problems started to become apparent in the NIPAS system. Firstly, the necessary laws required for each protected area have proved to be rather cumbersome, bureaucratic and expensive. Similarly, the process by which many protected areas can access funding has been bureaucratic, something that has stymied their management and successful protection. Moreover, the NIPAS law has come increasingly into conflict with laws that control fisheries and the rights of the Philippines' many indigenous peoples. The Fisheries Code demands that 10 per cent of coastal areas should become protected areas, but there has been considerable disagreement as to who is responsible for establishing these, with the DENR and the Bureau of Fisheries and Aquatic Resources (BFAR) frequently being at loggerheads.

On land, the creation of protected areas and/or the attempted enforcement of conservation laws by the DENR has faced considerable opposition from the indigenous peoples who occupy many of the regions to be protected. Understandably, they resent the imposition of central government controls that often run counter to their own traditional ways of conserving their areas, not to mention the rights they have, guaranteed under the Indigenous Peoples Rights Act.

Worse still, from the biodiversity conservation perspective, the enormous amount of wildlife research that has been conducted since the 1990s has shown that there is a significant mismatch between the location of many protected areas and those regions richest in wildlife biodiversity – known as Key Biodiversity Areas (KBAs) – where protection is therefore most urgently needed. In fact, less than 50 per cent of all protected areas cover any KBA, and in total those 5.4 million hectares of protected areas cover just 35 per cent of all the KBAs so far identified.

The way forward

Recognition that all the above problems needed solving saw the launch in 2010 of a new programme,

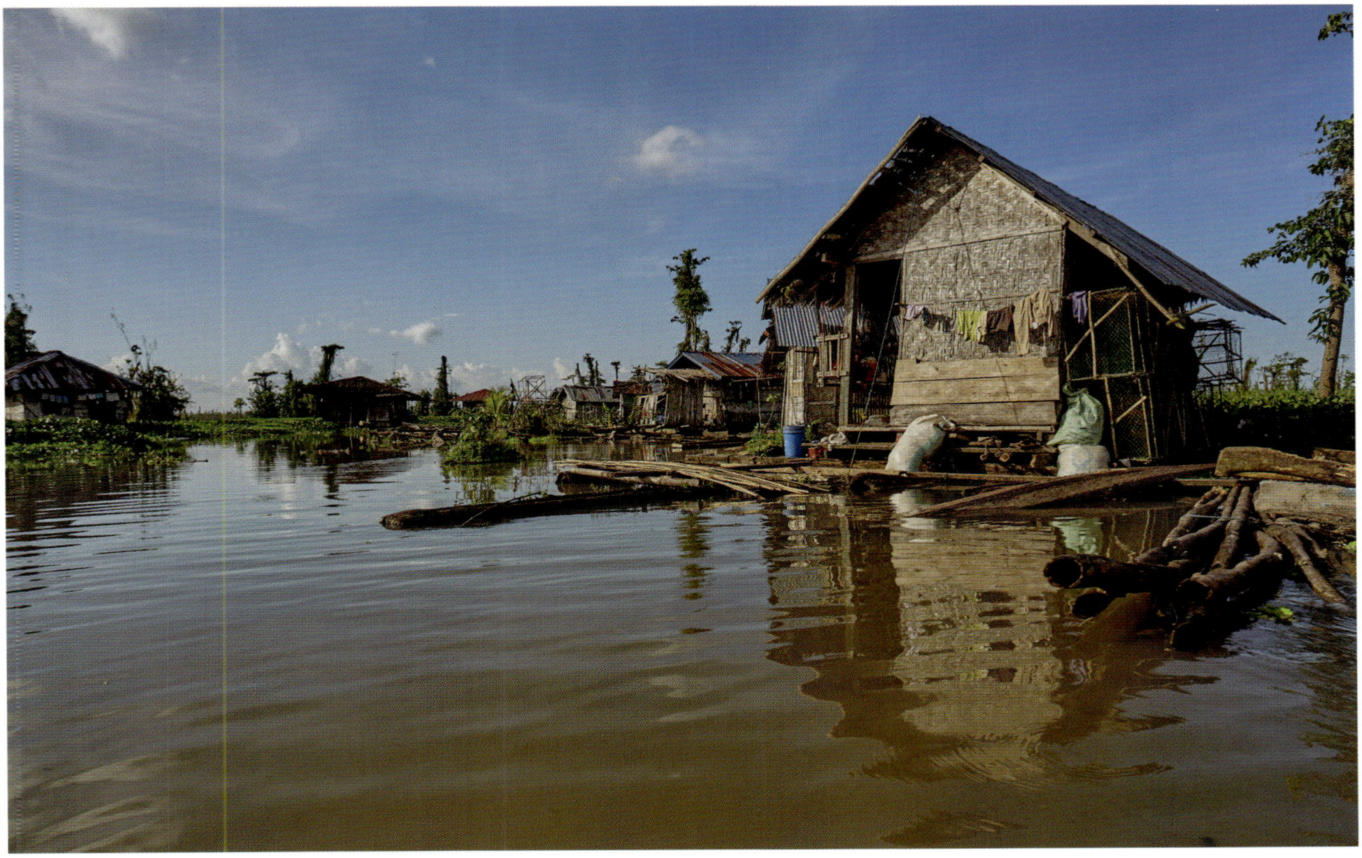

Above: *Many of the Philippines' cultural minorities live within protected areas, giving them a special role in environmental protection. Such is the case for the Manobo people, many of whom live in homes that float on the waters of Mindanao's Agusan Marsh.*

called the New Conservation Areas in the Philippines Project (NewCapp), which ran until 2015. It aimed to create a series of new protected areas that would cover 400,000 ha (988,000 acres), developed under the NIPAS umbrella but managed through a series of innovative solutions. Over the five years, the project worked in 14 sites across 10 KBAs throughout the Philippines, and has resulted in the creation of several new protected areas from northern Luzon to the Sulu Islands.

The result has been decentralization, moving control away from central government to local bodies and stakeholders, with two new categories of local protected area management: Indigenous Community Conservation Areas (ICCAs) and Local Conservation Areas (LCAs), the former allowing indigenous communities to protect their areas using largely their own traditional means, the latter giving management to a combination of local government and local conservation non-government organisations (NGOs).

Along with this has been the creation of a new category of protected area, the Critical Habitats (CHs), which allows important KBAs, or parts of them, to be taken into protection using one of the above management methods. The largest such CH protected area created so far is Cleopatra's Needle, one of Palawan's highest mountains, whose slopes are cloaked in rich primary forest, a home to an estimated 85 per cent of Palawan's mammal and bird species. This new protected area was declared in 2017 and will protect 41,350 ha (102,134 acres).

Coupled with this, a plethora of NGOs have taken or been given responsibility for conservation in a number of specific areas. A few examples would start with the some of the most prominent, such as the Haribon Foundation and the Philippine branch of the World Wide Fund for Nature (WWF), both of which work throughout the country. Some smaller bodies include

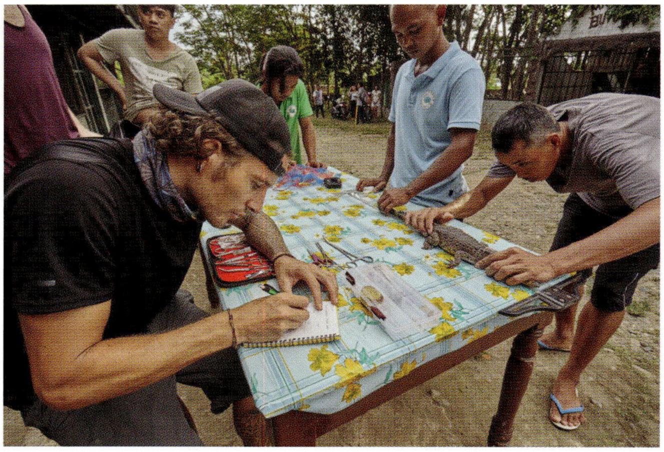

Above: Mabuwaya Foundation staff measure a two-year-old Philippine Crocodile before releasing it into the wild in the Northern Sierra Madre Natural Park.

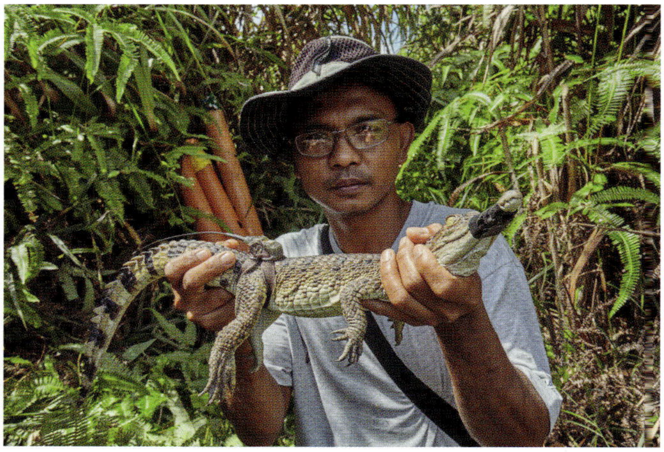

Above: A Mabuwaya Foundation member holds up a juvenile Philippine Crocodile minutes before releasing it into a lake in the Northern Sierra Madre Natural Park.

the Katala Foundation, originally set up to protect the critically threatened Philippine, or Red-vented, Cockatoo (*Cacatua haematuropygia*) in Palawan, though it has since broadened into wider conservation projects in the province; the Mabuwaya Foundation, largely responsible for work to save the Philippine Crocodile principally in the Sierra Madre mountains of north-east Luzon; the Talarak Foundation (which now incorporates the previously separate Negros Forests and Ecological Foundation Inc), which runs a number projects in Negros and Panay; and of course the Philippine Eagle Foundation, which works to save the Philippine Eagle from its base in Davao.

Below: A ranger station inside Mt Pulag National Park, on the shore of Lake Tabeyo, an area where the park's forest has long since been converted to farmland.

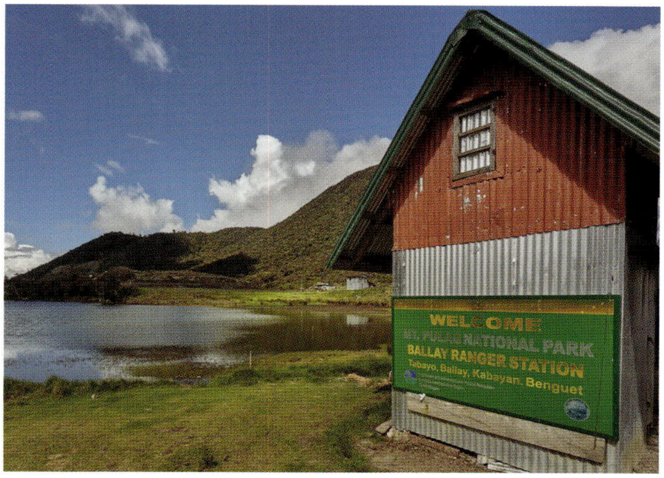

In the marine environment, literally hundreds of protected areas have been created, mostly as local initiatives and often by fishing communities that have realized the benefits of creating even quite small no-take zones that help as fish breeding grounds to maintain a supply to neighbouring fishing areas. As a result, the Philippines has more marine protected areas than any other country in Southeast Asia. Because these reserves are usually very small the total area protected is still rather restricted, and there is quite some way to go before the combined efforts (and hopefully cooperation) of local communities, conservation NGOs and the various government bodies can create a significantly effective marine protected area network.

That said, the impetus provided by the Coral Triangle Initiative, along with the large marine protected areas already in existence, such as Tubbataha Reefs Natural Park, Apo Reef Natural Park, El Nido-Taytay and Malampaya Sound Managed Resource Reserves, and a number of very popular scuba-diving areas such as those around Bohol and Puerto Galera, will give viable models for future marine protected areas, as well as strong economic

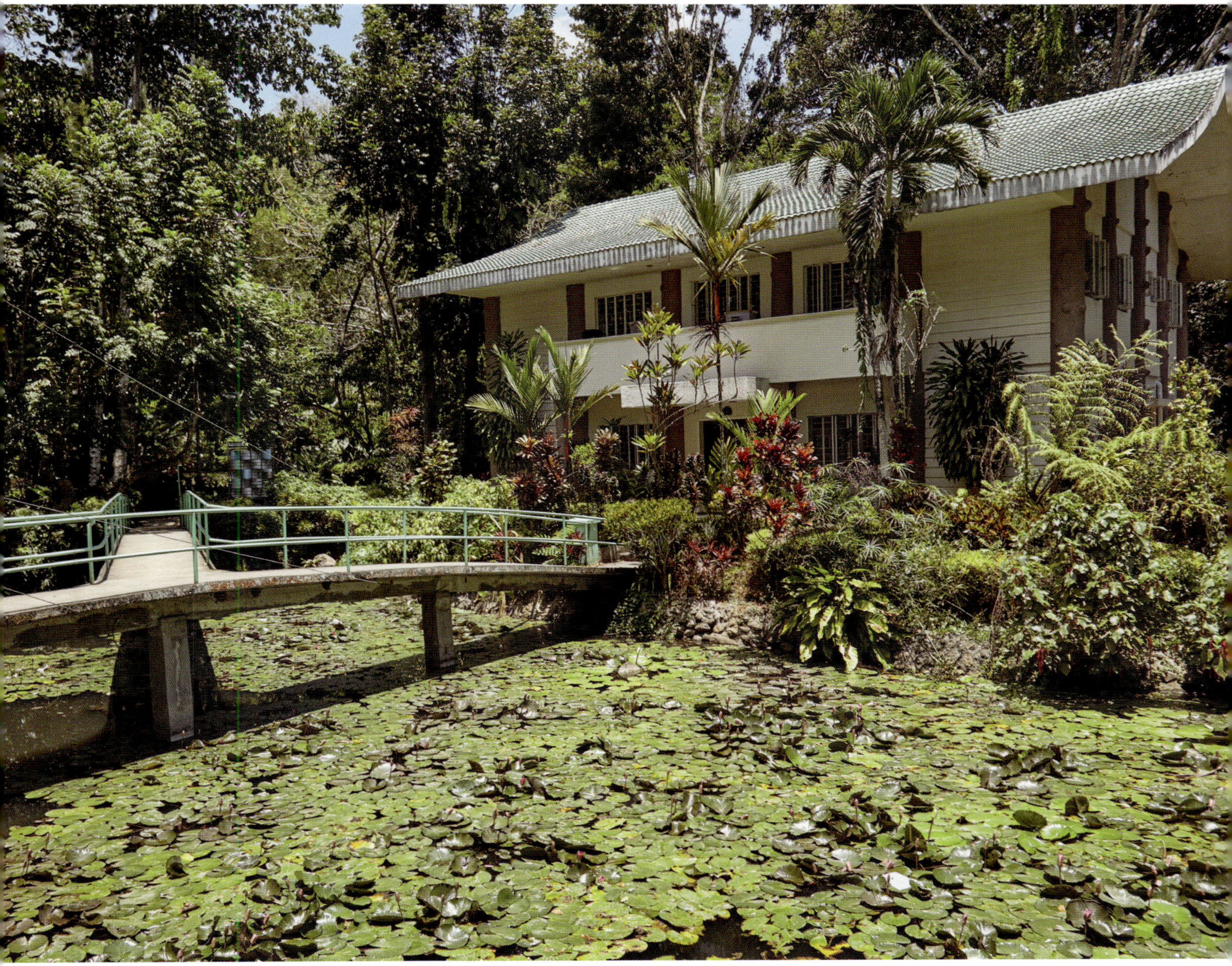

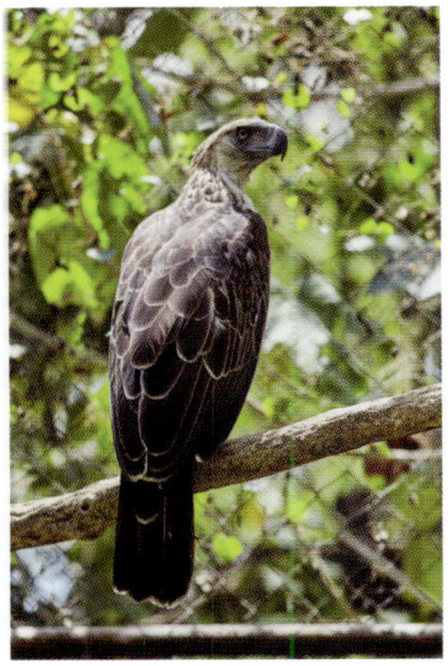

Above: Headquarters of the Philippine Eagle Foundation, near Davao in Mindanao. PEFI is one of the country's leading conservation NGOs, working to rescue the Philippine Eagle from extinction.

Left: One of the Philippine Eagles, sitting inside a flight cage, that can be seen at PEFI's Davao headquarters.

and conservation reasons for enacting greater protection. Ultimately, the most successful conservation measure would probably be the lifting of millions of coastal and rural poor out of their poverty trap. There are signs that the huge numbers of overseas Filipino workers may be starting to achieve just that, with the income they are able to send back home (and which today is by some margin the country's largest earner of foreign currency) helping to give family members some choices. One of those choices may be to move back down those rain-soaked mountain slopes, leaving the forests to regenerate.

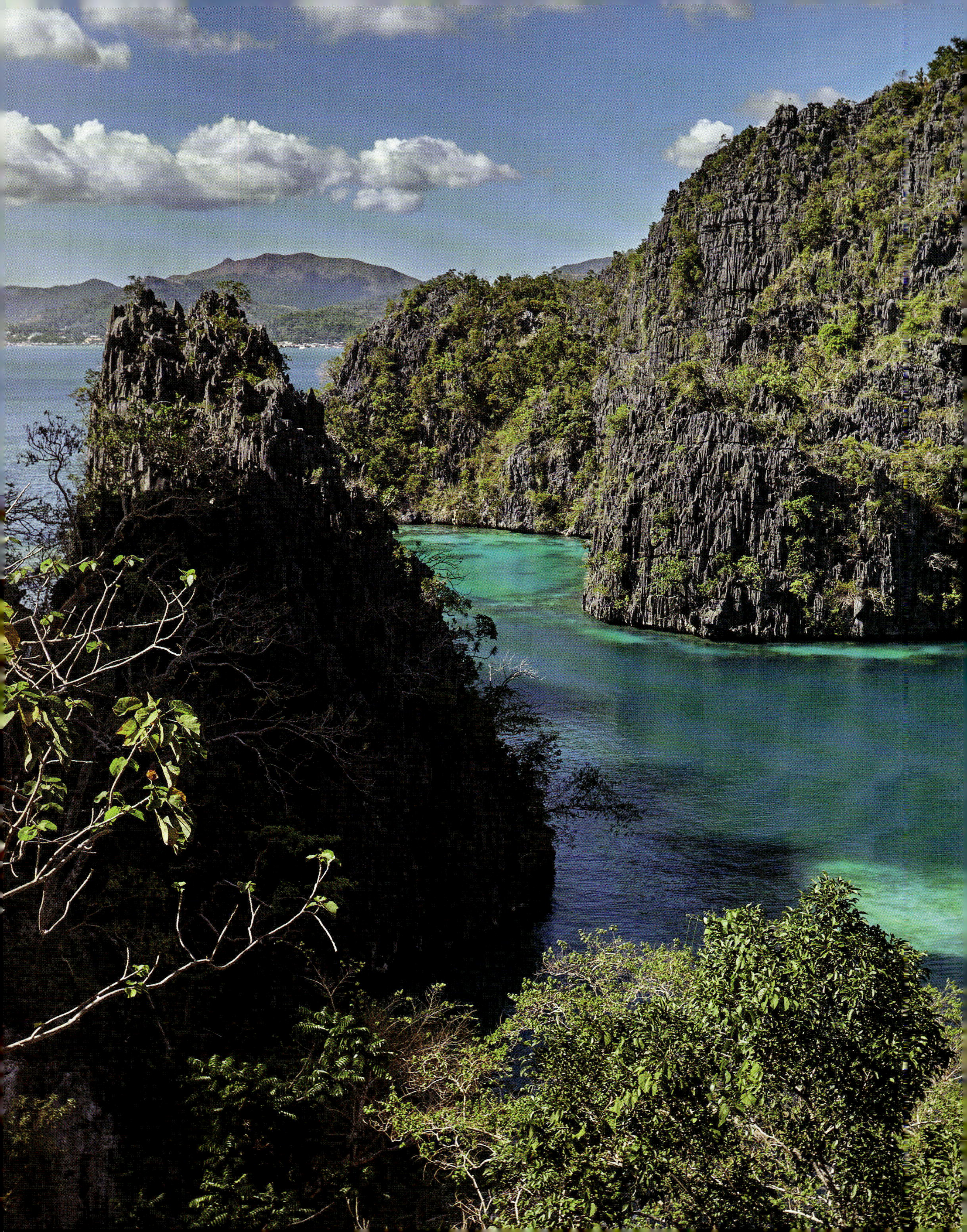

3

From Ocean to Mountain

Natural environments and habitats

One of the most striking impressions of even just a cursory glance at the Philippines' natural environment is that this is a land of blue and green. The two colours seem to envelope, surround and exude from just about every natural scene both on land and at sea: the blue-green of the marine environment, as well as the vibrantly rich greens of the country's terrestrial vegetation, not to mention that great over-arching dome of tropical blue sky.

This blue-green intensity masks a great wealth of habitats, from the depths of the seas to the tips of the tallest volcanoes, which have given the Philippines one of the world's greatest biodiversities, much of which is found nowhere else on Earth. In an intact environment the habitats merge one into another, starting with the open seas off the country's coasts, then continuing with the coastal coral reefs. These in turn merge with sandy or muddy sea beds in the shoreline shallows, which then give way usually to either sandy beaches or mangrove forests. On their landward sides these are backed by beach forest, a type of lowland forest that is specialized to a sandy and salt-laden environment.

We are now firmly into the terrestrial environment, the beach forest quickly giving way to lowland rainforest, which if underlain with deep, moist soils creates a tropical rainforest of giant trees. If the substrate is porous limestone, however, then the resulting forest will generally consist of smaller drought-tolerant trees. At the other extreme, in areas that are extensively flooded with fresh water, there is a specialized form of swamp forest.

As we head inland and climb the mountains, falling temperatures and higher rainfall lead to changes in the forest composition until it becomes montane forest, a place of small trees, which at the highest elevations become quite stunted and gnarled, draped with huge amounts of moss. Finally, at the highest altitudes, around the country's tallest summits, the forest peters out and is replaced with grassland.

This chapter describes each of these environments and habitats in turn, treating them as discrete ecological niches. It should be remembered, however, that these habitat classifications are no more than convenient pigeonholes that make it possible for us to understand each one. In nature, though, the habitats exist as a continuum, each one seamlessly merging into the next, local environmental conditions often making the dividing line separating one habitat from another a very hard one to be exact about. Nevertheless, treating each of these habitats as separate entities makes the hugely complex tropical natural world understandable and explicable.

Opposite: *The dramatic karst limestone coastal landscape of Coron Island, in the north of Palawan, has become iconic both for the country's tourism industry and the beauty and value of its natural environment.*

Open seas and coral reefs

The Philippines is the world's second largest archipelago, with more than 7,000 islands, ranging from tiny sandbars that disappear at high tide, to the large islands of Luzon and Mindanao, topped by smouldering volcanoes and home to millions of people. As already mentioned, all those islands add up to a land area of 300,000 km^2 (115,800 mi^2), an area dwarfed by the amount of sea surrounding them: the Philippine government claims over 2 million square kilometres (0.8 million square miles) of water as its own, though some areas are disputed with neighbouring countries. Much of the surrounding sea consists of relatively shallow coastal regions, no more than 100 m (330 ft) deep, but it also contains some of the world's deepest ocean trenches. These include the Philippine Trench, lying barely 120 km (74 miles) off the country's Pacific coast, and dipping down to more than 10,000 m (33,000 ft) deep. This is the deepest trench of all – at least around the Philippines – but there are other deep ones around the coast; in the Sulu Sea (lying between Palawan and the main body of the country) and in the Celebes Sea (to the south of

Below: *A* Heteractis *species sea anemone, with attendant* Amphiprion *species clownfish on a reef in Bacuit Bay, El Nido, Palawan.*

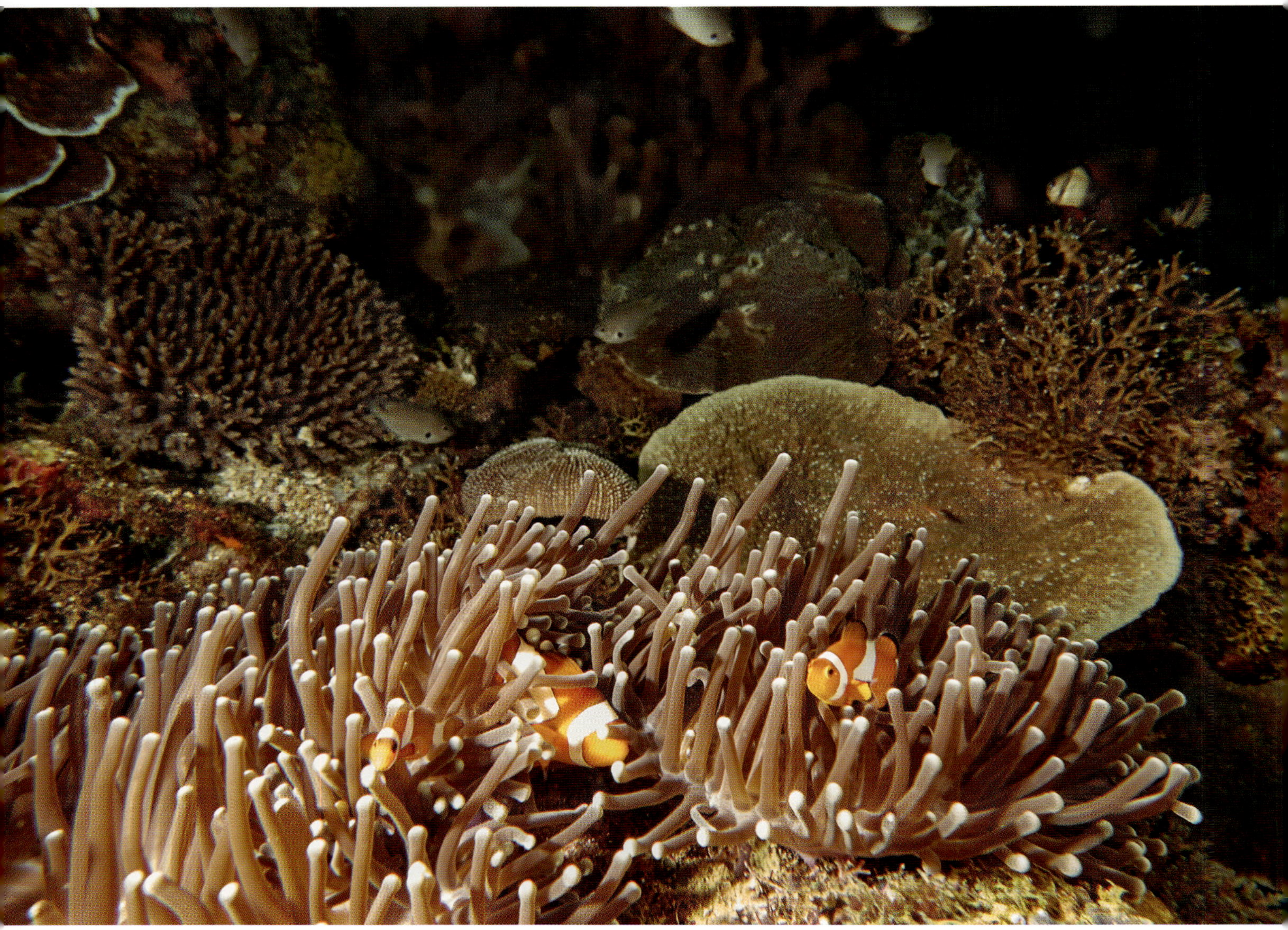

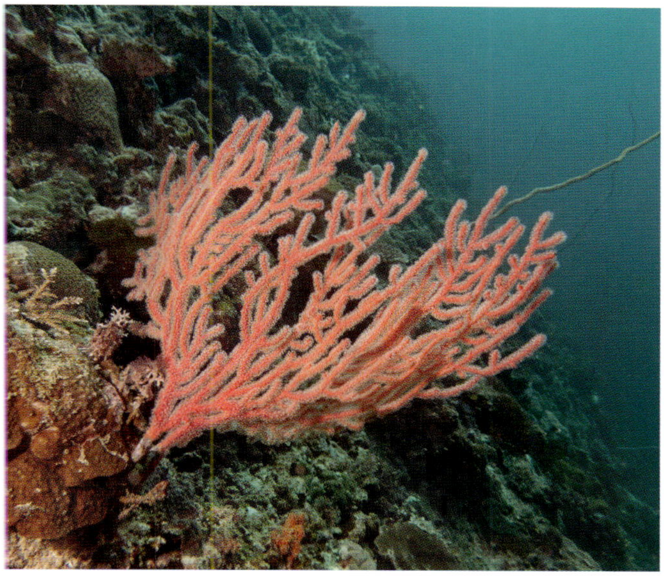

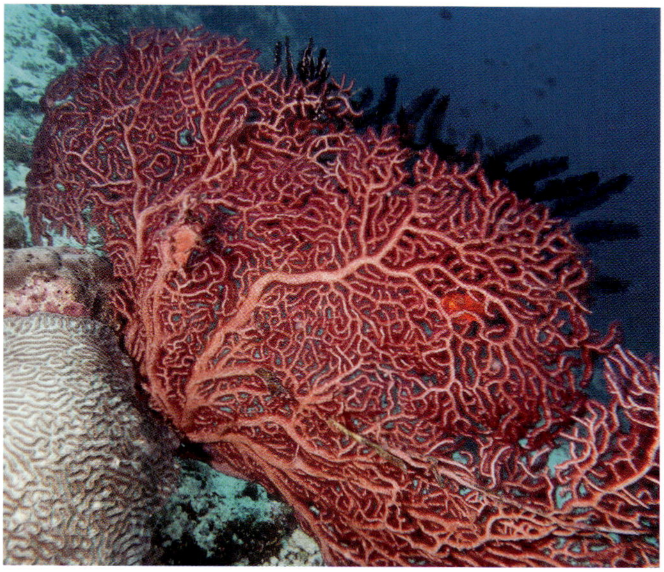

Gorgonian sea fans are arguably the most spectacular reef corals, creating enormous, fan-like shapes that are frequently vibrantly coloured. With so many very closely similar genera and species, they can be hard to identify definitively. **Above left:** *A gorgonian sea fan, probably* Acalycigorgia *species, on a steep wall in Bacuit Bay, El Nido, Palawan;* **Above right:** *A spectacular gorgonian sea fan, probably* Subergorgia *species, off Moalboal, Cebu.*

Mindanao and the Sulu islands), for example.

Such deep oceanic seas give plenty of opportunity for Philippine waters to be home to some of the world's great oceanic animals, such as whales and commercially important fish like tuna. In general, however, the warm seas surrounding the Philippines are nutrient poor, leading them to contain less life than might otherwise be expected.

The explosion in life for which the Philippines is renowned only really happens when you come to waters that are shallow enough for coral reefs to form, mostly of course along the coasts, in water less than about 50 m (164 ft) deep. In an archipelago of over 7,000 islands there is a lot of coast; roughly 35,000 km (21,700 miles) of it, much of this providing the possibility of environments suitable for corals to form.

Philippine coral reefs

The Philippine coast is surrounded by a staggering 25,000 km² (9,650 mi²) of coral reefs, spread around large parts of the coastline, though not surprisingly concentrated especially into those areas that offer the best growing conditions. The latter consist essentially of clear saltwater, with plenty of light and a temperature range of roughly 20–28°C. Areas with high sediment, where clarity and light levels are reduced, such as river estuaries and harbours, do not make for good coral reef habitats.

The great majority of the Philippines' coral reefs are fringing reefs; they literally fringe the coast, following it closely just a short distance off the shore. That shore is often a sandy beach, but frequently also a rocky shoreline or sometimes a mangrove forest. Typically, the first part of the reef, the area closest to shore, is shallow – perhaps only a few metres deep – and only very gently sloping; the water here is calm and protected by outer parts of the reef. As you reach deeper water – perhaps about 10–15 m (33–49 ft) deep, a reef 'crest' develops. This is a ridge running along much of the reef, built up towards the sea's surface and consisting of large, strong hard coral structures. Once over the crest, the sea bed usually falls away rapidly in a drop-off that is either a slope or a sheer wall, down into deeper water, where the corals peter out.

Different types of coral inhabit different parts of the reef. In the drop-off zone, in water too deep to be affected by wave action, you will find some wonderfully delicate corals such as sea fans, sea whips

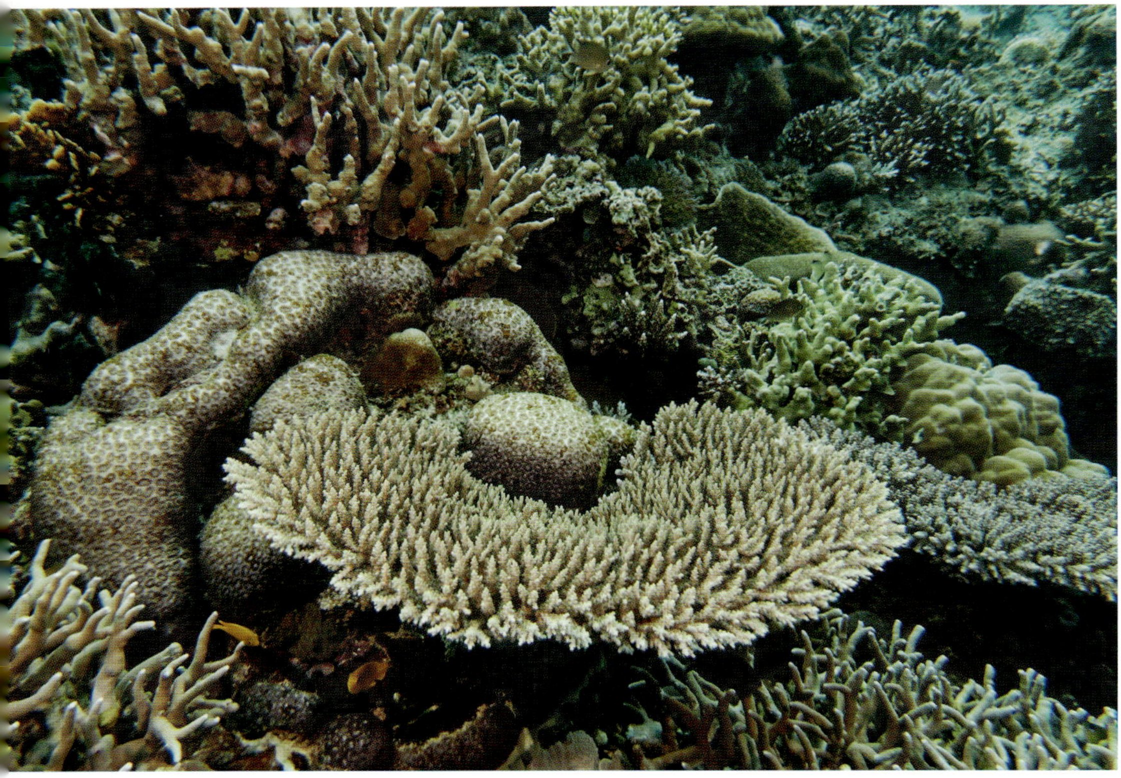

Left: A mass of different hard corals, mostly various *Acropora* species, jostle for space on the crest of a healthy reef off Moalboal, Cebu.

Below: Although resembling a soft coral, with its elongated polyps swaying in the currents, this is actually a *Turbinaria* species hard coral; seen off Moalboal, Cebu.

Bottom: A digitate leather coral (*Sinularia* species), a soft coral typified by finger-like lobes clustered together on stalks, seen here on a shallow reef flat off Panglao Island, Bohol.

and some forms of soft corals. The sea fans are some of the most beautiful corals imaginable, delicate fan-like lattices designed to ensnare passing plankton, sometimes up to 2 m (6.6 ft) across, and having a range of colours from white to red or orange.

The reef crest is the area exposed to the greatest wave action, where the corals have to be strong, solid structures. Here you will find the biggest corals, such as the large, boulder-like brain corals, or the spreading table or staghorn corals produced by a number of *Acropora* species of hard coral. In the calmer waters of the reef flat, behind the protective reef crest, is a host of soft corals that are able to wave around in the currents and small waves, as well as smaller outcrops of hard corals. Many of the species here are able to tolerate the extremely high light levels found in very shallow waters, and some can even put up with being exposed to the air for a few hours at low tide.

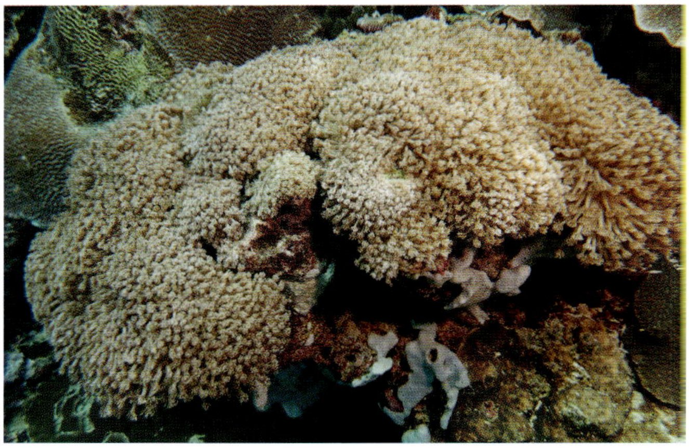

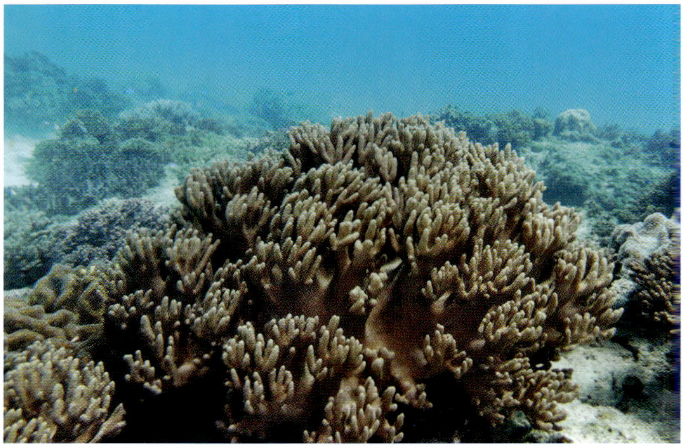

The Philippines at the heart of the Coral Triangle

When scientists started to study the distribution of coral species around the tropical oceans, one finding

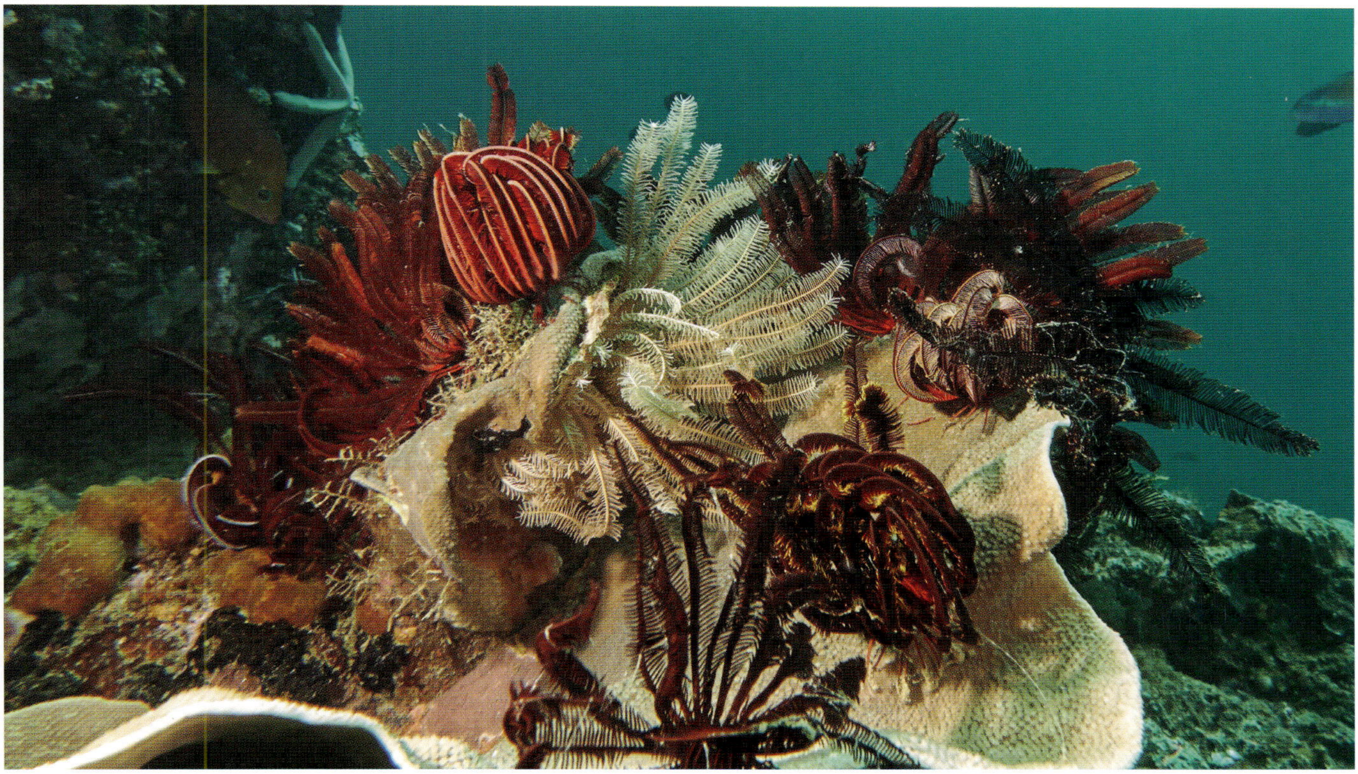

Above: A host of very different and colourful feather stars, or crinoids, crowds onto the surfaces of a Turbinaria *species plate coral; seen off Miniloc Island, El Nido, Palawan.*

that kept coming up was that species diversity was at its greatest in a roughly triangular region around the great mass of islands that make up the archipelgoes of Southeast Asia and the western Pacific. It became clear that this region is the central 'hub' of global coral biodiversity, with a rapid drop-off in variety when heading eastwards into the Pacific and a more gradual decline westwards into the Indian Ocean. This region has become known as the Coral Triangle, a title formalized in 2007 by the governments of the six countries affected, namely Indonesia, Malaysia, the Philippines, Timor Leste, Papua New Guinea and the Solomon Islands, which agreed to work together to improve coral reef protection throughout the Triangle.

Out of a total of 720 coral species worldwide, the Coral Triangle alone is home to just over 600 species. Contrast this with less than 500 for Australia's Great Barrier Reef and a mere 61 for for the whole of the Caribbean. The Philippines hosts about 95 per cent of the Triangle's coral species (and 80 per cent of the entire world's), putting the country firmly at its heart. Indeed, it has been argued that the Visayan Sea and/or Tubbataha Reefs are the very epicentre of the Coral Triangle, acting as the main source of coral larvae – as well as the young of many other reef species – to be fed out by oceanic currents to the Indian Ocean and western Pacific.

Not surprisingly, it is not just coral species that reach their maximum numbers in the Coral Triangle, but also species of many other reef animal groups, including sponges, crustaceans, echinoderms and fish, which together add up to several thousand coral reef animal species in the region.

Non-coral invertebrate animals

A biology of the corals is given in Chapter 4 (see page 85), but of course they are far from being the only invertebrates (animals without backbones) on a coral reef. Huge numbers of jellyfish, sponges, worms, molluscs, crustaceans, echinoderms and sea squirts call a reef home too, many of them conspicuous by size and/or colour, others requiring a sharp eye to spot.

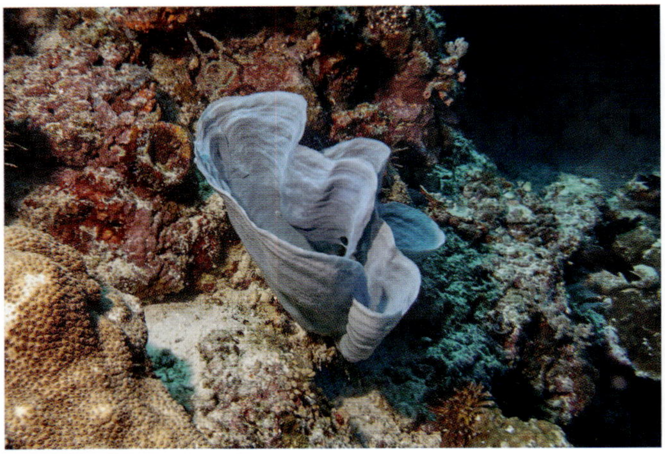

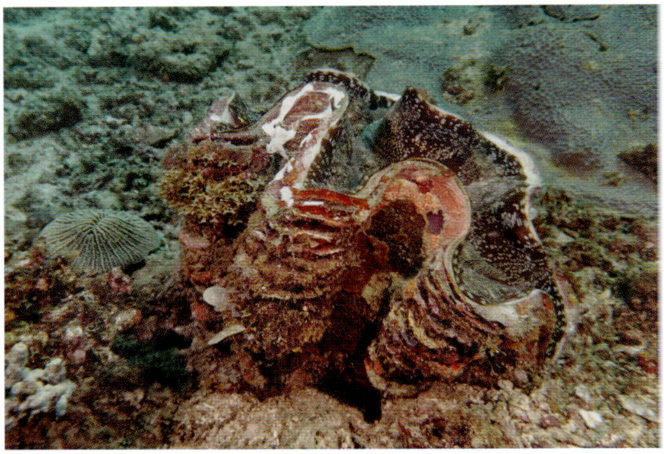

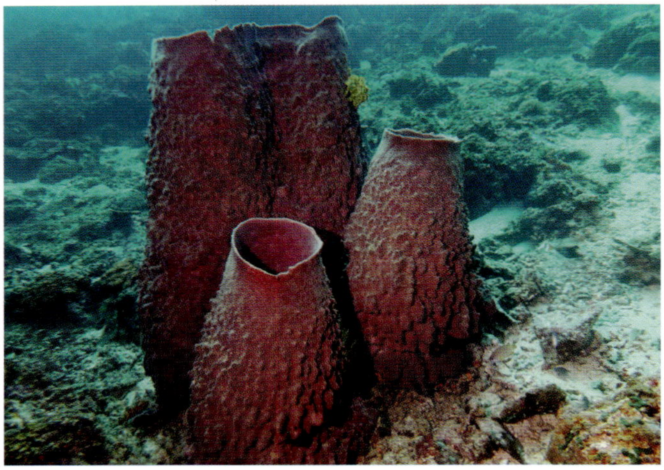

Above: *The giant clam (*Tridacna *species) is the largest mollusc around the Philippine coast, reaching over 1 m (3.3 ft) in length. Over-collection has caused it to become quite rare. This individual was seen in Bacuit Bay, El Nido, Palawan.*

Above left: *A magnificent blue vase sponge (*Strepsichordaia *species), off Balicasag Island, Bohol.*

Left: *A cluster of enormous barrel sponges (*Xestospongia *species), over 1 m (3.3 ft) high, in Bacuit Bay, El Nido, Palawan.*

Jellyfish are a close relative of the corals, being also coelenterates, but much larger and free swimming. Well known for their translucency and ability to give bare skin a nasty sting, their presence in waters around Philippine reefs is quite variable, though not uncommon.

Sponges are among the most common members of the reef community. Consisting of thousands of species, they are a very simple animal made up from layers of cells separated by a gel-like substance that gives the sponge its rigidity. It has no skeleton or nervous system and does not move, spending its life anchored to the reef. It feeds by filtration of the water, its body filled with pores and channels that direct the water into its central cavity before being expelled. The most striking of the sponges seen on many Philippine reefs are the huge barrel sponges (*Xestospongia* species) that can grow to more than 1 m (3.3 ft) tall. Others commonly seen include various vase and tube sponges, their English names accurately describing their shapes, plus many that simply form hard-to-describe reef encrustations.

Molluscs consist of a vast diversity of animals, ranging from snails, cowries and limpets, through to giant clams, sea slugs and the cephalopods. The sea slugs include the wonderfully colourful nudibranchs, a common, though very small, hard-to-spot feature of every Philippine reef. These can sometimes be confused with the worms, many of which burrow into corals. Giant clams, which can easily be 1 m (3.3 ft) long, are quite rare these days due to over-collection, though reintroduction programmes in some protected areas may be helping to bring them back.

Cephalopods comprise the octopuses, squid and cuttlefish, as well as the rare Chambered Nautilus (*Nautilus pompilius*), some of the largest and most intelligent of the reefs' invertebrates. Seen only occasionally during daytime, they often rise up to the reefs from deeper levels at night.

Crustaceans are another vast group, taking in many of the mobile shellfish, such as prawns, shrimps, crabs and lobsters, a vast wealth of species that ranges from those large enough to be hugely important as seafood, to others that can barely be seen with the eye, resting on a coral polyp. Many of the crustaceans are nocturnal, spending the daytime hiding in the reef's nooks and crannies, and coming out to feed only at night.

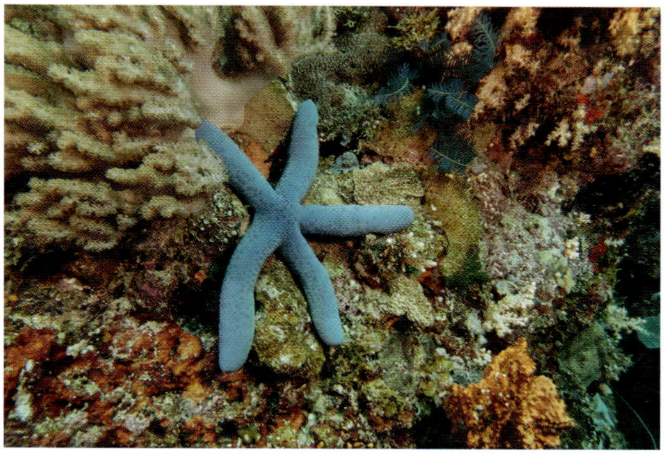

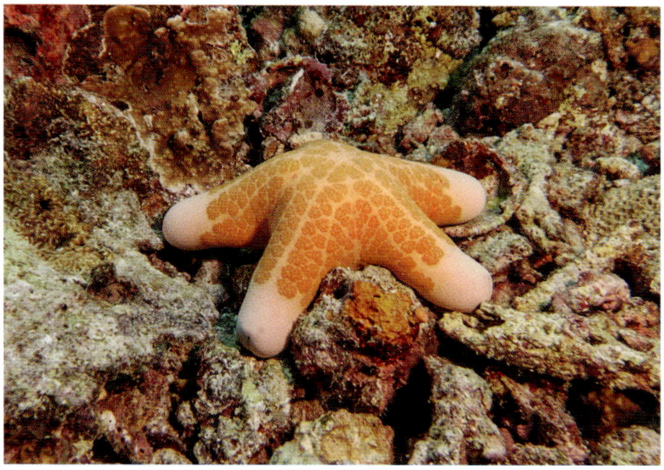

Top right: *Linckia species sea stars are among the most common of the Philippines' starfish, typically seen in the shallow waters of the reef flat. This individual was seen close to Miniloc Island, El Nido, Palawan.*

Right: *Another common starfish of the shallows,* Choriaster granulatus, *here found off Moalboal, Cebu.*

Below: *A spectacularly colourful predatory mantis shrimp (*Odontodactylus *species). Usually nocturnal, this one was spotted during the day off Miniloc Island, El Nido, Palawan.*

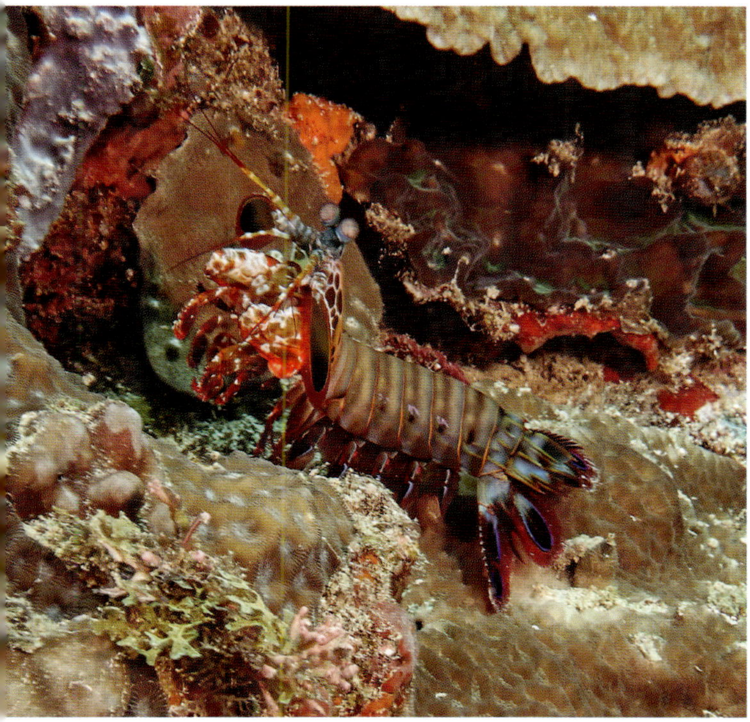

Echinoderms consist largely of sea urchins, sea cucumbers and starfish. Sea urchins are not frequently seen, but sea cucumbers are moderately common and starfish are hugely visible. Sea cucumbers vary from very fat, cucumber-shaped animals that can be a variety of colours, often seen crawling across sand or coral substrate, to tiny white 'alabaster sea cucumbers' that frequently decorate the surfaces of sponges. The highly variable starfish range in appearance from those with the typically simple five legs (often blue), as in the very common *Linckia* species, to the fantastically colourful and complex feather stars (often known as crinoids). These range in colour from yellow to red to black, and can often be seen forming large clusters on corals in all parts of the reef, though especially in the shallower areas.

Finally, there are the tunicates, or sea squirts. Rather delicate, often translucent, and immobile

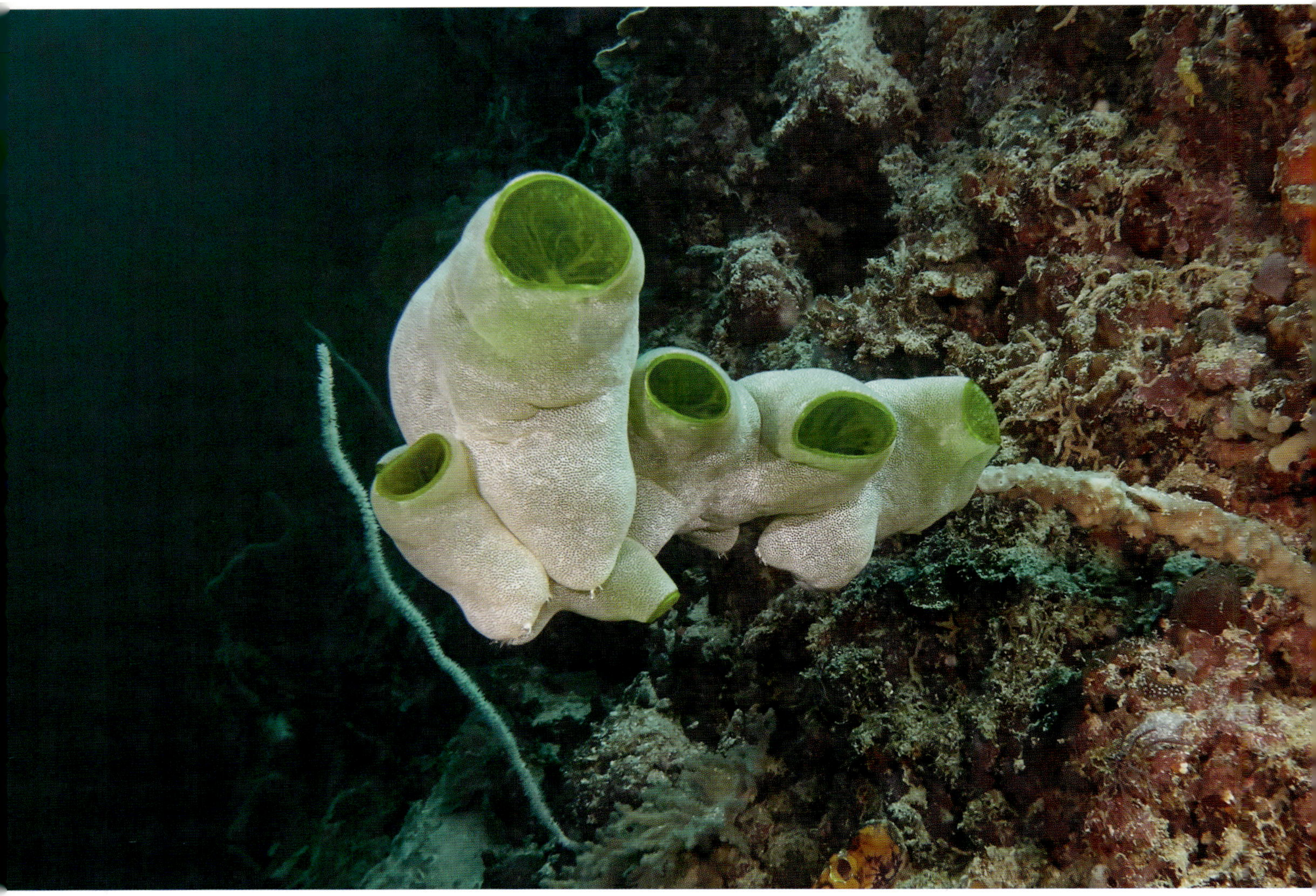

animals, some sea squirts superficially resemble sponges, in as much as they contain a large central cup cavity. This is actually part of their syphon mechanism for filtering food out of water as it is passed through the body. Many are very colourful. Common examples on Philippine reefs include *Didemnum molle*, a greenish, speckled, rather urn-shaped sea squirt that is opaque, and several *Rhopalaea* species, which are usually a stunning blue colour and can vary from opaque to completely transparent.

Above: A magnificent example of the sea squirt Didemnum molle, *one of the most common tunicates around the Philippine coast. Here seen in Bacuit Bay, El Nido, Palawan.*

Fish on the coral reefs

Philippine waters are home to more than 3,000 fish species, a truly staggering figure, divided roughly into reef and pelagic species. Pelagic fish are those that live in the open seas, though some do often visit the coral reefs. They range from the mighty Whale Shark (*Rhincodon typus*), the world's largest fish, through a host of other sharks, including thresher and hammerhead sharks (*Alopias* and *Sphyrna* species), to such fish as tuna, barracuda (*Sphyraena* species), sardines (*Sardinella* species), anchovies and the Big-eyed Scad (*Selar crumenophthalmus*).

The reef fish, not surprisingly, tend to live most of their lives on or around a reef, generally not travelling at all far. They make up a large part of the hugely varied and colourful spectacle awaiting anyone diving or snorkelling over a coral reef. Of those readily visible, some of the smallest are the clownfish, or anemonefish (generally *Amphiprion* species), which are mostly just a few centimetres long, and have a close territorial association with several sea anemone species. Rather larger are the beautifully named damselfish, angelfish, squirrelfish, soldierfish, cardinalfish, goatfish, gobies, groupers, triggerfish, parrotfish, pufferfish, batfish, sweetlips, snappers and wrasses, each consisting of multiple species, which are regularly seen swimming across almost any part of a reef, as well as using many of the crevices and holes that a reef has to offer as shelter.

Other fish include moray eels, various stingrays and a number of reef sharks. One of the most common rays is the blue-spotted (*Neotrygon* species), occasionally seen half-buried in the sand or rubble around a reef, while occasionally sighted at just a few reefs is the Reef Manta Ray (*Manta alfredi*). The most common sharks on Philippine reefs are Whitetip and Blacktip Reef Sharks (*Triaenodon obesus* and *Carcharhinus melanopterus*, respectively), neither of which grows larger than about 2 m (6.6 ft), plus the

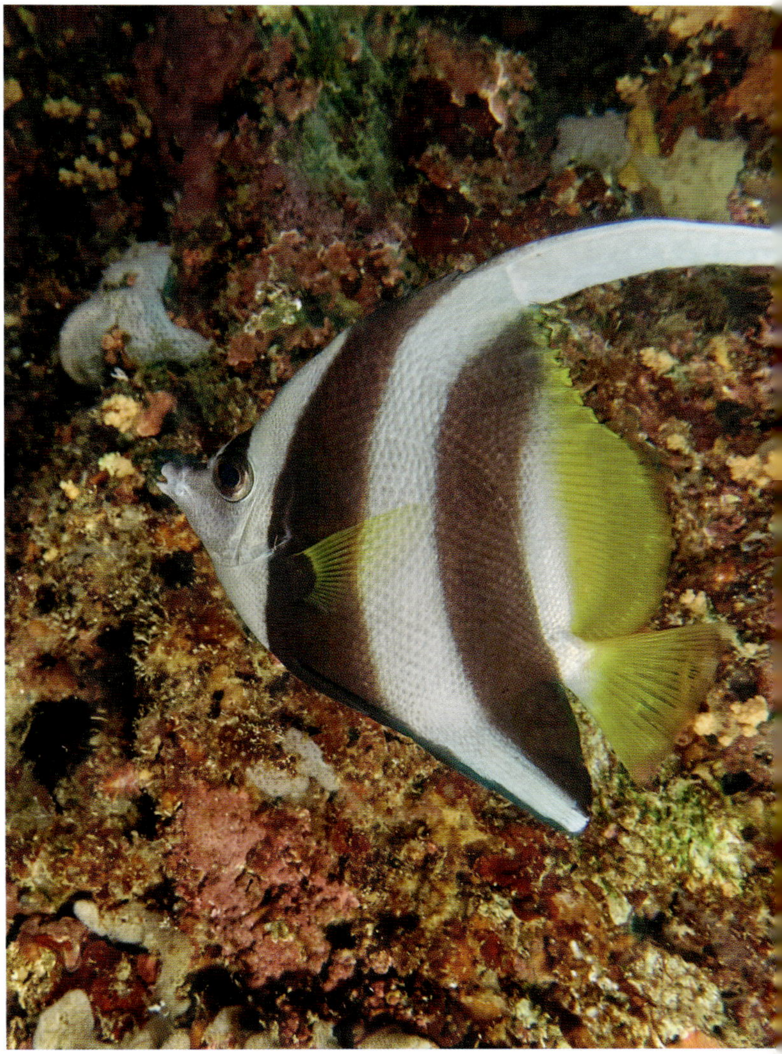

Right: *The Schooling Bannerfish (*Heniochus diphreutes*), a species of angelfish that is easily confused with the well-known Moorish Idol, seen in Bacuit Bay, El Nido, Palawan.*

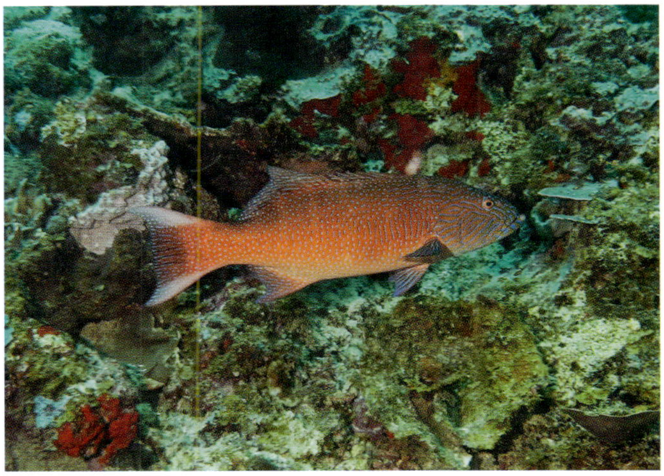

Below: *A coral grouper (*Cephalopholis *species), one the largest and most colourful of the Philippines' reef groupers, seen off Miniloc Island, El Nido, Palawan.*

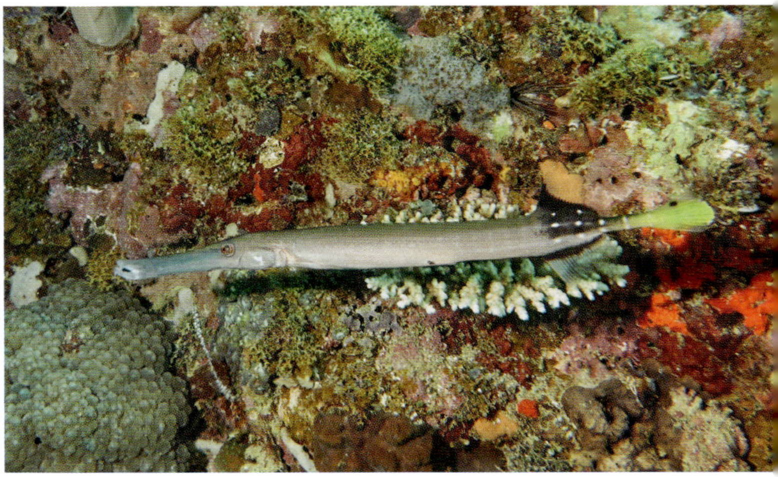

Below: *This rather unusual fish, the Pacific Trumpetfish (*Aulostomus chinensis*), is only occasionally seen on Philippine reefs, in this occasion off Miniloc Island, in Bacuit Bay, El Nido, Palawan.*

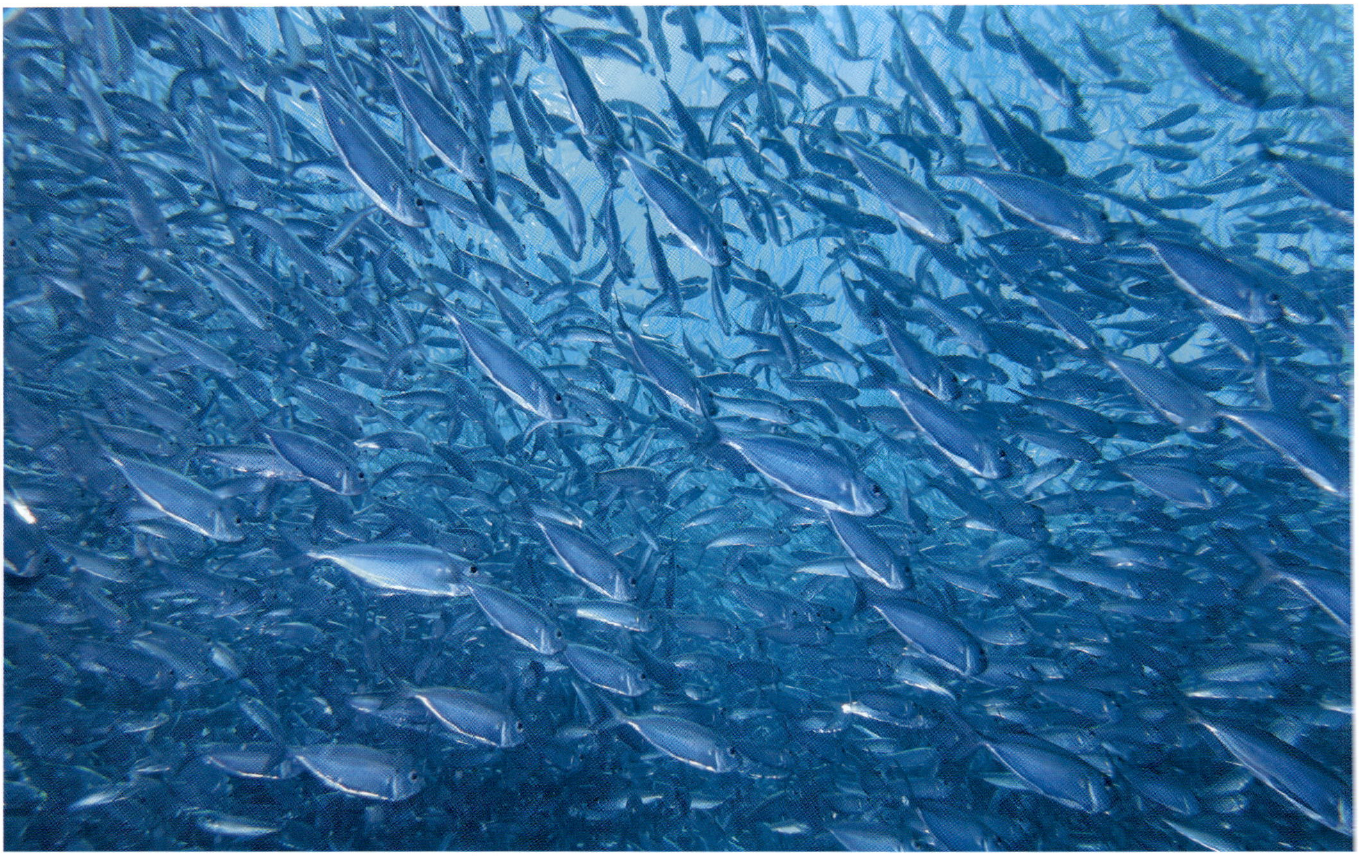

slightly larger Grey Reef Shark (*Carcharhinus amblyrhynchos*).

Although a few of the larger fish feed on other fish, for most of them food consists of algae, plankton, seagrasses, general detritus and the reef itself. Several species of reef fish, including some parrotfish, specialize in feeding on the coral polyps, grinding up the reef skeletons into tiny fragments and thus helping to create those wonderful white coral sand beaches. The Humphead Wrasse (*Cheilinus undulatus*, also known as the Napoleon Wrasse) is an important predator in that it feeds on the Crown of Thorns Starfish (*Acanthaster planci*), a species that when left unchecked by predators can proliferate to a point that its coral-eating diet can do huge amounts of damage to a reef.

Most of these reef fish are a spectacular rainbow of colours, with just about every colour imaginable. Complementary colours on a single fish are a regular feature, with blue and yellow particularly common. It is a bit of a mystery why colour should be so important in this submarine environment, where the only light colour to penetrate down as far as about 20 m (66 ft) is blue. It is possible, however, that such colour combinations as blue and yellow, while being highly visible close up, can render a fish almost invisible against the reef when seen from a distance.

Above: Every afternoon, a vast shoal of sardines congregates just metres offshore from Moalboal's Panagsama Beach, creating quite a spectacle.

Threats to the coral reefs

As already mentioned, the Philippines' coral reefs are under huge pressure, with at best about 5 per cent estimated in the most recent surveys to be in good condition, meaning that they have more than about 60 per cent healthy, live coral cover and abundant fish life. Almost all reefs have been damaged to varying degrees and by various man-made causes.

The most common form of damage is simply over-fishing, a result of the huge and growing coastal communities, many of which are still living in poverty and have little choice but to catch fish simply to feed their families every day. It is believed that fish catches for these people have declined by as much as 90 per cent in terms of both fish numbers and weight over the past 50 years. Linked to this are destructive fishing practices, particularly the use of explosives and cyanide (see page 32). A further locally caused factor is siltation as a result of deforestation on adjacent land. Once tree cover is removed from steep hills, heavy rain is able to wash the topsoil down into rivers and then the sea, clouding the waters and reducing light levels for photosynthesis. Similarly, chemical pollution from inland mines and factories is increasingly being swept into the sea, poisoning the reefs.

Factors beyond Philippine control are centred largely around global warming. Rising temperatures are causing an increase in both the number and severity of storms, including typhoons, resulting in rising levels of physical damage to many reefs. Furthermore, as the seas are getting warmer, so coral bleaching is becoming more of a threat. This occurs when the sea temperature rises above about 32°C, and consists of the coloured zooxanthellae algae being expelled from the corals, turning the corals white and leading them to starve. In recent years there have been two major coral bleaching incidents in the Philippines, in 1998 and 2010, both the result of El Niño events, when an abnormal reversal of sea currents drove sea temperatures up and rainfall down. On both occasions the corals recovered once sea temperatures fell again, but with sea temperatures around the Philippines frequently hovering around the 28–30°C mark – already above the optimal coral growth temperature – conditions are perilously close to the bleaching trigger temperature much of the time.

Conservation measures

Quite apart from the Coral Triangle Initiative, a large number of conservation measures is being put into place to try to overcome the problems, which should produce some positive results in the coming years.

Over the past 20 years, a huge number of Marine Protected Areas (MPAs) has been created; more than 1,700 in fact, making the Philippines the world's leader in the number of marine sanctuaries. Unfortunately, most exist on paper only, with little information available even on their boundaries and exact location, let alone any effective enforcement. Most are also very small, covering an area of less than 1 km^2 (0.38 mi^2), rendering them of limited value. Reliable information and management controls exist for rather less than 400, though this is still an impressive number.

The 400 MPAs consist of quite a mixture of sites. The largest include the vast Tañon Strait Protected Seascape, Tubbataha Reefs Natural Park and Apo Reef Natural Park, covering thousands of square kilometres, and created by the DENR as part of the NIPAS. At the other end of the scale – and much more numerous – are tiny coastal sanctuaries covering a few hectares, established by local communities and governments as part of the Fisheries Code.

Apart from size, one of the main differences between these two types of sanctuary is motivation. Most of the small, locally established sites have been created with the aim of boosting fish yields for local municipal fishermen, rather than for protecting biodiversity. There is growing evidence that small no-take zones (that is, the local MPAs) can have a very positive impact on fish yields in surrounding areas, provided there is no poaching within each MPA. The large NIPAS sites, on the other hand, are mostly aimed at protecting biodiversity, though there are a few that do have a dual role for both this and fisheries, including Tañon Strait, El Nido and Malampaya Sound, the last two in Palawan, the first in the Visayas.

Although a start has been made in protecting the marine environment, there is still a long way to go. The Philippines' fisheries and Coral Triangle Initiative laws call for at least 10 per cent of coastal habitats to be protected, but at present all the MPAs together still amount to less than 1 per cent, a much lower percentage of habitat than is protected on land. There is still a lot of work to be done to ensure preservation of the Philippines' coral reefs and their glorious biodiversity.

The littoral zone: the shallows and the shore

In the shallowest waters, at the uppermost limit of the coral reefs but below the low-tide line, you start to see an important plant habitat, consisting of seaweeds and/or seagrasses. Both are important as food for certain marine species, but also provide shelter for many more, helping to nurture the young of a wide range of fish, crustaceans and molluscs. They are, therefore, vital components of the marine coastal environment.

Seaweeds, of which there are about 1,000 species in the Philippines, are large multicellular algae, living mostly attached to a solid substrate, usually either parts of the reef structure or shoreline rocks. While many have to be immersed the whole time, some can survive temporary exposure above the water, so can be found on the shore in the intertidal zone, between the low- and high-tide lines.

Seagrasses, on the other hand, grow out of a sandy or muddy sea bed, usually in very shallow water, but

***Opposite:** Seagrasses, seen at low tide, sway in the shallows close to young mangrove trees in Olango Island Wildlife Sanctuary, Cebu.*

***Below:** Seagrasses seen just below the low-tide line, along a sandy beach on Cowrie Island, in Honda Bay, Palawan.*

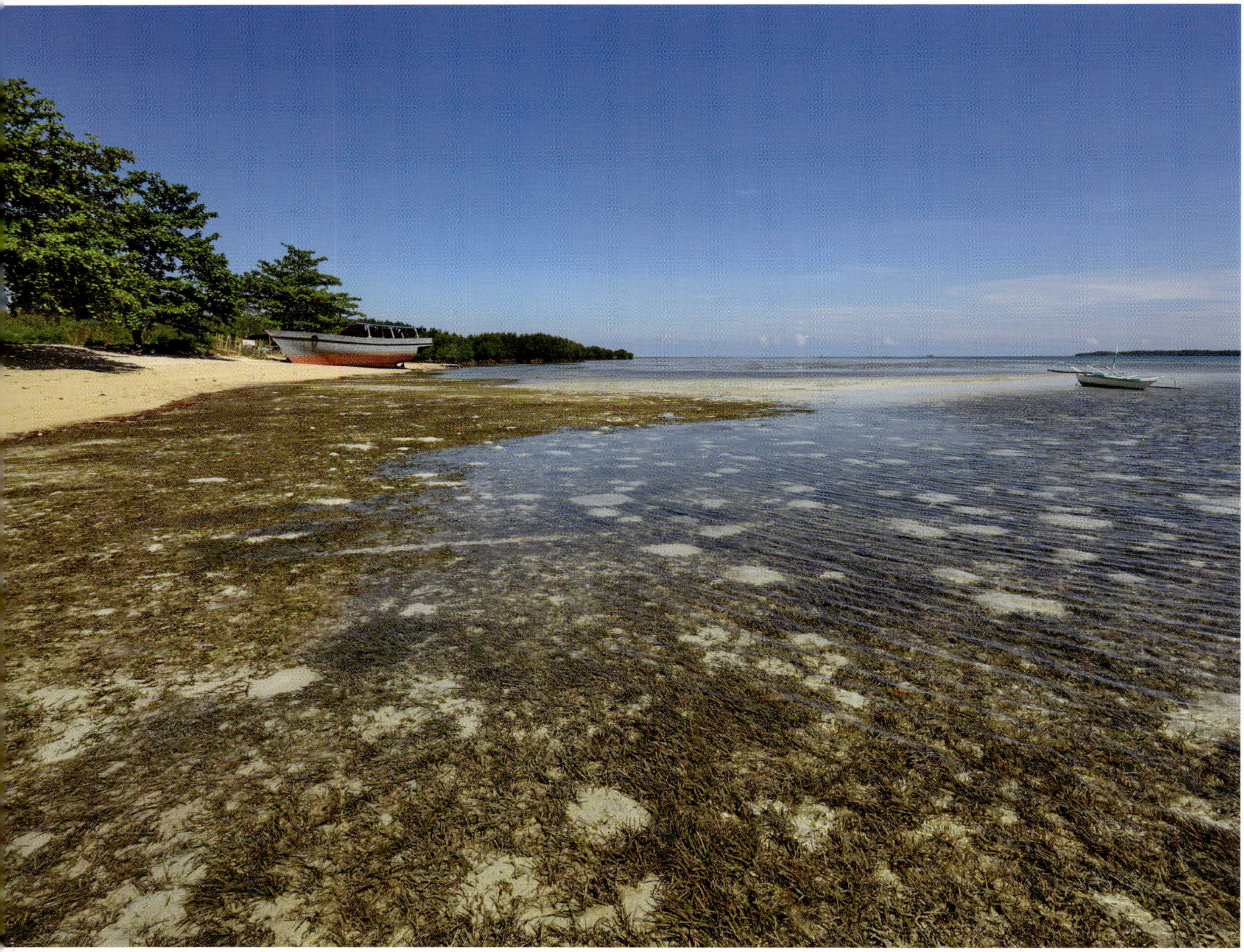

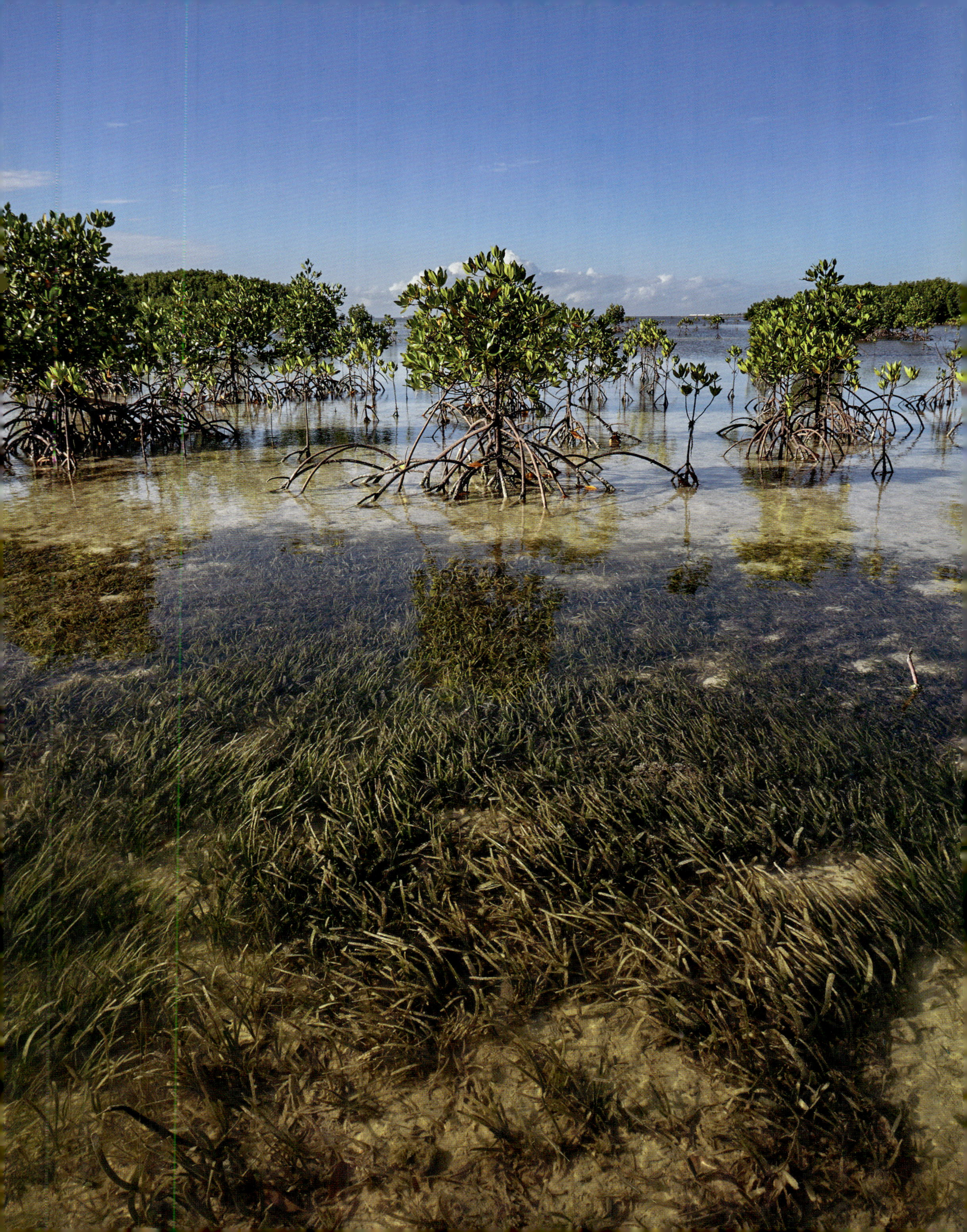

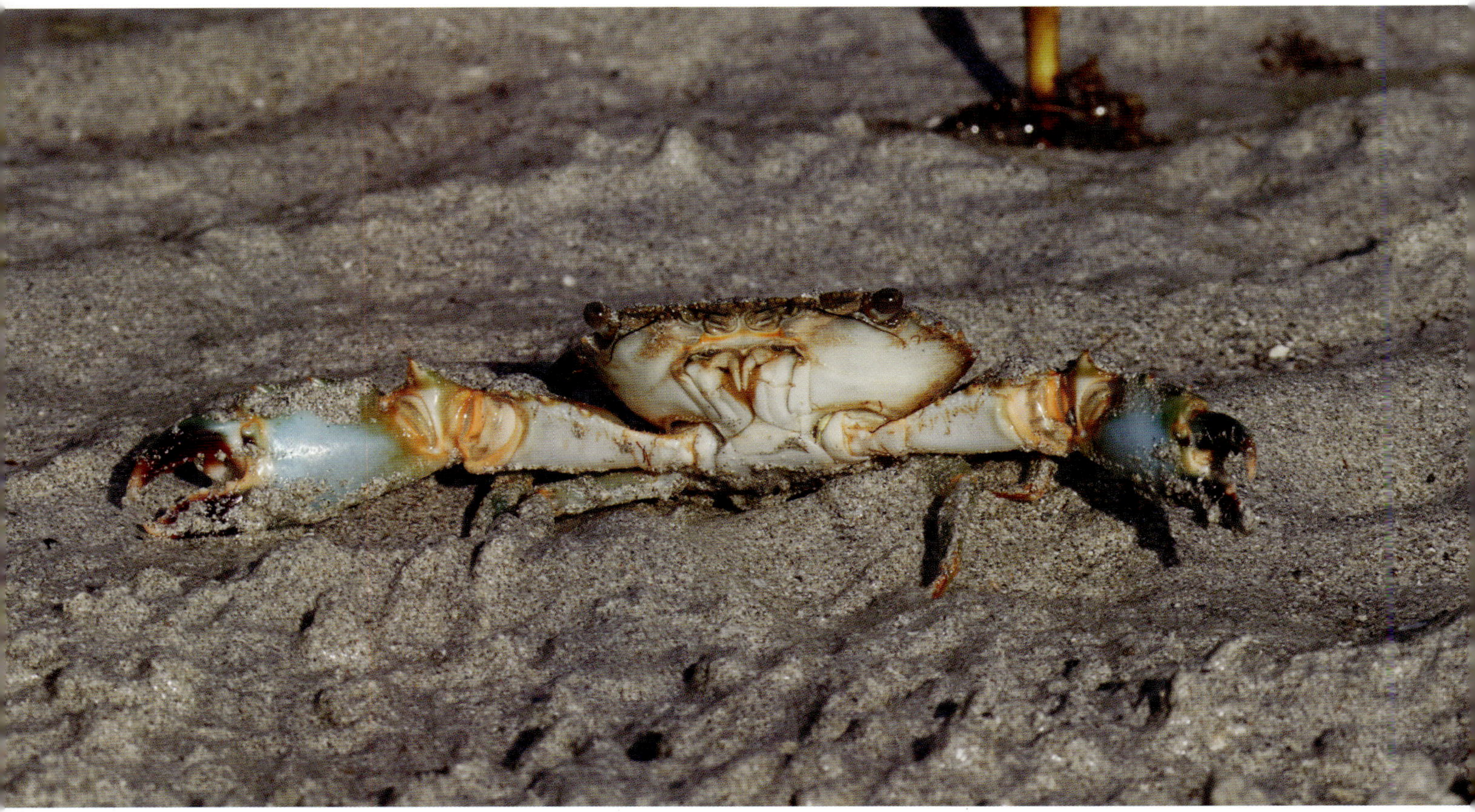

spreading above the low-tide line only to a very limited degree. They are the only flowering plants adapted to live fully immersed in seawater, and despite their name they are not grasses at all, but are related to lilies and gingers. Although found across the world

*Below: The tiny Light Blue Soldier Crab (*Mictyris longicarpus*), can be seen in large numbers, marching across the sands in mangrove forests, here seen in Puerto Princesa, Palawan.*

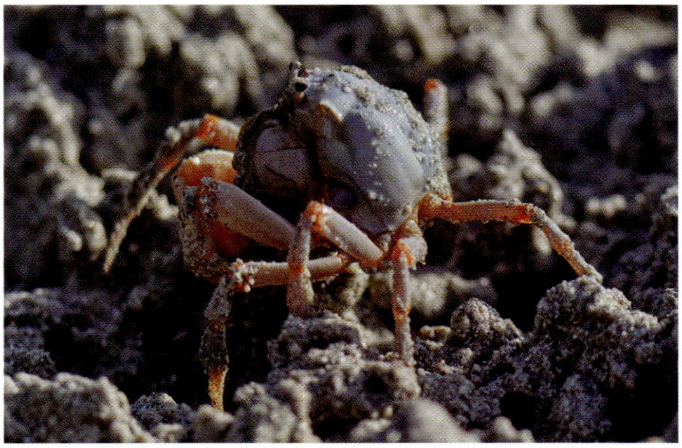

there are far fewer species of seagrass than seaweeds, with only about 60 worldwide. The Philippines is home to 19 species, giving it the world's second highest seagrass species diversity.

Seaweed and seagrass environmental roles

Seagrasses are important in binding together sandy or muddy sea beds, helping both to reduce levels of suspended sediment and indirectly to build more land. This is achieved through their extensive root and rhizome (underground stems) networks, which travel through the sea bed, physically linking together separate plants across entire seagrass beds.

Although both seagrasses and seaweeds provide shelter for a range of marine species, comparatively few see them as food. Seagrasses in particular are fed on by only a limited number of reef fish, molluscs and urchins. They are, however, hugely important to Dugongs (*Dugong dugon*) and Green Turtles (*Chelonia midas*), both of which regularly graze on the plants.

Although seagrass beds still cover 2.73 million

Opposite: The Spiny Rock Crab (Thalamita crenata), a typical resident of mangroves, seen in muddy sands close to mangrove trees at Puerto Princesa, Palawan.

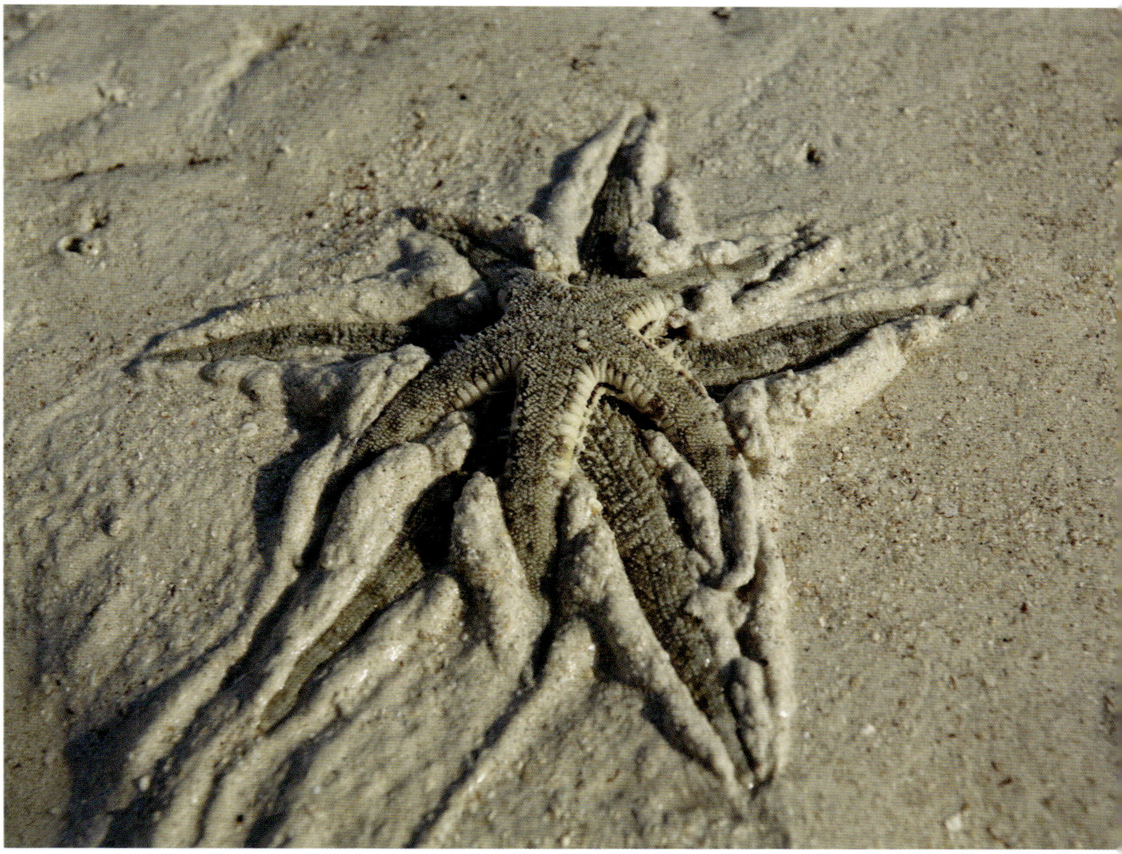

Right: A pair of Sand-sifting Starfish (Archaster typicus), a common Philippine starfish, mating on sand at low tide at Sandugan, on the north coast of Siquijor.

hectares (6.74 million acres) of Philippine shores, mostly in Palawan and to a lesser extent in the Visayas, this represents a 30–50 per cent drop from their levels 50 years ago. The decline is largely the result of coastal development, and conversion of mangroves and mudflats to fish ponds.

On and beneath the sand and mud

Quite apart from the plants, the sand and mud both above and below the low-tide line harbour a wealth of life, consisting of burrowing animals such as worms, molluscs and crabs. It is also not uncommon to see several species of starfish in the shallowest waters, though usually immersed, below the low-tide line.

The sand and mudflats are of great importance to both coastal communities and avian wildlife, providing rich pickings when exposed at low tide. Some of these areas can be quite extensive, and when also associated with adjacent mangroves can be particularly important to waders such as herons, egrets, curlews, whimbrels, stilts and redshanks, to name just a few. This is especially so at Olango Island, off Cebu, where mangroves and mudflats have been protected as a wildlife sanctuary due to the thousands of migrating waders that come here every year to rest and feed (see page 174).

Below: Turnstones (Arenaria interpres) are one of many bird species that thrive on the huge numbers of worms, molluscs and crustaceans in the intertidal sands and mudflats.

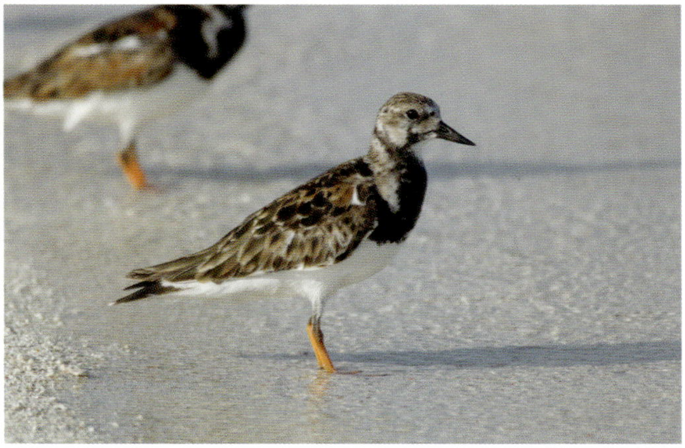

Mangrove forests

Mangrove forests flourish on the shore between the high- and low-tide lines, their trees a group of terrestrial plants that have evolved to tolerate alternating inundations with salt or brackish water at high tide, and dry periods when the tide is out. Generally, they prefer calm, shallow waters with a muddy substrate, conditions typical of river estuaries and many enclosed bays. In such places they frequently establish extensive forests several hundred metres deep, from sea to solid land, or occasionally 1 km (0.62 miles) or more. In less favourable conditions, such as along beaches, where the substrate is usually sandy and with a greater slope, they generally form only a narrow band the depth of a few trees, and even then only when behind the protective screen of a fringing reef. Unlike seagrasses, they cannot tolerate inundation in saltwater the entire time so they rarely grow below the low-tide line.

Mangrove biodiversity

Although an estimated 260 plant species can typically be found in mangroves, most of these, such as epiphytic orchids and ferns, are able to live in a range of different forest types, not just mangroves. The 'true' mangrove species, as they are called – plants that live only in mangroves – come to only about 70 species worldwide, all of them restricted to the tropical or subtropical zones. About 40 of the true mangrove species can be found in the Philippines, making this country one of the world's most important for

Below: Mangrove trees in the vast mangrove swamp along the shore of Siargao Island, Mindanao. Avicennia *species trees often dominate this leading, colonizing mangrove edge.*

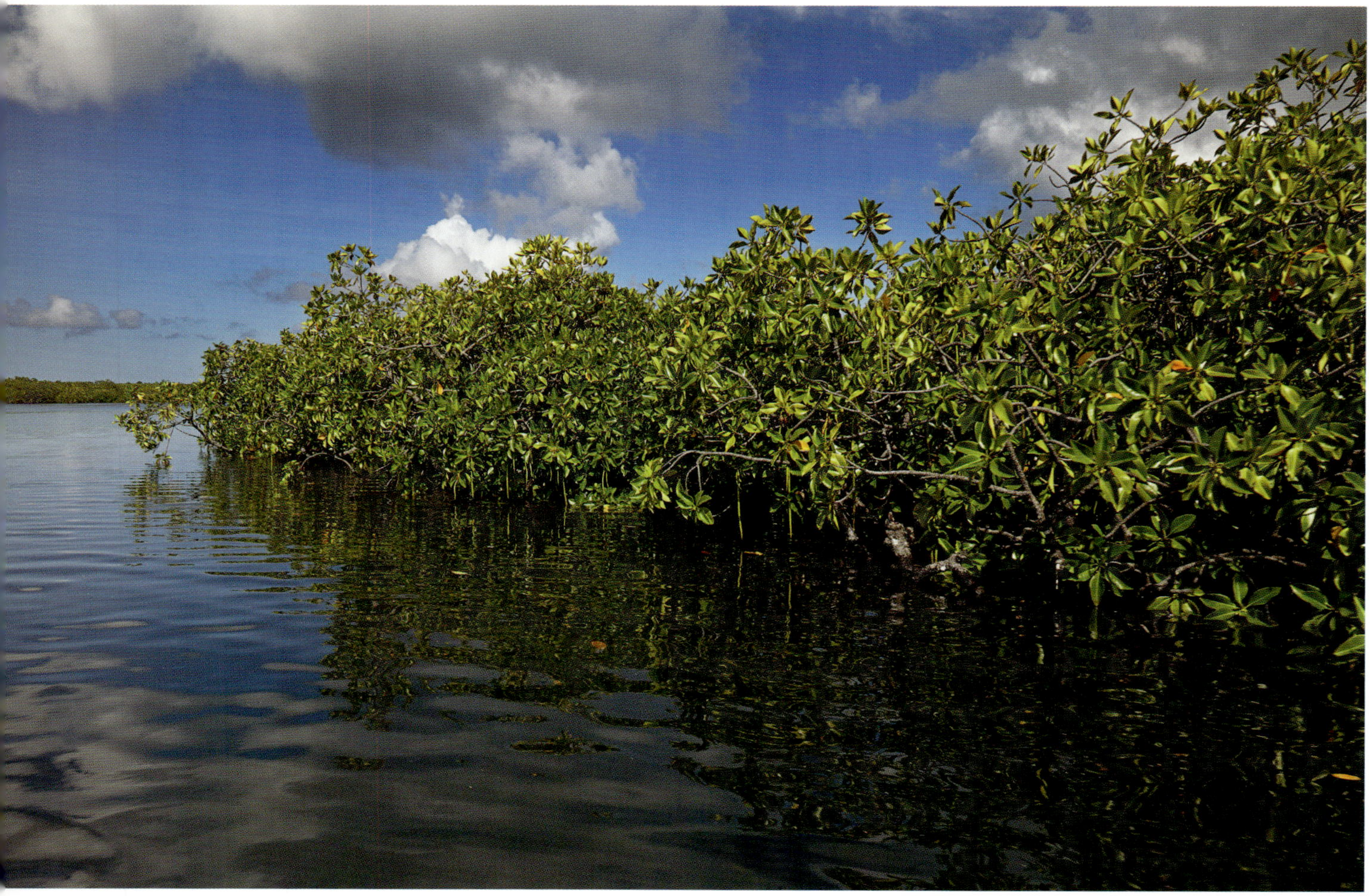

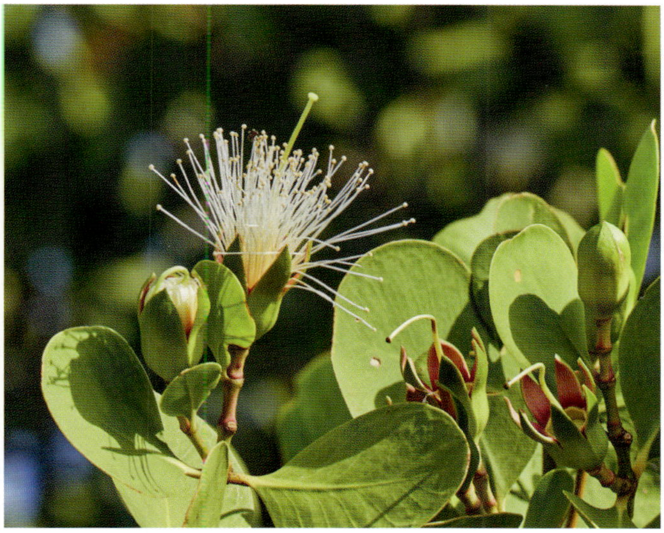

Above: The beautiful white flower of the mangrove tree *Sonneratia alba* (known locally as Pagatpat); Puerto Princesa, Palawan.

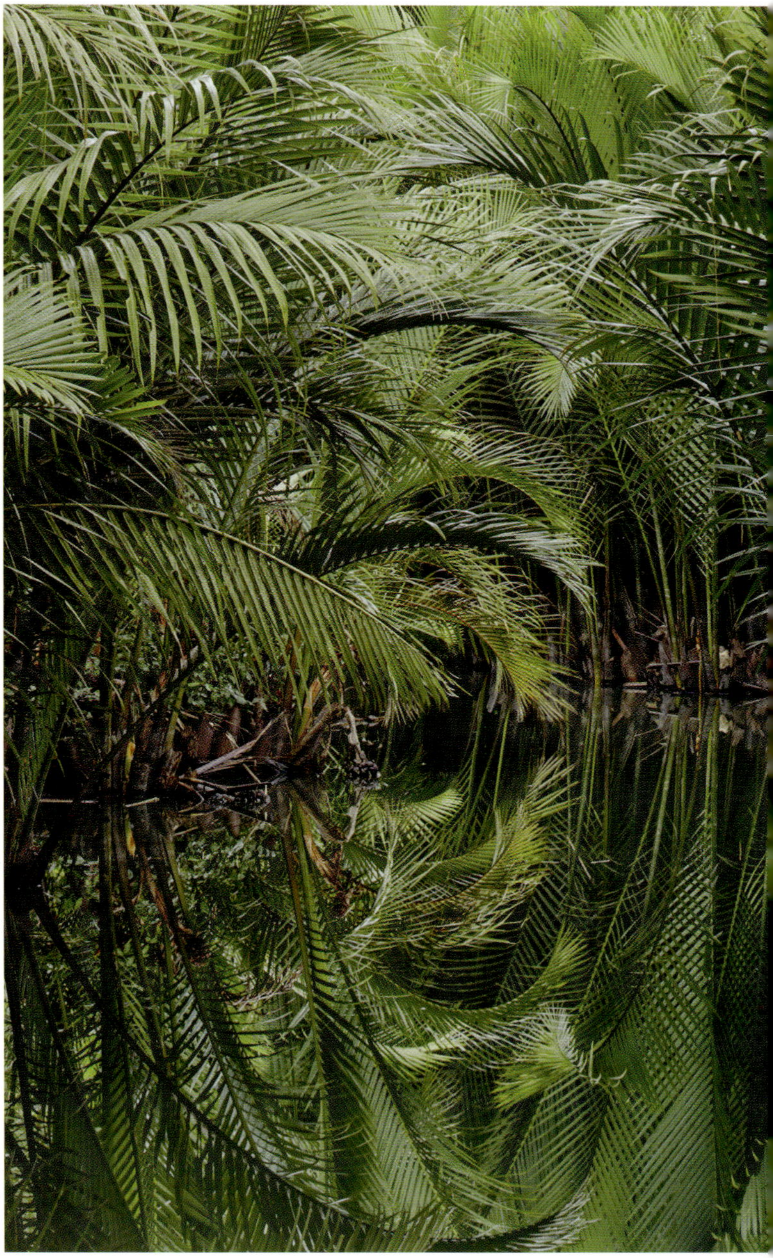

Right: A dense stand of Nipa Palms on the landward side of a mangrove forest, along the Tibag River, Baler, north-east Luzon.

mangrove biodiversity. Most of them we would recognize as trees, ranging in height from 3 to 20 m (10–66 ft), though there is also one very important palm, the Nipa Palm (*Nypa fruticans*), widely used in the Philippines in traditional construction.

The mangrove trees most commonly encountered in a Philippine mangrove include *Avicennia*, *Rhizophora*, *Sonneratia*, *Lumnitzera*, *Xylocarpus*, *Cerios* and *Bruguiera* species, which often occur in mixed species forests, particularly in areas closer to the sea, with species diversity decreasing towards the landward side. However, *Avicennia* species are often the main new colonizers, and often dominate the leading seaward edge of a mangrove.

Heading further inland, and particularly on passing inland along a brackish river estuary, the mangroves are increasingly dominated by Nipa Palms, to the point that in some areas – such as along the Loboc River in southern Bohol – vast expanses of saltmarsh are dominated by almost pure stands of Nipa Palm.

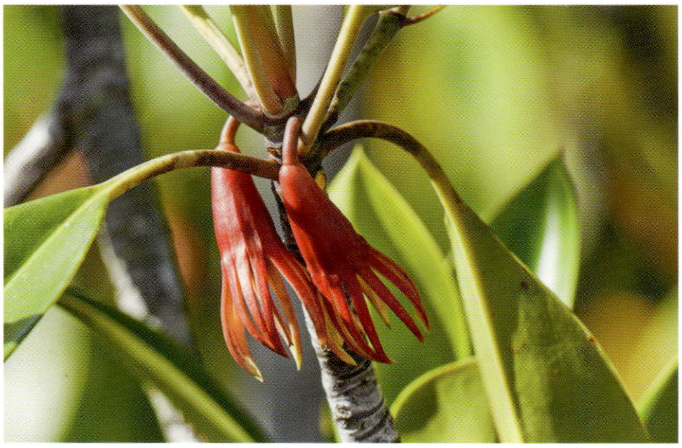

Right: The fruit of a *Bruguiera* species tree in the mangroves of Rasa Island, Narra, Palawan.

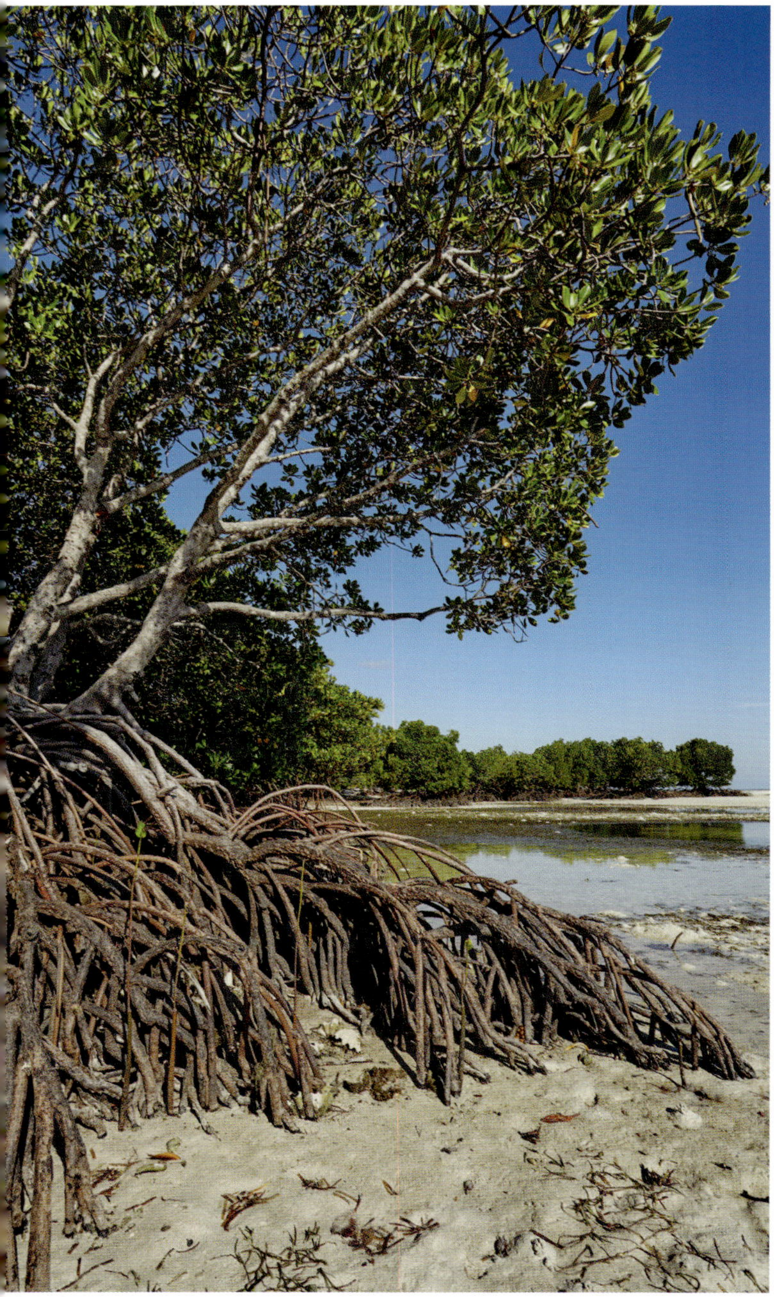

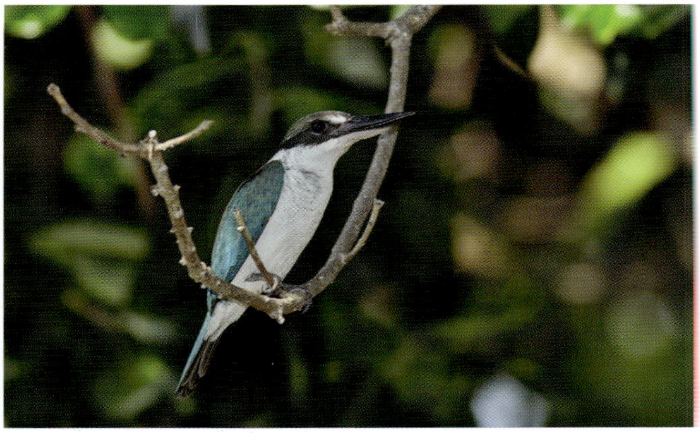

Above: The Collared Kingfisher, a common resident of mangrove forests.

Left: Thick aerial roots, or 'knees', seen here at low tide in a mangrove forest in Honda Bay, Palawan, help mangrove trees absorb oxygen even when submerged.

*Below: Mudskippers (*Periophthalamus *species), a type of amphibious goby, are commmon in mangrove forests.*

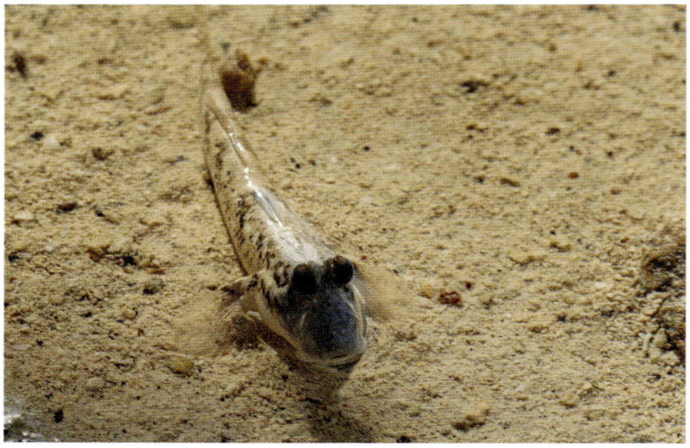

Interestingly, because this species is a palm and not a tree, it is sometimes not considered a mangrove species, even though it is clearly an essential part of the habitat as the forest progresses from tidal swamp, through saltmarsh, to solid land.

Two adaptations mangrove trees have to survive in their harsh environment are thick, waxy leaf coatings and masses of dense aerial roots. The former work to conserve hard-won fresh water, while the latter – consisting of either dense masses of curving 'knees' or vertical, upwards pointing pneumatophores that make mangroves such dense, tangled places – specialize in absorbing oxygen to enable respiration in the roots.

The importance of mangroves

Mangroves are crucial to both the natural environment and the human economy. Not only do they protect the shore from wave action, helping to reduce erosion of low-lying coastal areas, but by stabilizing the soil they also help to build new land. A number of studies have found that coasts lined with mangroves fare much better against both typhoons and tsunamis than those without.

Mangroves also provide protection to the young of a vast range of marine animals, including fish, prawns and crabs that are vital to the fishing industry. Related to this, the trees are a home to quite a range of birds, most especially kingfishers, egrets, herons, bitterns and other waders, which come to feed on the rich pickings in the water. Another unusual animal frequently seen in mangroves is the mudskipper (a member of the goby family), an amphibious fish that in evolutionary terms is an intermediate form between marine and terrestrial animals.

Reptiles find mangroves a good place to live in. They range from a number of snake species, through various water monitor lizards, to the biggest of them all, the Saltwater Crocodile (*Crocodylus porosus*), which is still not uncommon in some mangrove areas.

As well as the collection of fish and shellfish, human communities also use mangroves as a source of building timber and fuel, particularly charcoal, a vastly cheaper material than petrol or paraffin.

Destruction and conservation of mangroves

It is estimated that until the early part of the twentieth century, the Philippines had about 400,000–500,000 ha (988,000–1,235,000 acres) of mangrove forests. By 1994 this had dropped drastically, down to about 120,000 ha (296,400 acres), and the Philippines looked set to completely lose one of its most important but most taken-for-granted environments.

Since that low point, however, things have steadily improved, and by 2003 mangroves had recovered to about 247,000 ha (610,000 acres). The latest figures published by the DENR show mangrove cover up to 310,000 ha (765,700 acres), so there is plenty of reason to hope for the future.

The main reason for the huge decline in mangrove cover was its large-scale commercial conversion to fish and shrimp ponds, which were originally thought to be more productive than the mangroves, but many of which have since been abandoned. The mangroves' resurgence has been due to a tide of replanting projects, following a realization of just how much coastal communities depend on them, coupled with an acknowledgement that they represent an effective protection against global warming and rising sea levels.

Visiting mangroves

Mangrove forests are fascinating places to visit, if only because they are so utterly different from any other forest type. They are also increasingly popular among birdwatchers looking for kingfishers and migrating waders. However, the mass of aerial roots makes exploring a mangrove forest on foot quite a challenge. Usually, it is wise either to hire a boat to explore at least the navigable channels, or wait until low tide makes exploration on foot along the seaward side possible. Alternatively, growing numbers of local communities have installed raised walkways through mangroves, making exploration possible at any state of the tide, penetrating right into the deepest parts of the forest, and with admission fees helping to generate some local income.

Below: A raised boardwalk through a mangrove makes it possible to explore these forests at any state of the tide. Local communities are increasingly building such facilities to cash in on growing visitor interest in exploring the mangroves.

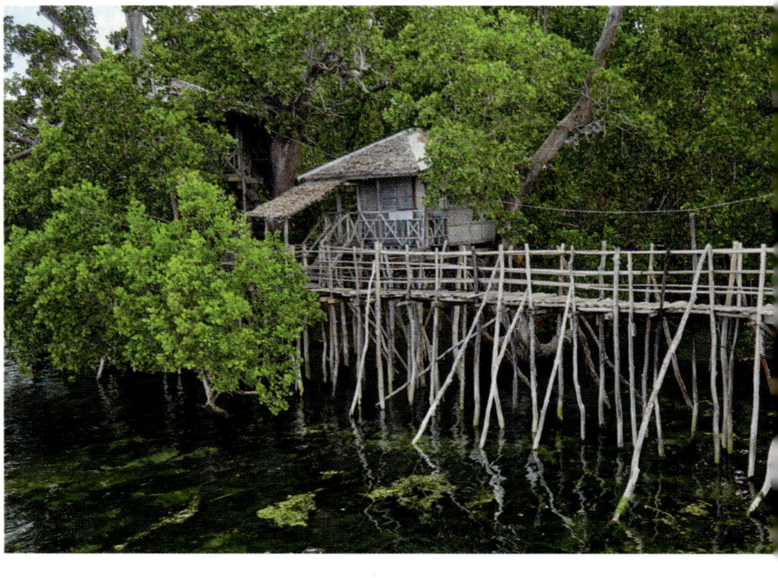

Beach forest

Beach forest consists of a narrow band of vegetation that grows along the landward side of many sandy or pebbly beaches, often also merging with the landward side of a mangrove forest. A short distance inland, once away from the beach, beach forest then merges with terrestrial lowland rainforest. This is recognized as a distinct habitat due to the very specialized nature of the vegetation needed to colonize such a harsh environment. Beach forest trees are adapted to survive the dry, high-salt content of the sandy beachside environment, as well as regular assault by strong, salt-laden winds.

Trees of the beach forest

Examples of trees typically found in beach forest include the Talisay (*Terminalia catappa*), Botong (*Barringtonia asiatica*), Agoho (*Casuarina equisetifolia*; known in English as Common Ironwood or Whistling Pine), Dapdap (*Erythryna variegata*; Indian Coral Tree or Flame Tree), and several pandan species (such as *Pandanus tectorius*; Screw Pine), though there are numerous others.

Terminalia catappa (sometimes known in English as the Indian or Tropical Almond) is a deciduous tree

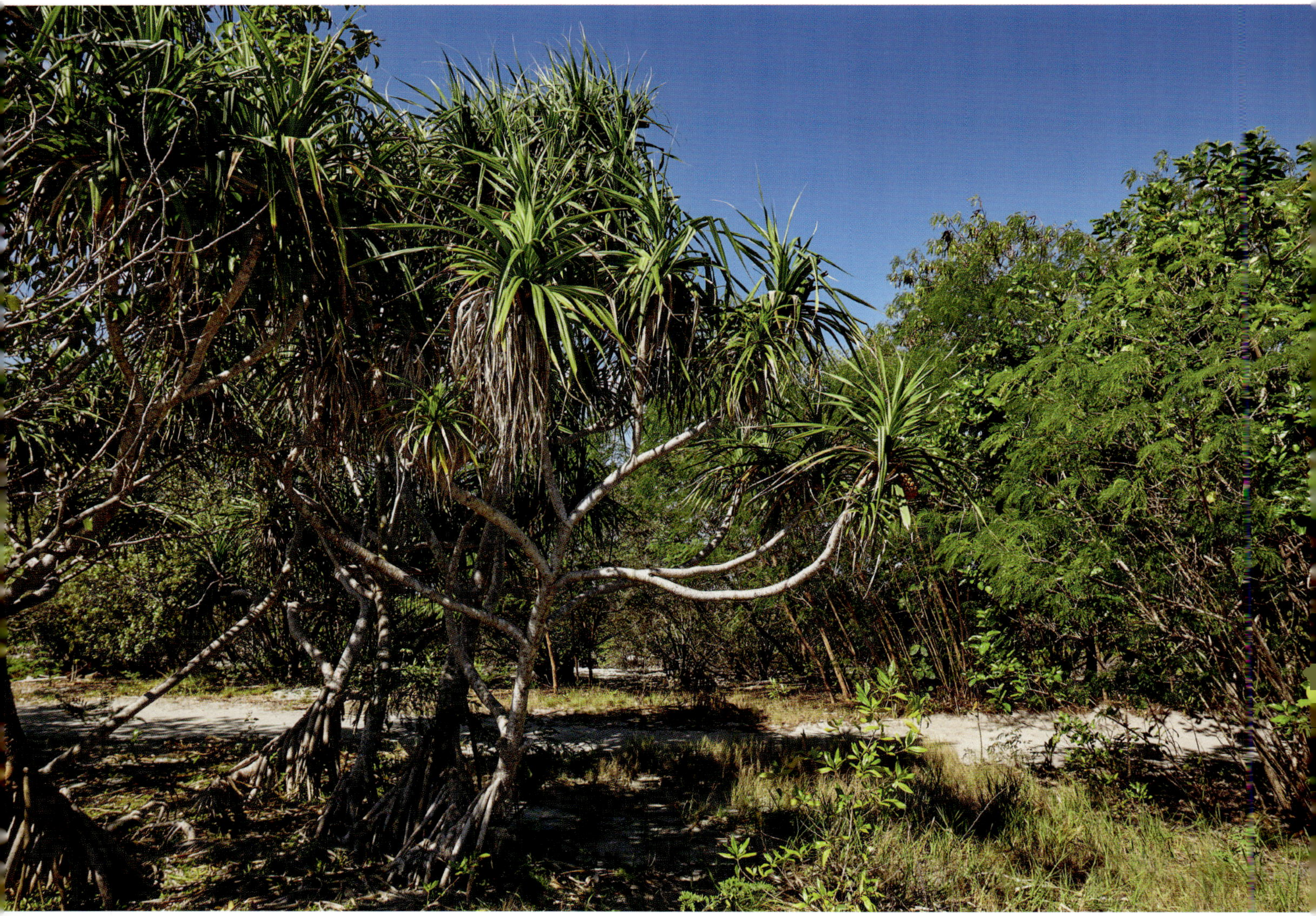

Below: *Beach forest is often quite scrubby and untidy, characterized by a mix of* Pandanus *and small leguminous trees. Here seen in Olango Island Wildlife Sanctuary.*

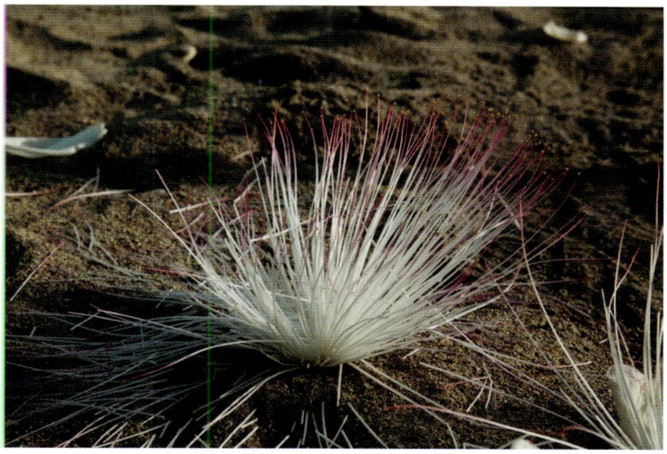

Above: In early morning the pompom-like flowers of the tree Barringtonia asiatica *are commonly seen lying on beaches that are still lined by beach forest.*

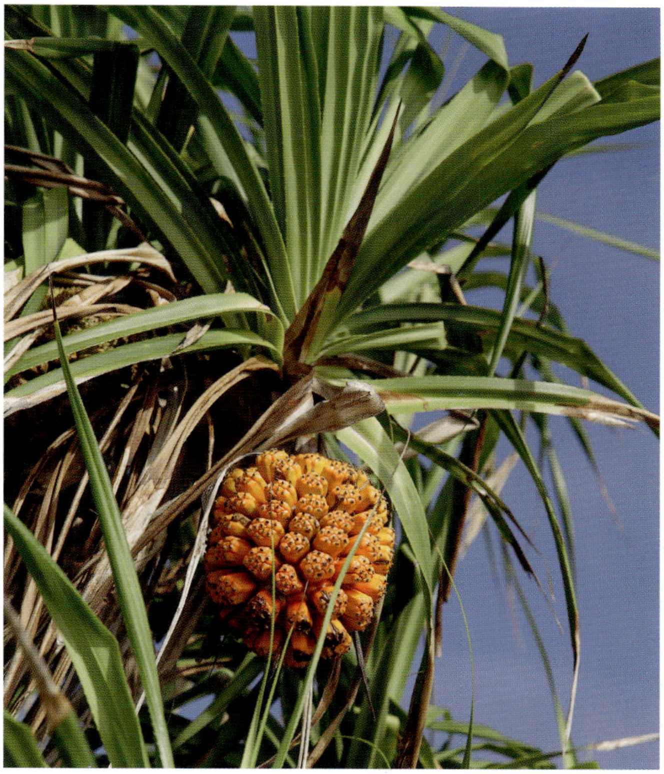

Right: The vividly orange fruit, vaguely resembling a pineapple, of the tree Pandanus tectorius*, is a common sight along forest-lined shores.*

that can grow to quite a large size. Its leaves are long, ovoid and glossy, turning reddish shortly before falling during prolonged dry periods. The evergreen *Barringtonia asiatica* is known mainly for its wonderfully delicate, pompom-like white but pink-tipped flowers. Opening mainly at night, the flowers fall into the sand each morning. Its fruit is rather box shaped, a little smaller than a coconut, and is able to float, a means by which it readily spreads. *Casuarina equisetifolia*, with its long, fine, needle-like leaves superficially resembles a pine, though it quite definitely is not one. Instead, the scientific species name *equisetifolia* describes the leaves' resemblance to a horse's tail. *Erythryna variegata*, a deciduous tree, is easily recognized by its large clusters of vibrant red flowers, which are particularly prominent when the tree is leafless, and which regularly attract nectar-hunting birds such as the Olive-backed Sunbird (*Nectarinia jugularis*).

Several pandan species are particularly common along sandy beaches, and in areas merging with the landward side of mangroves. Resembling palms, they are in fact trees. One of the most common shoreside pandans is *Pandanus tectorius*, the most notable feature of which is its bright orange, pineapple-like fruits.

Beach-forest conservation

Naturally restricted to quite a small environment anyway, in recent years beach forest has been decimated right across the tropical world, including the Philippines. This is at least in part because its frequently rather patchy, scrubby appearance does not conform well to the perceived idea of how a tropical beach paradise should look. Native beach forest has freqently been removed in favour of resort buildings and/or coconut palms, along with many other types of coastal development. However, the foolhardiness of this has started to dawn in some quarters, along with the realization of just how effective the native beach forest is in stabilizing beaches and in resisting high winds and storm waves. This was highlighted by the damage inflicted in 2013 by Typhoon Haiyan (known as Yolanda in the Philippines), the most powerful typhoon ever to have hit the country. Since then, there have been growing numbers of beach forest replanting programmes.

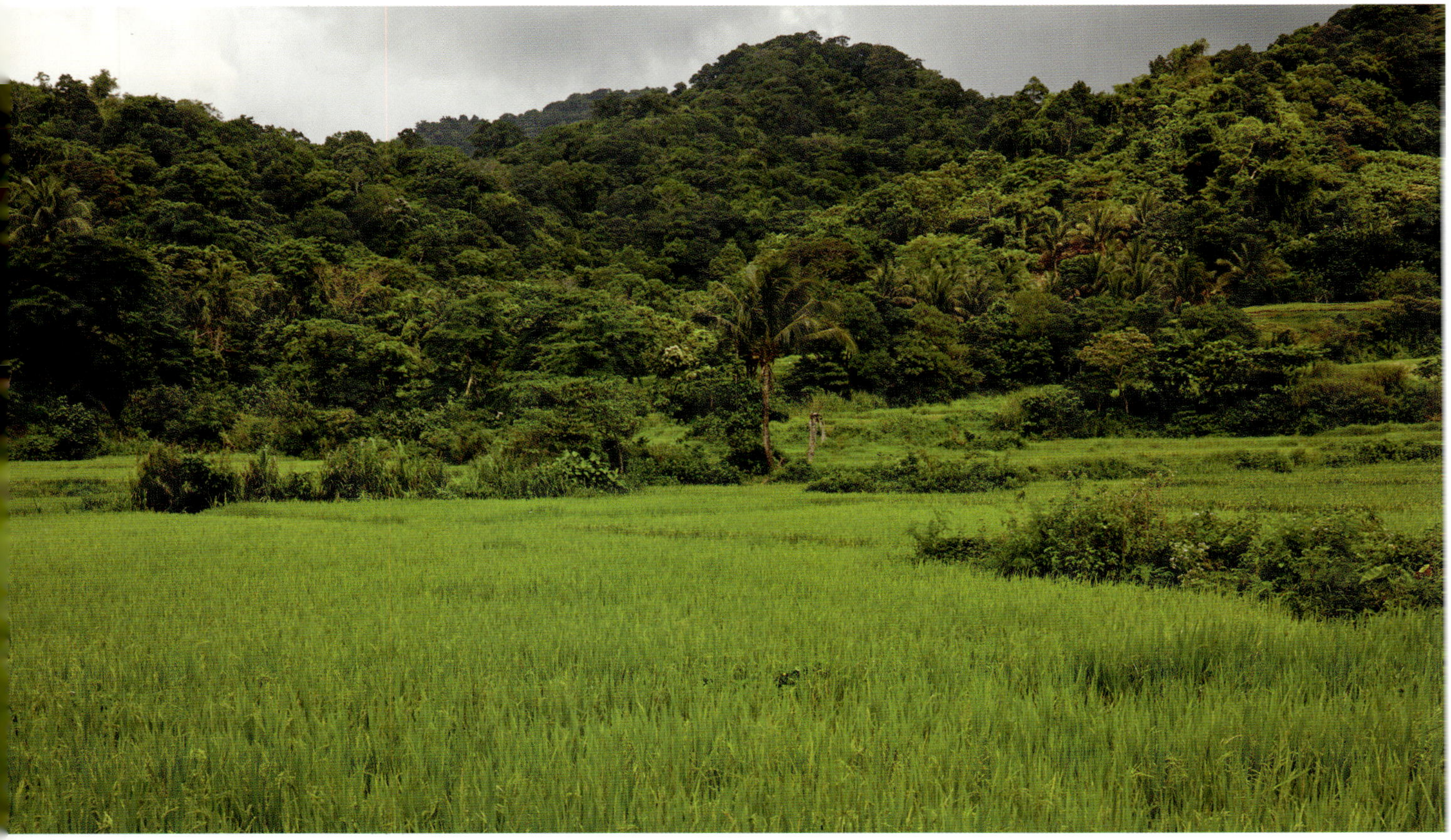

Lowland rainforest

As you move inland away from the shore, so the beach forest rapidly gives way to lowland rainforest, the domain of many of the world's biggest trees, the archetypal tropical rainforest for which the humid tropical regions of the world are so rightly famous. Here the forest has three layers, ranging from shrubs, small trees, palms and saplings at the lowest level, followed by mid-size trees in the middle and finally the tallest of the trees forming the canopy 40–60 m (130–200 ft) above the ground, with a few emergent trees standing above the canopy even higher, at up to 80 m (260 ft) high. Dominant among these giants, particularly in the Philippines, is a family of trees called the Dipterocarpaceae, so named for the winged structure of their fruits, which allow their seeds to drop gently to earth with a parachute effect.

Dipterocarp trees, as they are known, are spread throughout the whole tropical world, and consist of almost 700 species. Borneo is generally identified as the world's main dipterocarp hotspot in terms of species diversity, but the Philippines is not far behind, with almost 50 species, nearly half of them endemic to the country. Dipterocarp trees are divided into a range of genera, with different distributions around the tropical world. In the Philippines, the most common are *Dipterocarpus*, *Shorea*, *Parashorea* and *Hopea* species, though others that are regularly encountered include *Anisoptera* and *Vatica* species.

The majority of dipterocarp trees prefer deep loamy soils in lowland areas where temperature fluctuations are small, and in regions with little or no dry season. The greatest dipterocarp diversity occurs at the lowest altitudes, below about 600 m (1,970 ft), but a number of species are still able to thrive above this, creating 'upland dipterocarp forest'. Few dipterocarps are able to grow above 1,100 m (3,600 ft), though one that does reach into the lower montane forest region is *Shorea polysperma*, known locally as Tanguile.

Although the dipterocarps frequently dominate the Philippine lowland rainforests, there are many other tree types, including the Narra (*Pterocarpus indicus*), a host of fig trees (*Ficus* species), Ipil (*Instia bijuga*) and quite a range of forest pandans. The Ipil, Narra and many of the fig trees are quite large, with the Ipil reaching up to 50 m (164 ft) high and with a diameter of more than 2 m (6.6 ft). The Narra is well known for its beautiful yellow scented flowers, and is extensively planted in urban areas and parks both for this and for its shade value. It is so well regarded that it has become the Philippines' national tree.

The fig trees consist of a large number of species, many of them central to the ecology of lowland tropical rainforests for their seemingly endless supply of fruits. As a result, they are a major source of food for fruit-eating wildlife, especially the many fruit bats that live in Philippine rainforests, as well as birds such as hornbills and pigeons, plus ground-living mammals including rodents, deer and pigs. One of the most remarkable features of several fig species is their 'strangler fig' behaviour. Such a tree often starts life as

Opposite: *As the flat, low-lying farmlands give way to mountainous slopes, so the forest starts to take over, as here on the edge of the Cordillera Central, near Pagudpud, in northern Luzon.*

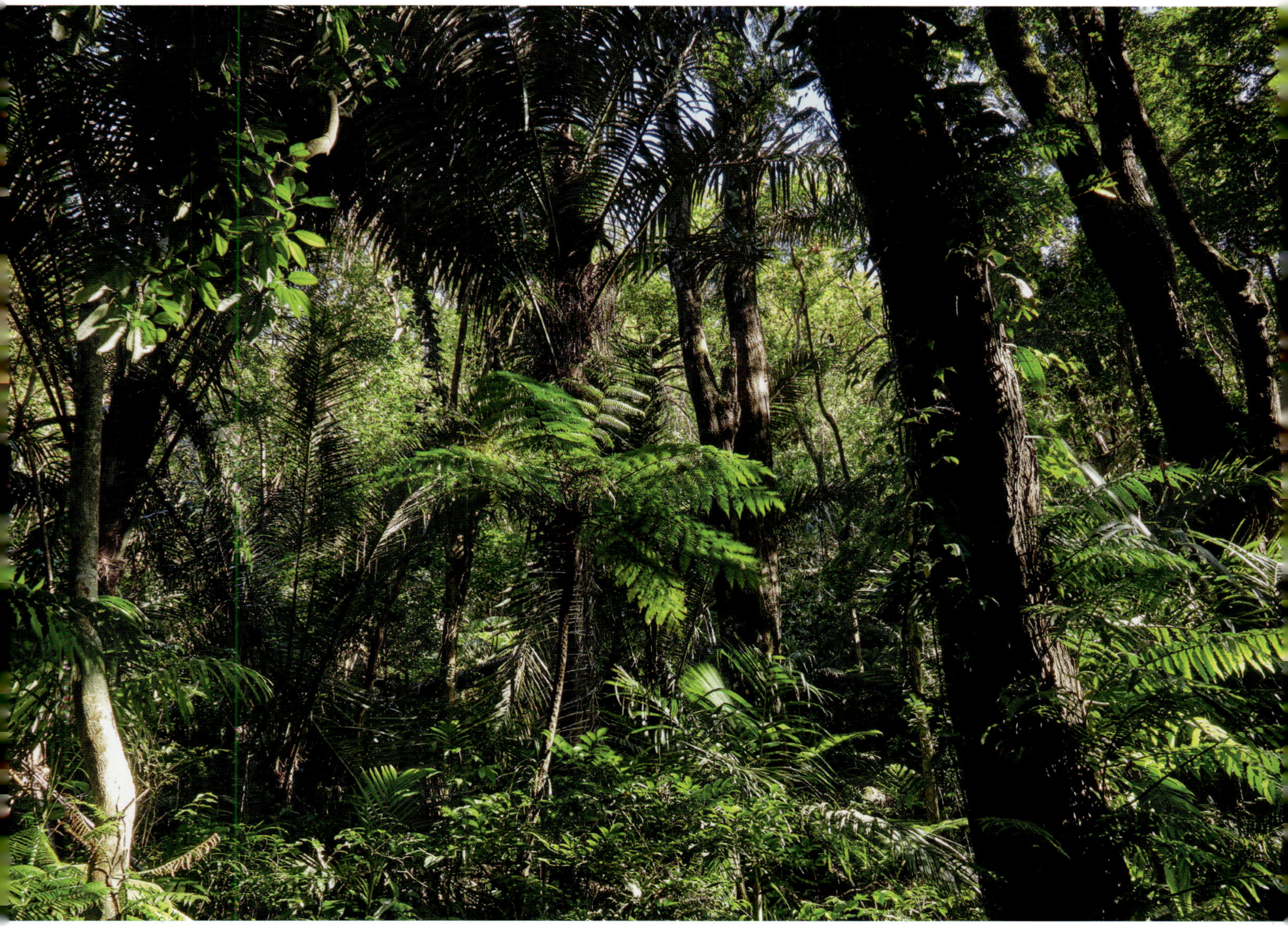

Below: *A mass of large trees and palms typifies lowland rainforest, the presence of a single tree fern revealing that this forest is at some altitude; in Bulusan Volcano Natural Park, southern Luzon.*

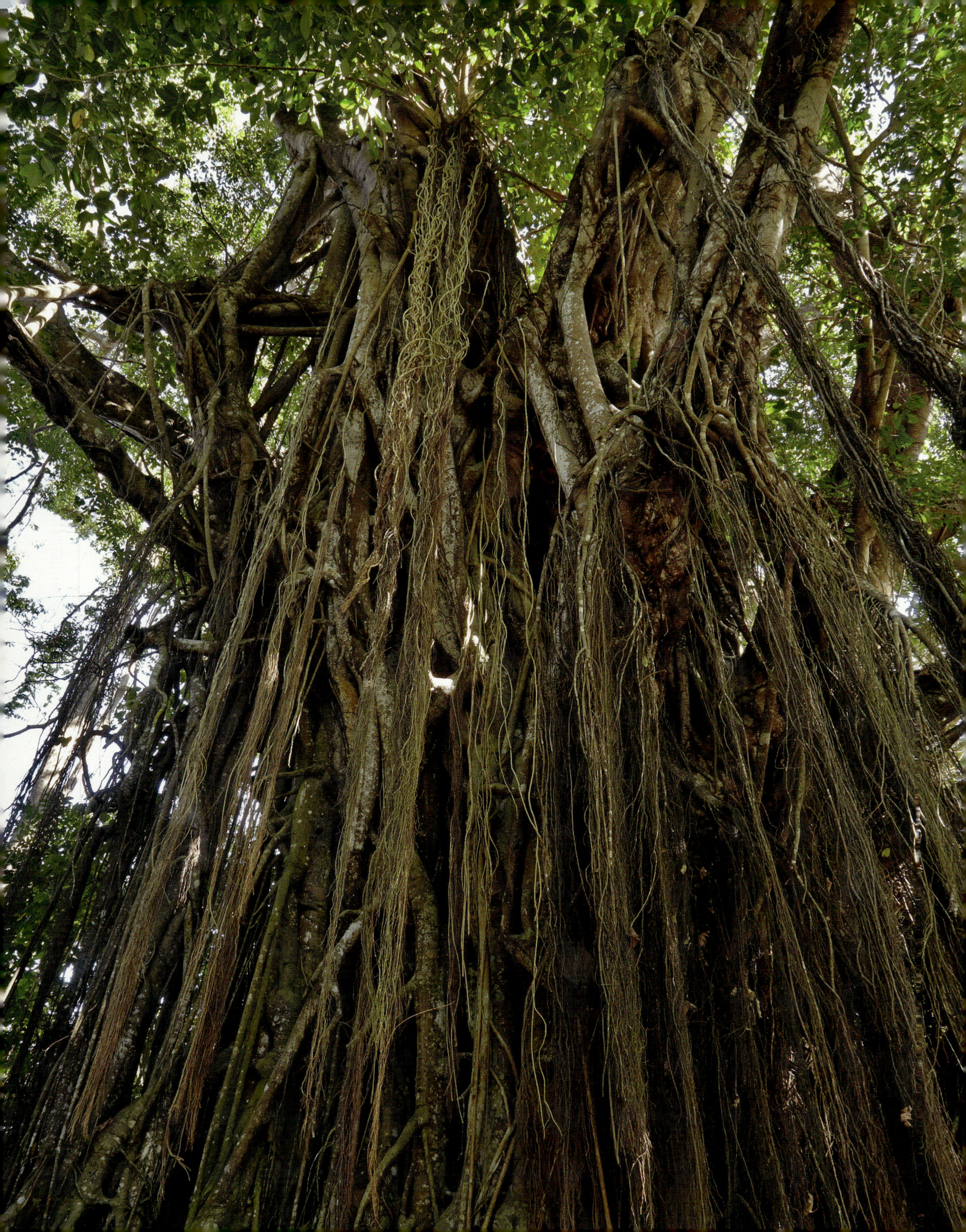

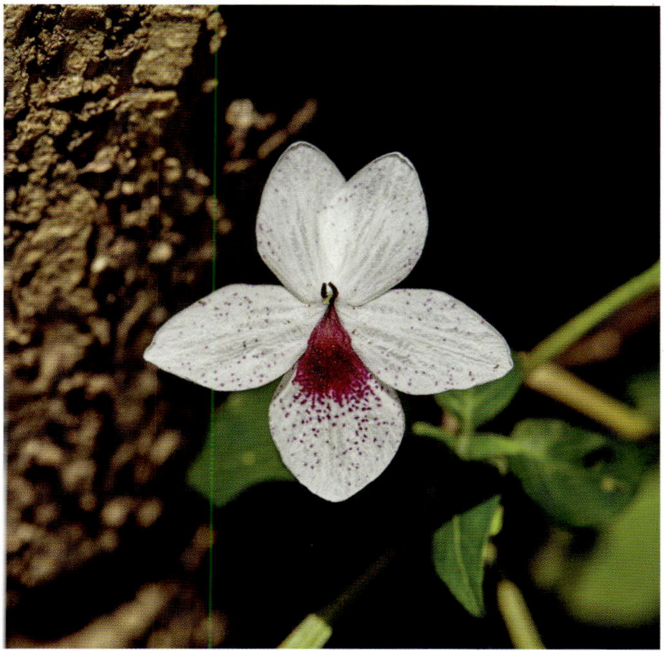

*Above: The beautiful flowers of the Shooting Star, or False Eranthemum (*Pseuderanthemum bicolor*), are sometimes seen in lowland rainforests, as here on Mt Talipanan, Puerto Galera, Mindoro.*

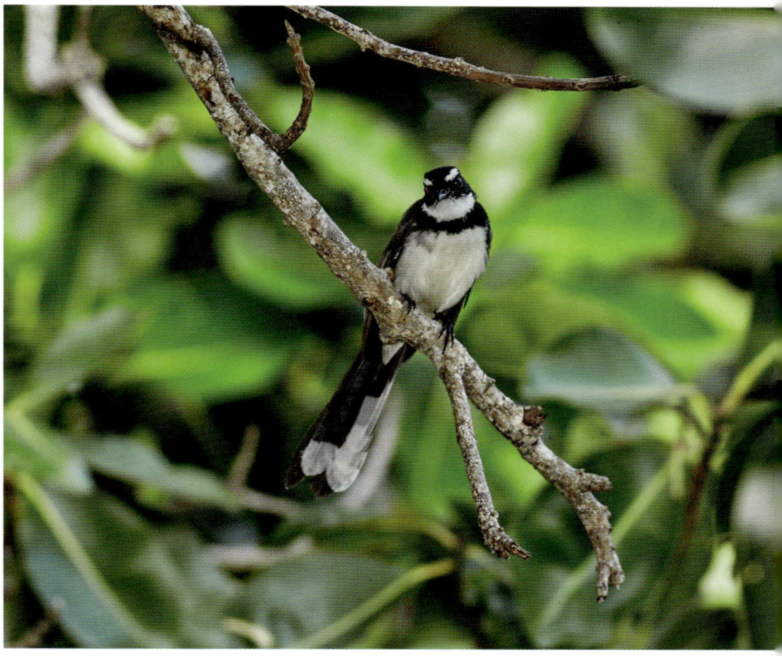

Above: The Pied Fantail is quite a common bird in lowland forests, parks and gardens. Beyond the Philippines, it is found in Malaysia and the islands of Indonesia.

several epiphytic saplings growing in the branches of another tree. These saplings send aerial roots to the ground, which then enable them to grow, using the host tree as scaffolding. Eventually, after a number of years, several saplings coalesce to completely encase, smother and kill the host tree, leaving only a large and often almost hollow fig tree.

Although trees dominate, of course they are not the only plants in the lowland rainforest. Ferns, orchids and epiphytic plants are quite common, but in closed canopy primary rainforest (that is, forest that is mature and has never been logged) lianas and bamboos are quite rare. They do become much more prevalent, however, in forests with a more open canopy (such as logged secondary forest), in which more light reaches the forest floor. Palms can also be quite common, including several species of rattan (*Calamus* species).

*Opposite: An enormous fig tree (*Ficus *species), characterized by a tangle of aerial roots reaching down from even the highest branches, the inner roots fusing with one another to create the tree's trunk.*

Evergreen and semi-deciduous lowland rainforests

There are, very broadly speaking, two types of lowland rainforest in the Philippines: evergreen and semi-evergreen (or semi-deciduous). The latter type of forest exists in parts of the country that experience a prolonged dry season, with some of the trees dropping their leaves to conserve water.

Evergreen forest predominates in regions where there is no rainless period, or at most only a short one. These include large parts of Mindanao, Samar, Leyte and Bohol, most of the Bicol peninsula of southern Luzon, the eastern side of northern Luzon (essentially along the Sierra Madre mountains, particularly the parts facing east), parts of Mindoro and the west coast of Palawan.

Areas of the Philippines with a prolonged dry season (usually running from October to May), and which therefore are most likely to host semi-evergreen lowland rainforest, include the western side of northern Luzon, much of the western Visayas, parts of Mindoro and the east coast of Palawan.

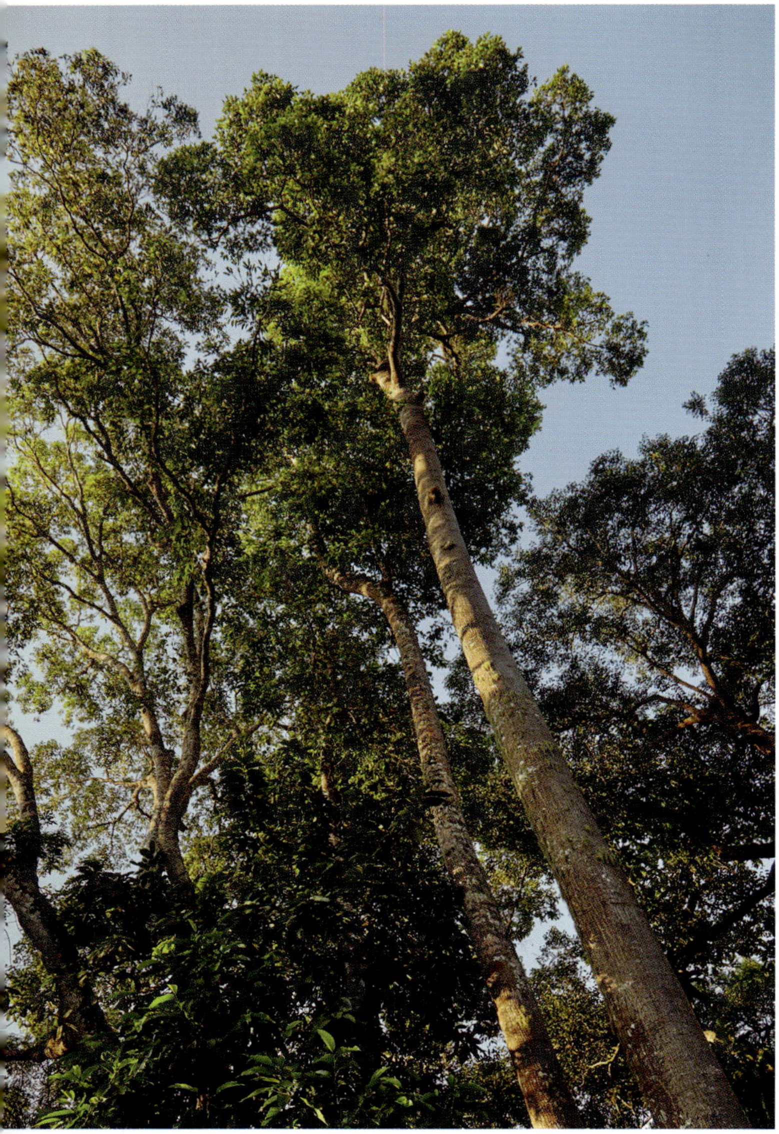

In general, dipterocarp trees greatly favour the evergreen rainforests, though some are able to tolerate the dry periods in the semi-evergreen forests. Indeed, at least two endemic species – the White Lauan (*Shorea contorta*) and Palosapis (*Anicoptera thurifera*) – are briefly deciduous when growing in semi-evergreen forests.

Destruction of the lowland forests

One of the characteristic features of many of the lowland trees is that, in growing very tall, they also grow very straight and free of any side branches, which makes them very attractive as timber. The wood of most is also quite hard and durable, further enhancing their usefulness. As a result, vast areas of lowland rainforest have been cleared just over the past 45 years, so that today little remains of the great forests that covered much of the Philippines less than a century ago.

Just about all the lowland trees well known to Filipinos for the quality of their wood (often known rather generically as Philippine mahogany) are now highly endangered. They include the Apitong (*Dipterocarpus grandiflorus*), White Lauan, Red Lauan (*Shorea negrosensis*), Guijo (*S. guiso*), Ipil (*Intsia bijuga*) and Narra (*Pterocarpus indicus*), to name just a few. Another tree worth adding here is the Philippine Ironwood (*Xanthostemon verdugonianus*), which has one of the world's hardest woods and is highly prized because of it. This Philippine endemic is now so rare that it is found in only a few locations in Palawan, Leyte, Dinagat Island off northeastern Mindanao and Homonhon Island off Samar.

The areas where lowland rainforest survives are primarily in isolated pockets, like the Sierra Madre mountains of north-east Luzon, on Samar Island, parts of Mindanao and along much of the length of Palawan.

Left: Healthy lowland rainforest is characterized by tall, straight, branch-free dipterocarp trees, some reaching more than 60m (198 ft) tall, the prime target of loggers.

Below: A classic feature in many of the large trees of lowland rainforest is the huge bird-nest ferns that grow epiphytically on the trees' branches high up off the ground.

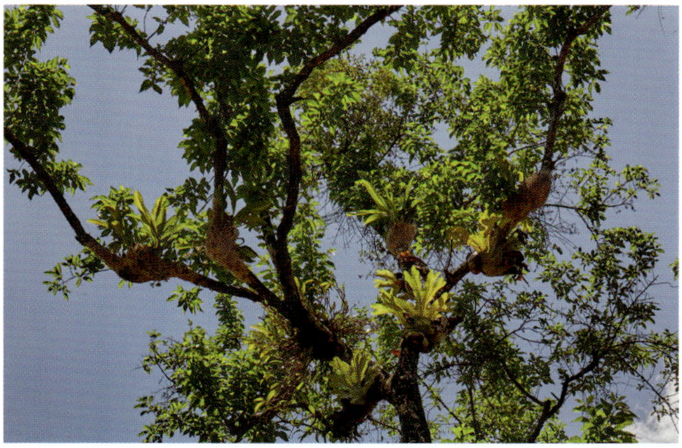

Calculating the decline in lowland rainforests

The destruction of the Philippines' lowland rainforest is little short of catastrophic, and despite bans on commercial logging forests are still being lost, though today it seems to be at a slower rate, and is usually through relatively small-scale illegal felling and relentless pressure to convert forest land to agricultural use.

The DENR's own figures, published in its Biodiversity Strategy and Action Plan 2015–2028, estimate that in 2010 (the last year for which figures seem to be available) the Philippines still had about 24 per cent natural forest cover, equating to a little less than 70,000 km² (27,000 mi²), about 10 per cent less than the forest cover in 2003. As if a 10 per cent decline in forest cover in seven years is not worrying enough, the DENR's figures also show a big drop over this period in the amount of closed canopy forest (that is, mostly primary forest that has never been logged), only partially offset by a rise in the amount of open canopy forest (secondary, logged over and hopefully recovering forest). These numbers indicate that primary forest is still being logged and converted to open canopy secondary forest.

These figures are for all forms of forest cover. When combined with numbers quoted by the Convention on Biological Diversity (CBD), the organization charged with overseeing international commitments to biodiversity protection, then the amount of lowland rainforest surviving across the entire country comes to a little over 43,000 km² (16,600 mi²), plus 12,600 km² (4,900 mi²) for montane forest (see page 74). This sounds as though there is a lot of lowland forest still surviving across the Philippines, but it represents only just over 14 per cent of the country's total land area, and a tiny fraction of the amount present just 50 years ago.

Over the past 20 years, most of the surviving areas of lowland rainforest have been included in large protected areas such as the Northern Sierra Madre Natural Park, Samar Island Natural Park, Northern Negros Natural Park, Mt Kitanglad Range Natural Park, and the latest, Cleopatra's Needle Critical Habitat, which of course is an essential step towards saving these remaining forests. What still remains to be done, however, is serious effective enforcement of existing laws to stop further logging and conversion of forest land to agriculture, combined with extensive restorative measures to recreate lost forest.

Below: The magnificent lowland rainforest of the Southern Sierra Madre mountains as seen from the Pacific Ocean south of Baler, and just outside the present-day boundaries of the Aurora Memorial National Park.

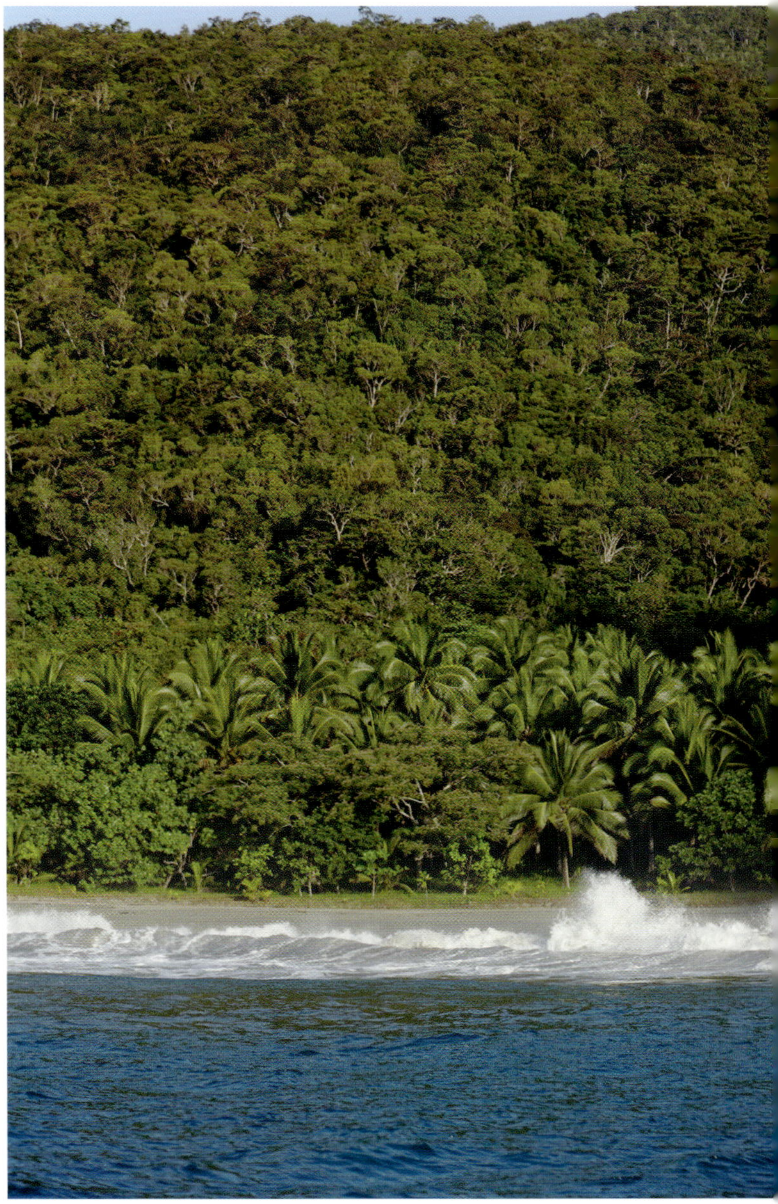

Limestone rainforest

An estimated 10 per cent of the Philippines' land consists of karst limestone, built up over millions of years by untold generations of corals and their calcium carbonate skeletons, then lifted above sea level – often many times over – by waves of tectonic uplifting, and alternating high and low sea levels. Being highly porous, rainwater tends to run through this limestone, eroding the rock as it filters through. The erosion, however, is uneven, resulting in caves, crevices, crags, outcrops and sheer cliffs, giving us the spectacular karst limestone landscape that is so well known across much of Southeast Asia, including the Philippines. Some of the largest areas of karst limestone in the Philippines are on Samar Island and in parts of the Cordillera Central mountains of north-west Luzon. However, the most famous from a visitor's perspective are probably those in northern Palawan, particularly the magnficent karst islands of Bacauit Bay around El Nido, off Coron Island in the Calamian group, and the rugged landscape of Puerto Princesa Subterranean River National Park.

Limestone and its vegetation

The fact that karst limestone is so porous to water creates a challenge for vegetation growing on it: there is simply very little water (or dissolved nutrients) at the surface for the plants to feed on, resulting in the constant threat of drought, even during quite short

Below: *The rugged cliffs of Palawan's Coron Island represent one of the Philippines' best-known limestone rainforests. The poor soil covering results in a forest made up largely of stunted, scrubby trees.*

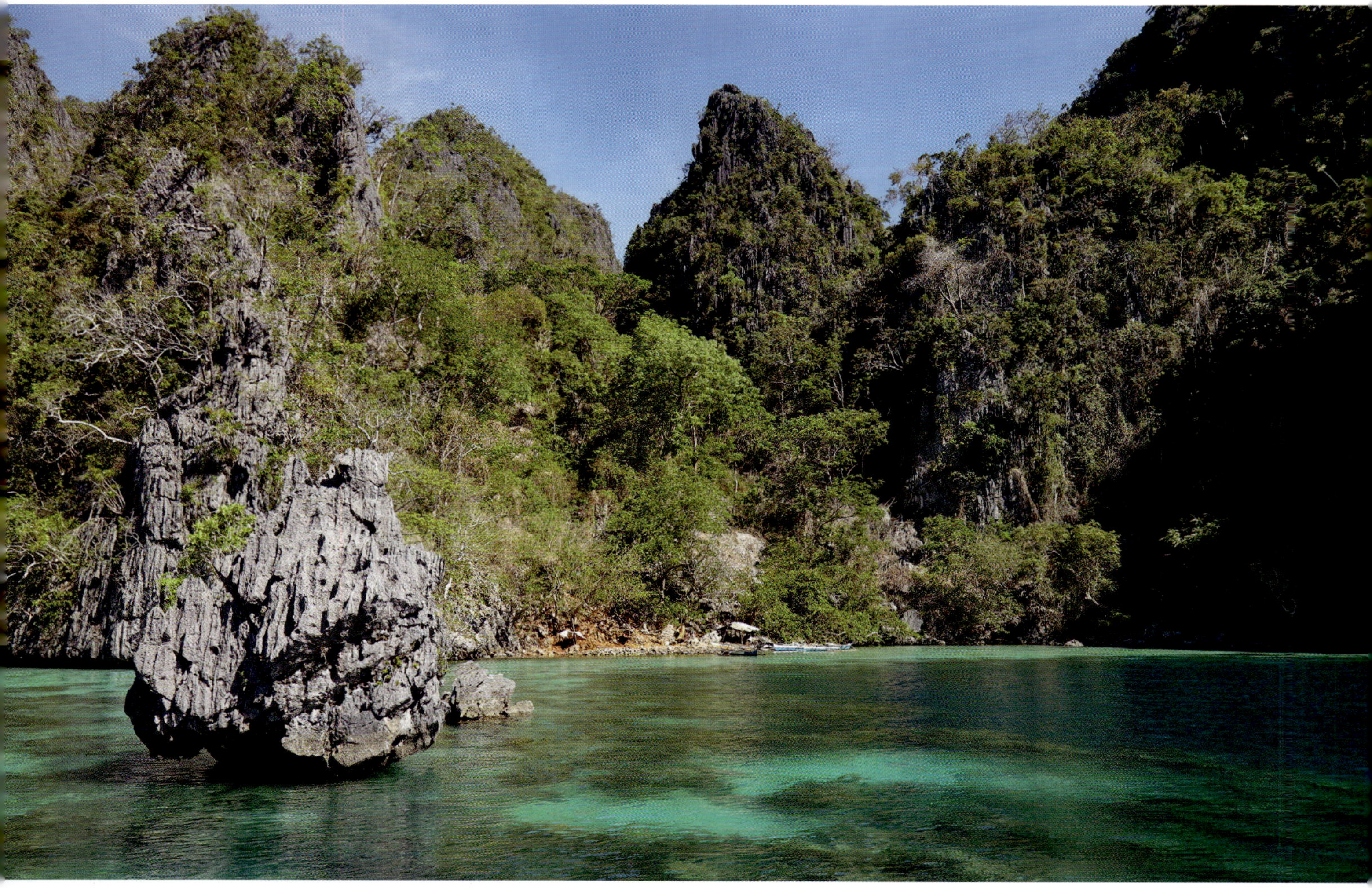

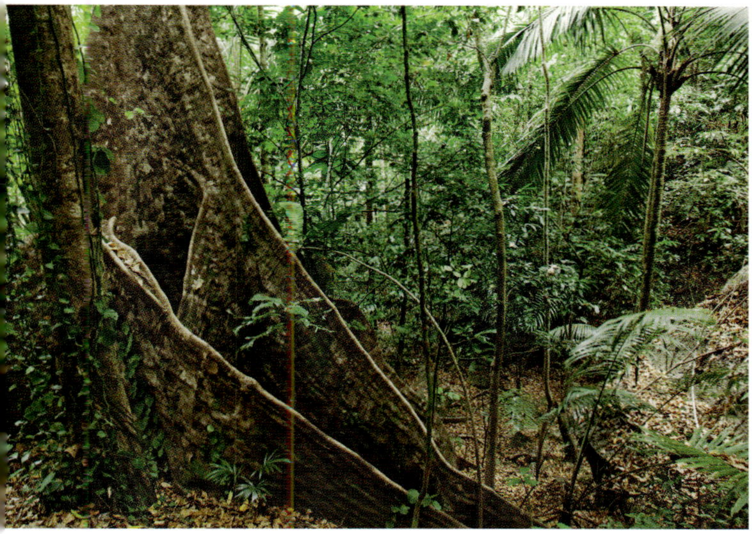

Above: In limestone areas with gentle slopes and deep soils trees can grow to significant sizes, including some dipterocarp trees as seen here on Lagen Island, in Bacuit Bay, Palawan.

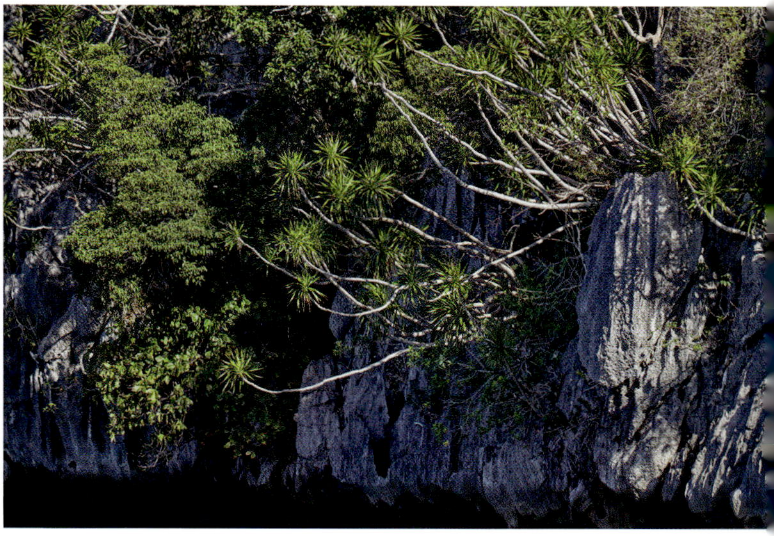

Above: The Dracaena *species, an extreme specialist, is able to colonize even bare rock, tolerating serious drought conditions, as shown here growing out of a cliff on Coron Island, Palawan.*

rainless periods. Consequently, many limestone areas require very specialized plants that are able to cope. That said, there is a wide variety of limestone forest conditions and hence species able to grow, ranging from areas where not only is the limestone covered with quite dense soils but also the climate has very little or no dry season, to places with prolonged rainless periods and little if any covering soil. The former conditions can result in dense forests, rich in dipterocarp trees, rattans and other trees typical of lowland forests. Examples of this type of limestone forest include those of Samar Island Natural Park and Bohol's Rajah Sikatuna Protected Landscape.

The other extreme is typified by the rocky cliffs of El Nido and Coron Island, where extensive areas of bare rock are vegetated with drought-resistant plants typified by having thick succulent leaves, such as *Dracaena multiflora* and various pandans.

Between the two extemes is what might be termed the 'typical' limestone rainforest, often found in areas with distinct dry seasons, and dominated by deciduous trees that lose their leaves during the (mainly) March–May dry season. The classic tree of this habitat is the deciduous Molave (*Vitex parviflora*), which under good growth conditions can reach nearly 40 m (131 ft) in height. Other plants include figs (*Ficus* species), the Wild Almond (*Sterculia foetida*), bamboos, small leguminous trees and pandans.

In terms of the dipterocarp trees found in limestone forests, a very few may be limestone specialists but most also live in lowland evergreen rainforests, forming a limestone-tolerant subset of the dipterocarp family. They include species that are well known in the Philippines, such as the Tanguile and White Lauan, both Philippine endemics.

Below: Fruits of the Wild Almond, a typical limestone forest tree specialist, in the Calamian Islands, Palawan.

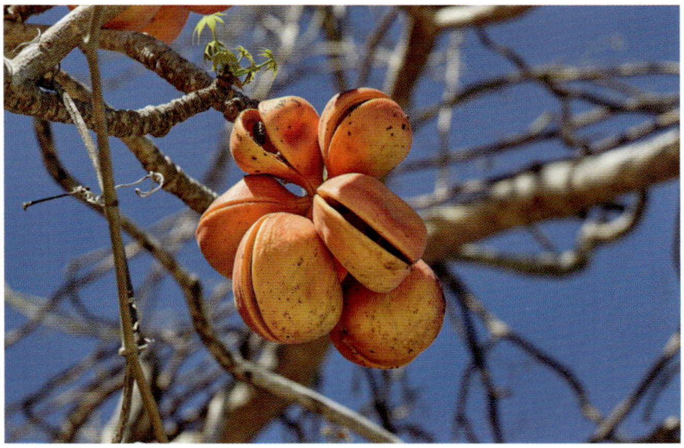

Freshwater marshes and swamp forests

Freshwater marshes and swamp forests cover a relatively small area of the Philippines and have not been as extensively studied as the lowland and montane rainforests, even though the swamp forests especially are important for Philippine biodiversity. They range from small slacks a short distance inland from a seashore, or riverside swamps, through to vast areas of lake and swamp-forest habitats covering thousands of hectares.

The first two occur in a number of lowland rainforest environments, including some of the coastal forest in El Nido, and along the banks of the Loboc River in southern Bohol. The large marsh environments, however, are restricted primarily to Mindanao, though there is also a smaller area in the Sab-a Basin of northern Leyte. Other areas that would have been marsh at one time, such as the Candaba Swamp in the lowlands of northern Luzon, were almost entirely converted to farmland some time ago.

The Mindanao marshes

The remaining large Mindanao marshes are Liguasan Marsh, which covers 280,000 ha (691,600 acres) of southern Mindanao, and the 90,000-ha (222,300-acre) Agusan Marsh in the east of the island. Both occupy low-lying basins surrounded by mountain ranges that feed rivers down into the basin, from where the water flows very slowly to the sea, the excess water backing up near the tributary confluences to create the extensive marshes.

Agusan Marsh has received the most attention, with a little less than half of its area protected as a wildlife sanctuary (see page 190). The much larger Liguasan Marsh (which is actually an aggregation of three smaller marshes) remains mostly unprotected and rather less studied – though this may be changing – problems that are at least in part due to long-running insurrections in the area. The two marshes are actually very different environments, Liguasan consisting largely of seasonally flooded land that is dry for parts of the year, during which it is farmed, and is generally devoid of forest. A relatively small proportion consists of a permanent lake, Lake Buluan. The marsh's biodiversity value lies mainly in its animal wildlife, which includes a large number of freshwater fish species, some of them endemic, extensive birdlife (in particular waders, many of them migratory), and a very rare home for both the Philippine Crocodile (*Crocodylus mindorensis*) and Saltwater Crocodile. Agusan Marsh, on the other hand, though also consisting of large lake areas, contains extensive swamp forests that are almost unique in the country. Furthermore, the marsh remains flooded the whole year round, though water levels do rise and fall considerably, and some areas become exposed.

Agusan Marsh has two types of swamp forest: freshwater and peat-swamp forest, the former consisting of trees growing in a flooded landscape, rooted in the marsh bed, the latter comprising forest growing on ground that contains water-saturated peat.

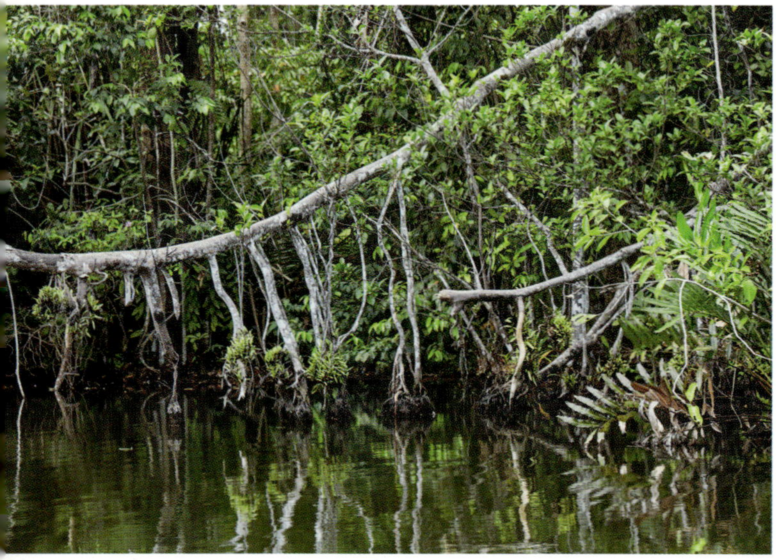

***Left:** A tangle of flood-resistant vegetation typifies this coastal freshwater swamp forest in El Nido-Taytay Managed Resource Protected Area, in Palawan.*

***Opposite:** Swamp forest consisting largely of* Terminalia copelandii *trees, growing out of the waters around the edges of Panlabuhan Lake, in the south of Agusan Marsh Wildlife Sanctuary.*

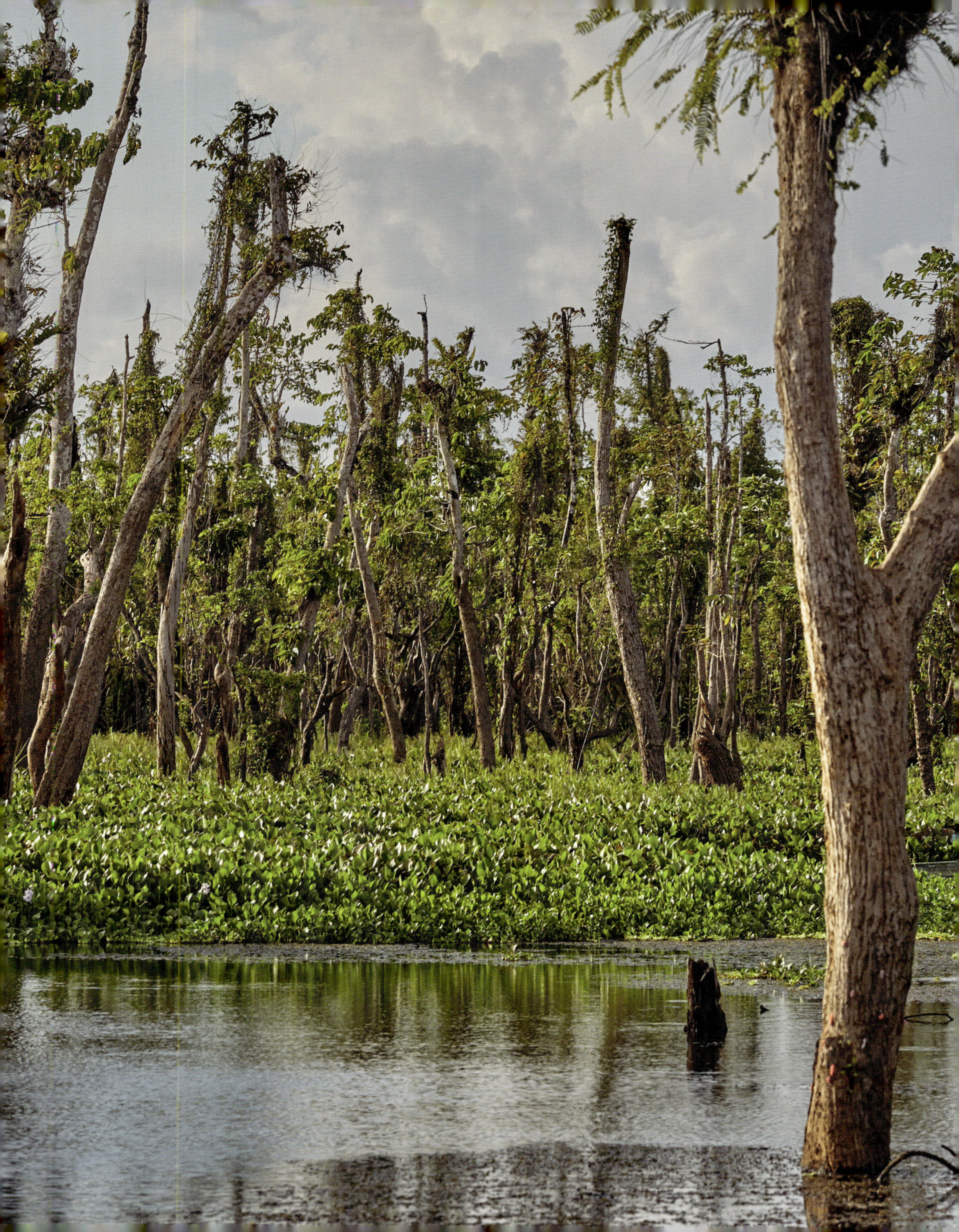

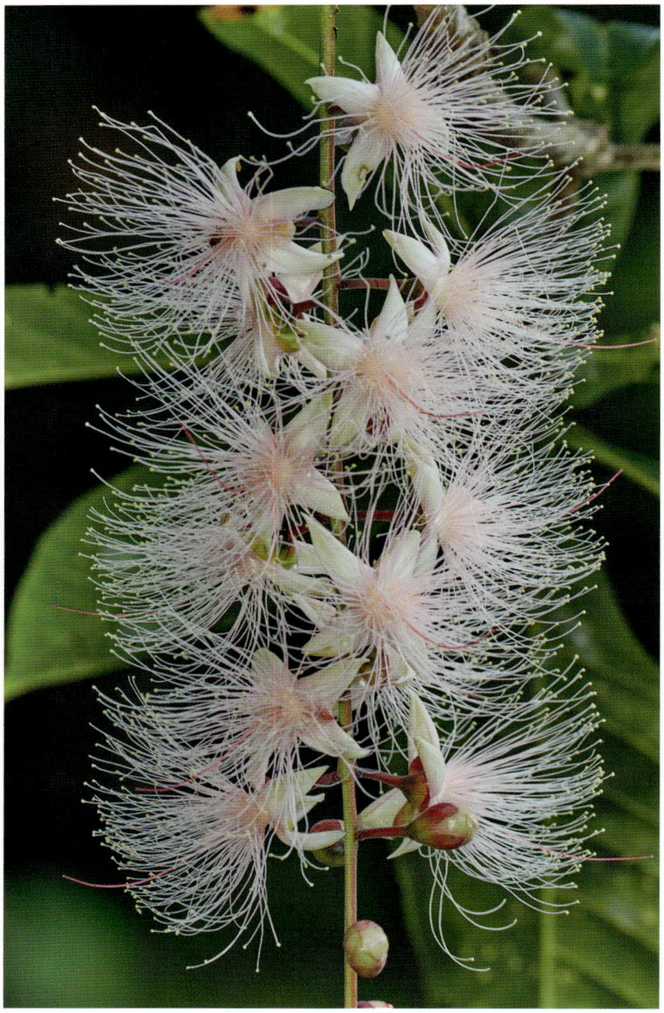

Above: The magnificent flowers of Barringtonia racemosa*, a freshwater swamp tree that flowers at night, dropping its flowers into the water in the morning.*

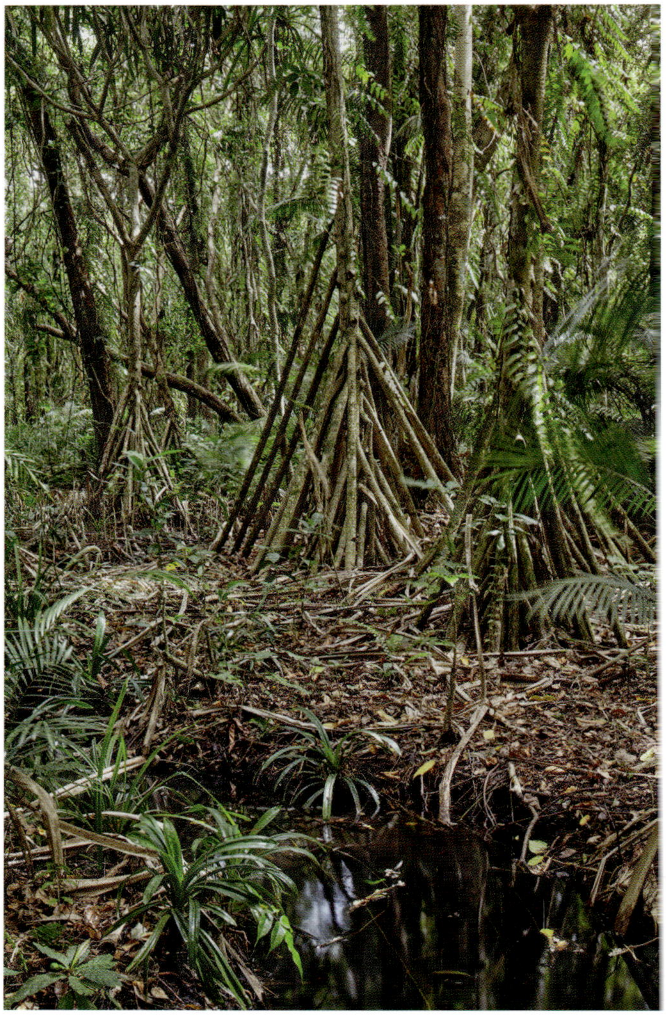

Above: Forest Pandanus *trees with buttress roots supporting the tall, thin trunks growing from swampy ground, in El Nido-Taytay Managed Resource Protected Area.*

Freshwater swamp forest

The freshwater swamp forests, whether in the vast Agusan Marsh or the smaller forest swamps elsewhere in the Philippines, have a mix of specialized tree species able to tolerate and thrive in permanently inundated environments. These include *Terminalia copelandii* (known locally as Lanipao), which grows extensively across Agusan Marsh; *Nauclea orientalis* (known as Bangkal in the Philippines, and in English as the Leichhardt Tree), which can grow to 30 m (100 ft) tall in ideal growing conditions; and *Barringtonia racemosa* (sometimes called in English the Powder-puff Tree), a small tree well known for its beautiful hanging streams (or racemes) of white flowers. The flowers are produced mostly at night, and are shed in the morning. The Sago Palm (*Metroxylon sagu*) is also sometimes found in the swamp forests, as are some prop-rooted *Pandanus* tree species.

Peat-swamp forests

Until recently, peat-swamp forests were largely overlooked in the Philippines, as a result of which their study is still very much in its infancy. Although they are relatively common in some other parts of Southeast Asia, particularly in Borneo, only a handful of sites has so far been identified in the Philippines, by far the largest of which is the Caimpugan peat forest in Agusan Marsh. Beyond Agusan, the next important peatland identified to date is in the Sab-a Basin in Leyte, an east–west elongated depression near the north coast, covering a little over 3,000 ha (7,400 acres). Unfortunately, nearly half of this has already been lost to agriculture, though plans are under way to protect and restore at least the remaining 1,740 ha (4,298 acres).

Peatlands found so far across the whole country add up to a total of only a little over 17,000 ha (41,990 acres), with about half of that coming from Sab-a and Caimpugan together. Other peatlands that contibute include further sites within Agusan Marsh, plus areas in Luzon's Quezon province, and in Mindanao's Agusan del Sur province.

Trees found in the peat-swamp forests include *Tristaniopsis micrantha* (called Tiga locally), a tree that is a member of the myrtle family, known for its orange bark; *Calophyllum sclerophyllum*, a stilt-rooted tree specialized for swamp environments; and *Pandanus* species trees.

In recent years the DENR has increased efforts to find more peatland forest areas and then start the process of bringing them into protection and restoration. This is in large part due to the Philippines' commitment under the Paris Agreement to bring its carbon dioxide emissions down. With the relatively recent recognition that peatlands are highly efficient storers of carbon (the International Union for the Conservation of Nature, or IUCN, estimates that the world's peatlands store about a third of the world's carbon), the Philippine government seems to be hoping that protection and restoration of its few peatland swamp forests will help. It remains to be seen just how much peatland and associated forest can be found across the country.

Below: A peaceful sunset scene on Lake Panlabuhan, in Agusan Marsh. Many of the swamp trees have become rather skeletal, possibly due to stress caused by rising temperatures.

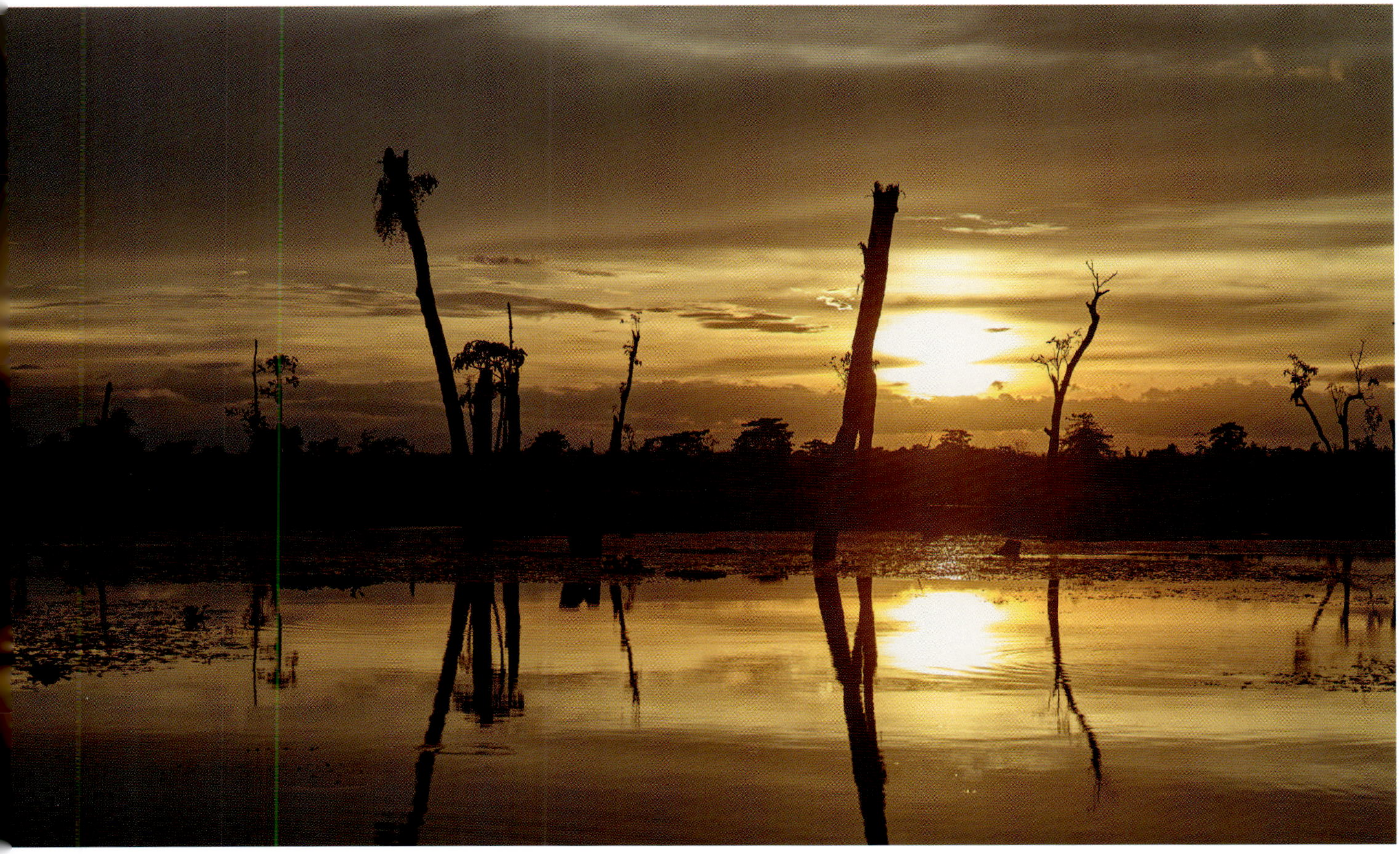

Montane rainforest

From an altitude of about 800 m (2,600 ft) and upwards the lowland rainforest starts to go through changes, so that by about 1,000 m (3,300 ft) it has become clearly quite a different kind of forest. The three vegetational storeys of the lowland rainforests have given way to just two storeys, in which even the canopy trees are, mostly, a good bit smaller. There are certainly fewer dipterocarp trees, the forest floor is increasingly crowded with ferns and many other

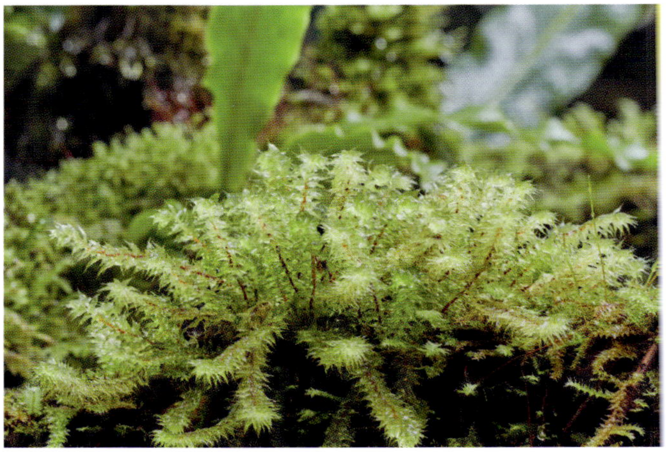

Below: *A view of typical upper montane forest on Mt Apo at an altitude of nearly 2,000 m (6,600 ft).*

Above: *A bed of sodden moss growing on a tree branch in the upper montane forest of Mt Apo.*

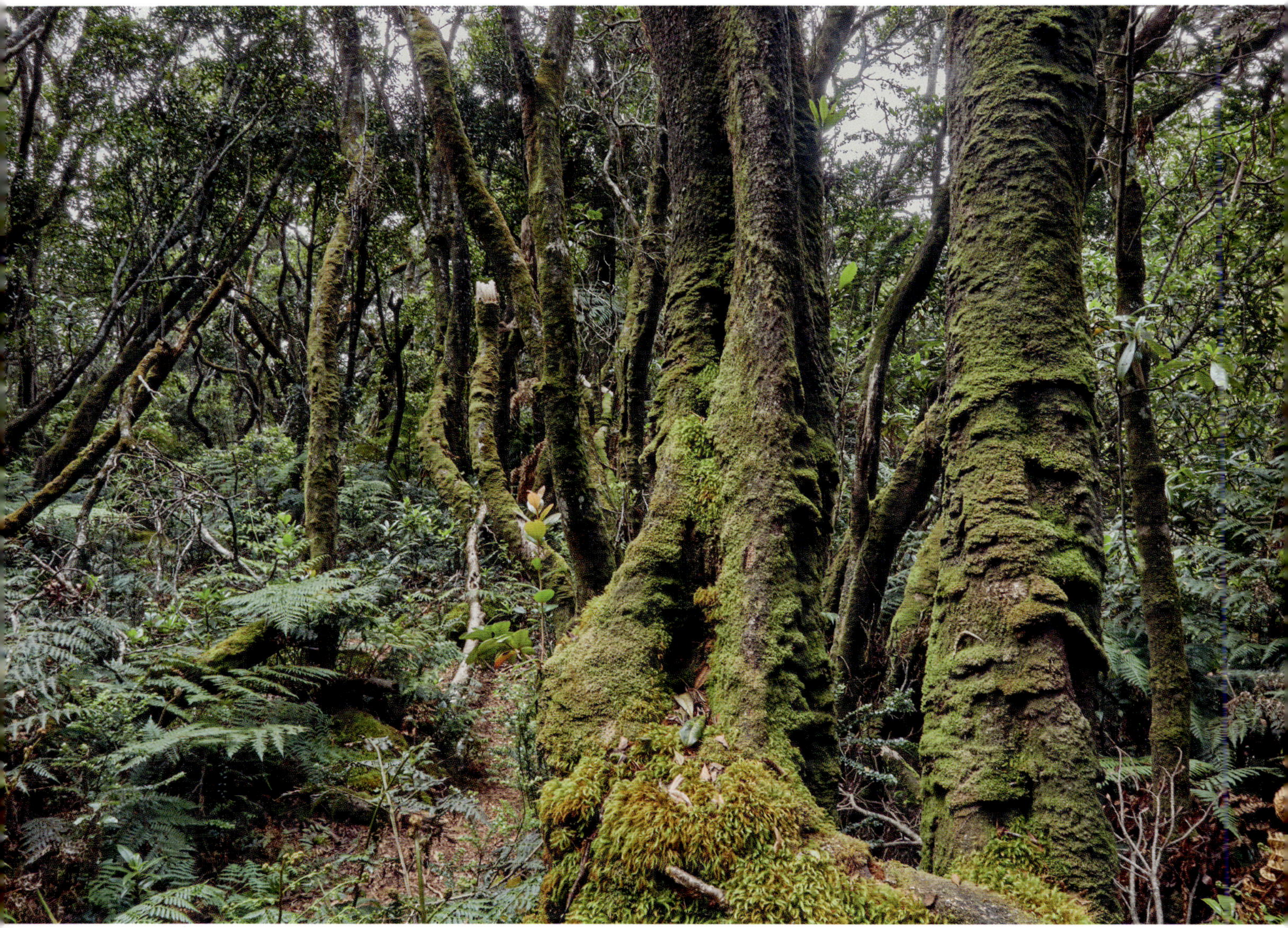

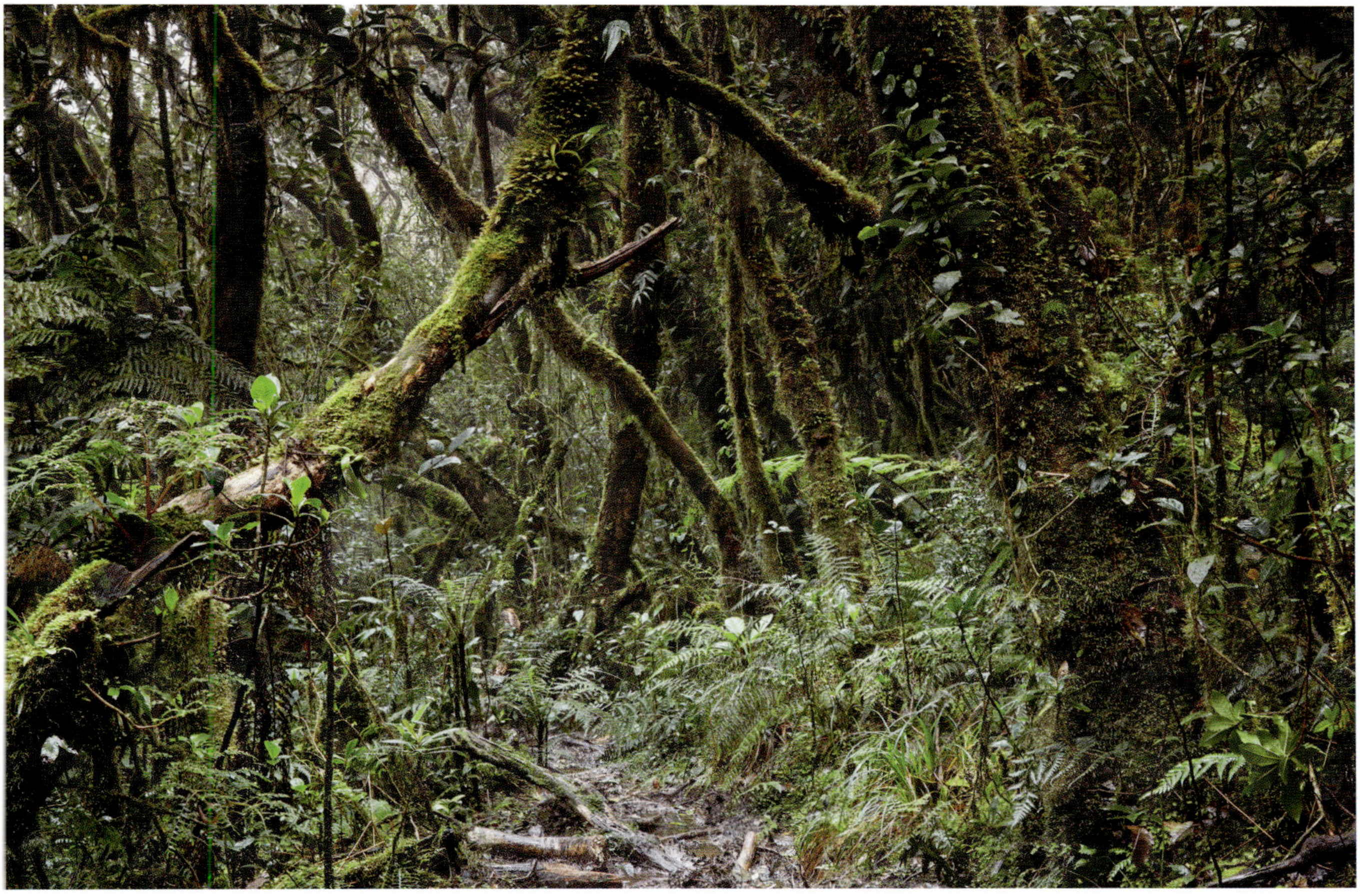

*Above: Rain-soaked upper montane forest on the north-west slopes of Mt Pulag, at about 2,000 m (6,600 ft), showing dense stands of stunted trees, mostly stone oaks (*Lithocarpus *species).*

*Below: A detail of a tree fern (*Cyathea *species), large, tree-like ferns that form a major component of the Philippines' montane rainforests, largely replacing the palms found mainly at lower altitudes.*

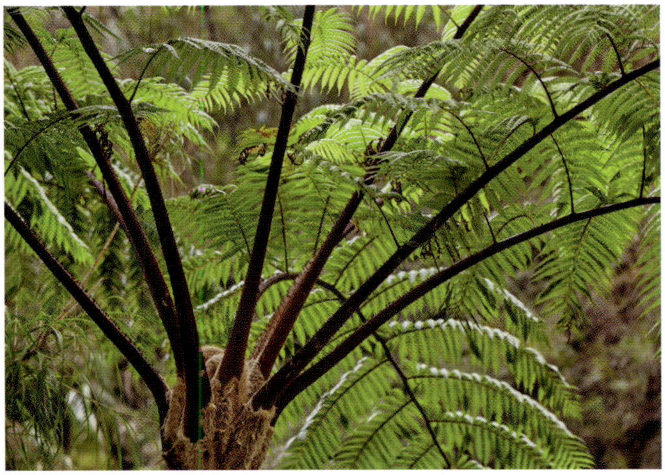

smaller plants, tree ferns (*Cyathea* species) are abundant, and the trees are festooned with many more lianas and epiphytic ferns and orchids. This is the montane rainforest, found on the higher slopes of mountains right across the Philippines.

The montane forest too undergoes changes with increasing altitude, the lower montane forest introduced above metamorphozing into higher montane, or mossy rainforest by an altitude of about 1,200–1,500 m (3,900–4,900 ft), or rather higher than this in a few areas, such as on Mt Apo on Mindanao. The two forms of montane rainforest are really quite different. For one thing, lower montane rainforest usually still feels and looks quite tropical. Upper montane/mossy rainforest, on the other hand, is a very different world altogether. This forest contains just one storey, the trees of which are stunted, gnarled and twisted into all kinds of weird and wonderful shapes. Epiphytic plants crowd many of the branches, and everything is festooned with moss, hanging in strands

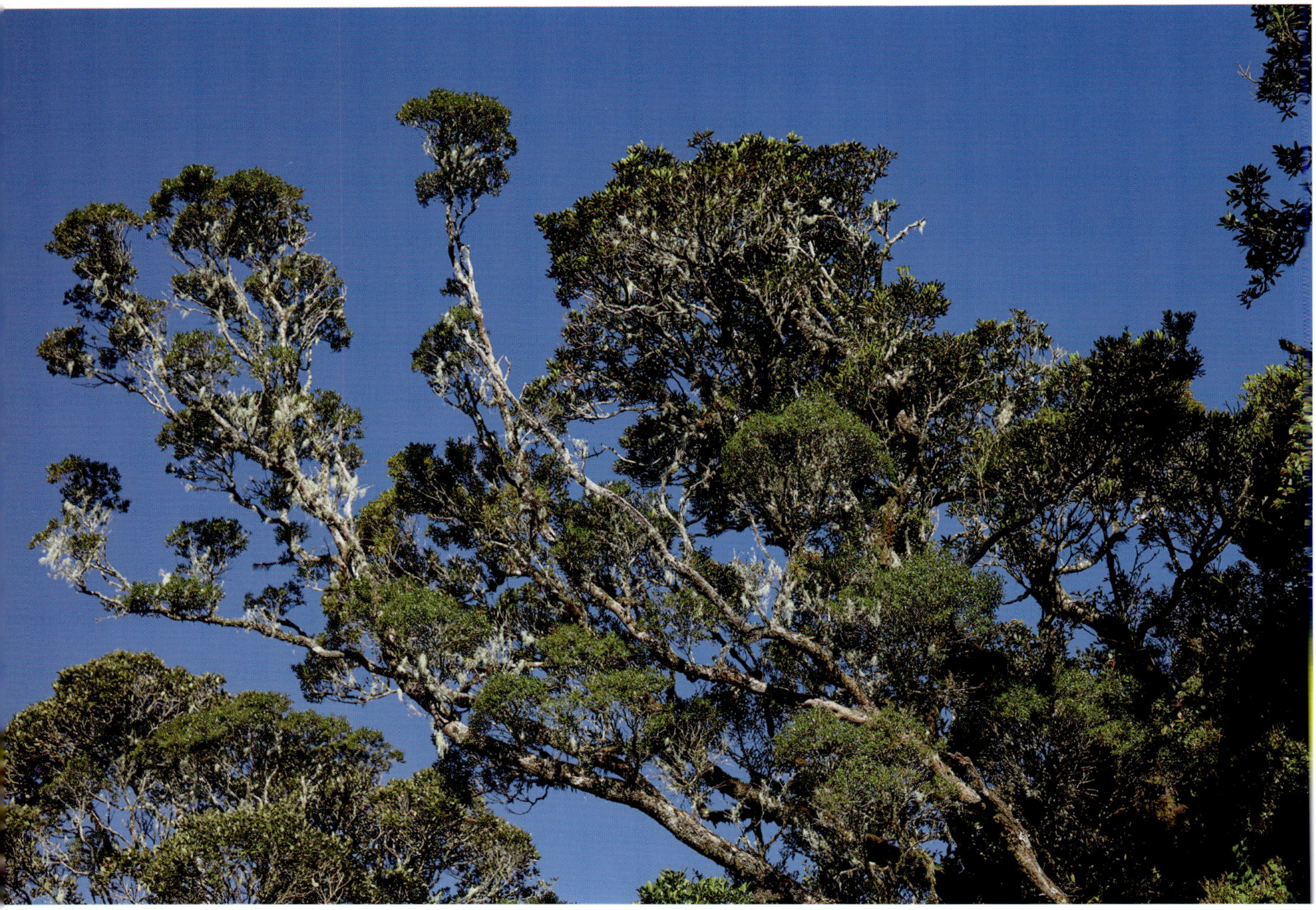

*Above: Fine weather and a break in the tree cover reveal the structure typical of the stone oak (*Lithocarpus *species) trees in Mt Pulag's upper montane rainforest, at about 2,600 m (8,550 ft).*

off just about every trunk, branch and stem, permanently dripping wet. With the temperatures so much lower at this altitude, dead matter does not decay well, and instead builds up deeply around the trees, often forming layers of semi-rotted leaves and wood a metre deep, or even more, above the ground. This is a very primeval place and it feels every bit as such.

As is the case with mountains everywhere, rainfall is a lot higher than it is down at sea level, but here in the Philippines it is really quite exceptional. Annual rainfall at sea level is typically about 2,500–3,500 mm (100–140 in), rising to about 6,000 mm (240 in) in the lower montane rainforest, and reaching a staggering 10,000 mm (400 in) or even more in the upper montane forests. Under such wet conditions the vegetation is permanently drenched, and the forest and highest mountain slopes are cloaked in thick fog for all but the first few hours of each morning. Even in parts of the Philippines with a prolonged dry season, at these altitudes there really is no such thing: rain can occur at just about any time, though it is true that during the dry season the montane forest is rather less drenched.

Plant and animal wildlife of the montane rainforest

Although dipterocarp trees quickly disappear in the montane forest altitudes, as already mentioned a few do reach into the lower levels of this habitat, particularly the Tanguile, though they peter out by an altitude of about 1,200 m (3,900 ft). However, they are joined by, and overlap with another giant of the Philippine rainforest, namely the Almaciga (*Agathis philippinensis*), a huge conifer that grows to as much as 65 m (213 ft) tall and a diameter of up to 2 m (6.6 ft). It usually starts to appear in the higher levels of lowland rainforest, but thrives in the lower montane forests. Even it cannot cope with the demands of the higher montane realms, however, though it can reach an altitude as high as 2,000 m (6,600 ft). It used to be found in lower montane forests throughout the country, but its high timber value has given it the same fate as the dipterocarp trees, and it has been logged almost to extinction. However, some of the mountainous protected areas, such as Mt Apo Natural Park, still have wild stands of Almaciga, while plantation forestry is increasingly growing it as a timber crop.

Despite these few giants, most trees in the montane forests consist of the much smaller stone oaks (*Lithocarpus* species), laurels (such as *Litsea* and *Cinnamomum* species) and in some areas Red Tea Trees (*Leptospermum flavescens*, known locally as Tinikaran). A handful of conifers can also be found in the montane forest, including *Podocarpus pilgeri*, *Dacrycarpus cumingii* and *Taxus sumatrana*. Shrubs often include such genera as *Medinilla*, *Melastoma*, *Syzigium*, *Clethra* and *Viburnum*.

On some of the highest mountains, such as Mt Pulag and Mt Apo, several *Rhododendron* species are common in the upper montane forest, along with a number of other plants more typically associated with the colder temperate regions of mainland Asia. Dwarf shrubs often found growing in the highest, most exposed locations, recalling a subalpine environment, include *Vaccinium myrtoides*, a plant related to cranberries and bilberries, and easily recognized by its purple hanging fruits.

The montane forests are increasingly recognized for their huge wealth of plant diversity that includes many orchids and ferns which simply do not grow at lower, warmer altitudes, as well as the literally hundreds of moss species. With many high mountain areas quite isolated from one another, it is not surprising that over the millennia they have evolved their own unique species, as a result of which some montane forest plant species are endemic not just to the Philippnes, but actually to a single mountain.

It is a similar story when it comes to animals, particularly those that cannot fly. Although a number of bird species are restricted to montane forests, such

Below: *In the Philippines a few rhododendrons do well on the highest mountain slopes. Here* Rhododendron subsessile *flowers on the edge of an upper montane forest clearing on Mt Pulag.*

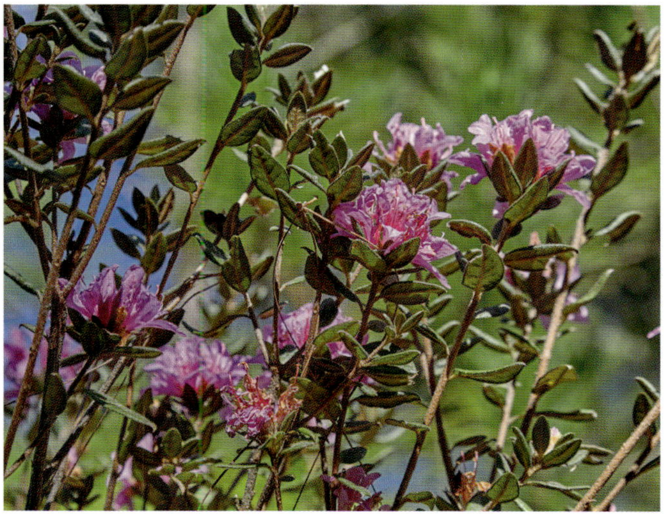

Below: *Another temperate weather specialist,* Vaccinium myrtoides, *a relative of the cranberries, thrives at the boundary between Mt Pulag's upper montane forest and its summit grassland.*

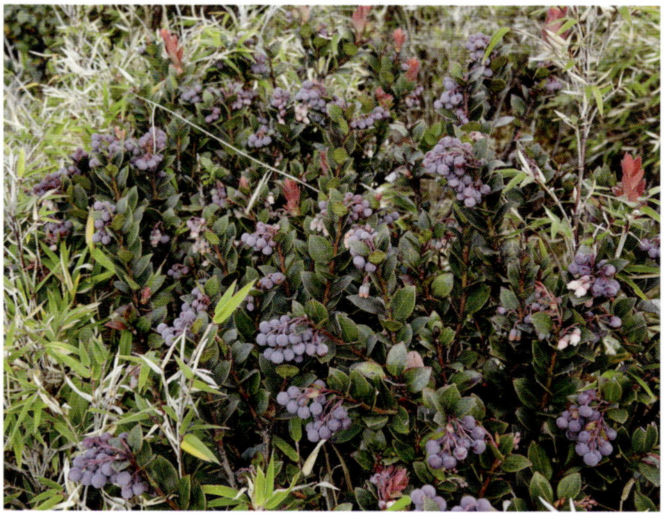

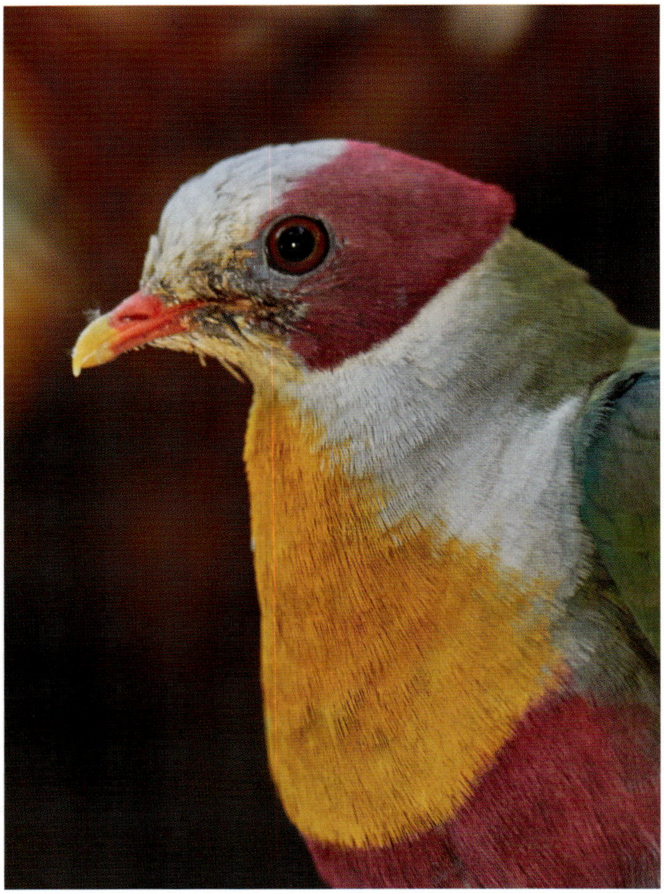

Above: *The colourful Flame-breasted Fruit-dove is a pigeon that is endemic to the mountains of Luzon, rarely coming down below 1,000 m (3,300 ft).*

as the Flame-breasted Fruit-dove (*Ptilinopus marchei*) and Whiskered Pitta (*Pitta kochi*), they can generally be found across a widely spread series of upland regions. This is much less the case for the rodent species that live at high altitude, some of which occupy very restricted areas, including single mountain ranges (see page 106).

Mountain summit grassland

Although upper montane rainforest covers many mountains right to their summits, on a few of the highest and additionally those with active volcanic crater summits, the trees give way to grasses. At these high altitudes the species of grass found are primarily temperate in origin, especially the Chinese Silver Grass (*Miscanthus sinensis*) and *Deyeuxia stenophylla*, among others. The summit grassland of northern Luzon's Mt Pulag, along with a few neighbouring peaks, is renowned for being dominated by the dwarf bamboo *Yushania niitakayamaensis*, which grows to about 40 cm (16 in) tall.

Species diversity is generally rather low in these grasslands, but other plants that can be found here include gentians (*Gentiana* species), small temperate-region flowering herbs more usually found in northern Asia.

Protection of montane rainforests

There appear to be about 12,600 km^2 (4,860 mi^2) of montane rainforest remaining in the Philippines. This is considerably less than the amount of lowland rainforest remaining, though this is largely due to the rather restricted high mountain locations in which it exists. In terms of percentage lost to logging and agriculture it has fared much better than the lowland forests, largely due to its remoteness, extreme weather, steep slopes and preponderance of trees less useful as timber. This does not mean, however, that montane rainforests are safe; far from it. Even most of the country's mountainous protected areas have very little lowland rainforest surviving, so they now consist almost entirely of montane rainforest. With the lowland rainforest gone, the montane areas have become more exposed and accessible, and in recent years even they have been increasingly eaten away by agriculture, even at high altitudes, with of course the larger trees in the lower montane areas lost some time ago to logging.

Some of the protected areas in Luzon's Cordillera Central mountains are illustrative. Mt Data, which is where much of the earliest research into Philippine biodiversity was carried out, has now lost almost all its forest cover, with all but a small patch of mossy rainforest converted to vegetable farms right up to its summit at 2,310 m (7,577 ft), despite being a national park. Mt Pulag, Luzon's highest mountain at 2,922 m (9,584 ft), has received rather better protection, and much of its upper montane rainforest remains intact, though even this, at an altitude of 2,400 m (7,900 ft) is

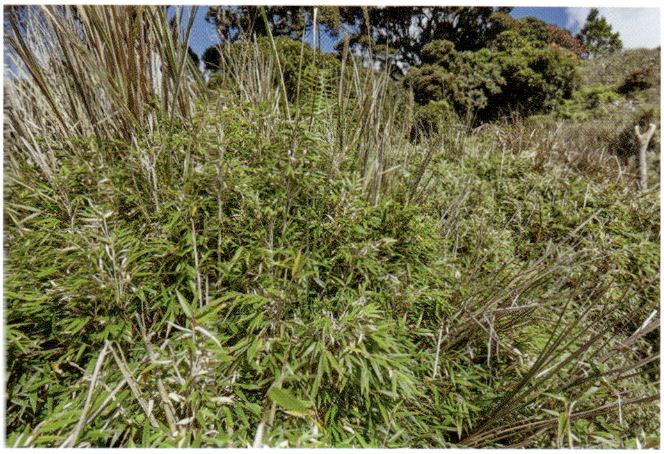

Above: Mt Pulag's summit grassland is characterized by dense stands of dwarf bamboo, a grass rarely found in the Philippines.

Below: Mt Pulag's summit grassland at about 2,850 m (9,380 ft), and looking across to montane rainforest on adjacent Mt Tabeyo.

being eaten away by farms well within the park's boundaries. Mt Pulag has no lower montane rainforest, it being replaced by a mix of vegetable farms and pine forest (see page 144) that transitions abruptly into upper montane rainforest at an altitude of about 2,400 m (7,900 ft).

It seems that the Philippines has an insatiable appetite for temperate vegetables, which in some parts of the country at least is fuelling this conversion of montane forest to farmland. If this can be brought under control with a combination of enforcement, better use of (and improved farmer access to) land at lower altitudes and the general unpleasantness of the high-altitude climate, then there is a reasonable chance that much of the country's montane rainforest can be successfully protected, along with its enormous biodiversity.

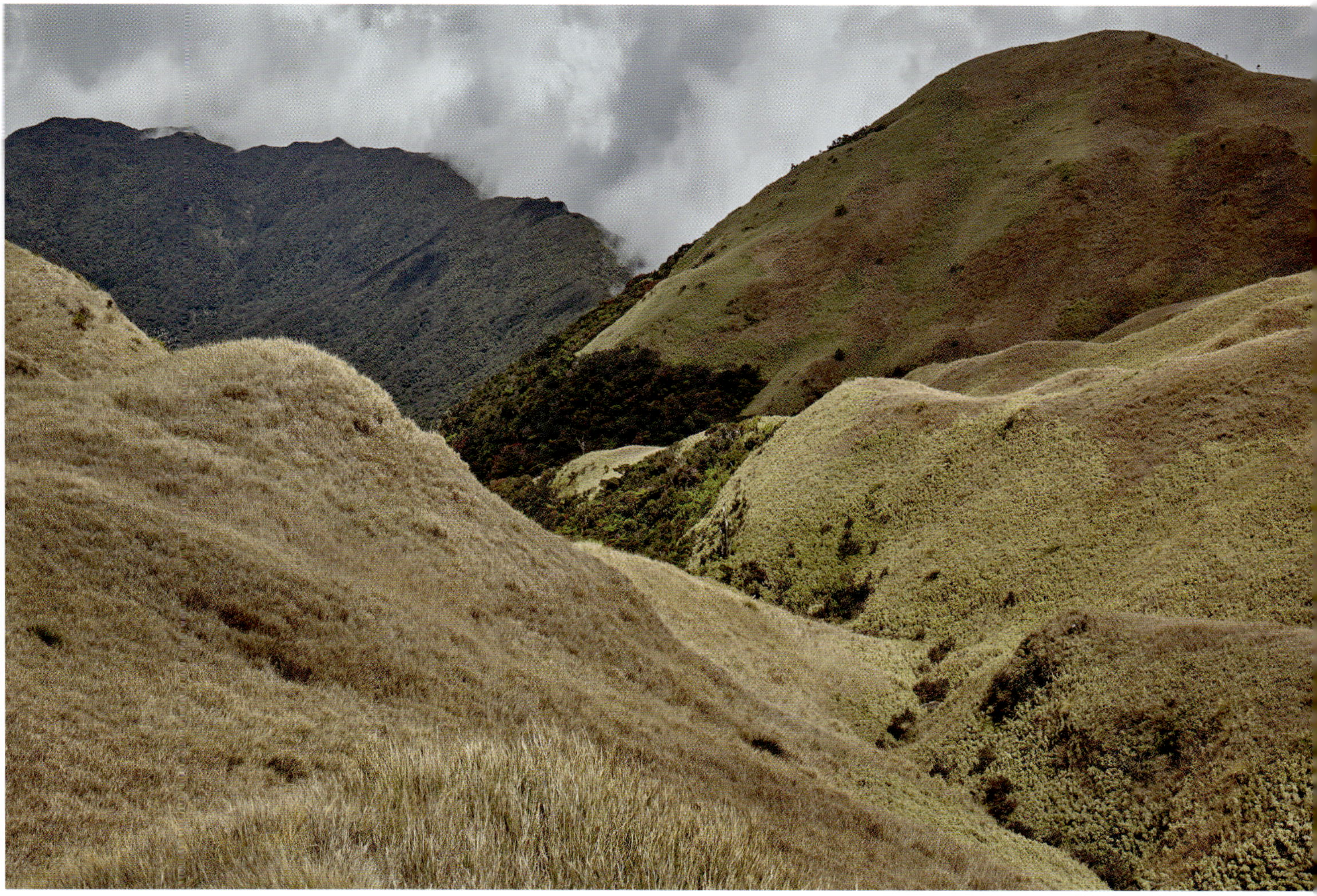

Pine forests

Pine forests are not the kind of habitat normally associated with a tropical country, but they do occur in three regions in the north of the Philippines: in north-west Mindoro, the Zambales Mountains of western Luzon and the Cordillera Central of northern Luzon. There are two species: Merkus's Pine (*Pinus merkusii*), which is found in Mindoro and Zambales, and Benguet Pine (*P. kesiya*, also known as *P. insularis*), which grows across large parts of the Cordillera Central. Both pines grow in areas, usually at altitudes of more than 1,000 m (3,300 ft), with a prolonged dry season and regular fires, both natural and man-made. There may also be a generally reduced rainfall, even during the wet season, due to a local terrain-related rain-shadow effect. In these conditions, pines are able to take hold, and even regenerate after a fire, whereas competing broadleaved trees, usually montane forest species, are unable to survive.

Below: *A forest of Benguet Pine, which covers many of the mid-altitude mountain slopes of northern Luzon's Cordillera Central, seen from the Halsema Highway.*

Merkus's Pine

In the Philippines this is found in small and fragmented stands in northwestern Mindoro, near the town of Santa Cruz, and in some rather larger forests in the Zambales Mountains, especially on Mt Tapulao (also known as High Peak, 2,037 m/6,681 ft), the highest peak in the range. Beyond these two locations, Merkus's Pine is found only in Sumatra, where its once widespread natural forests have been severely

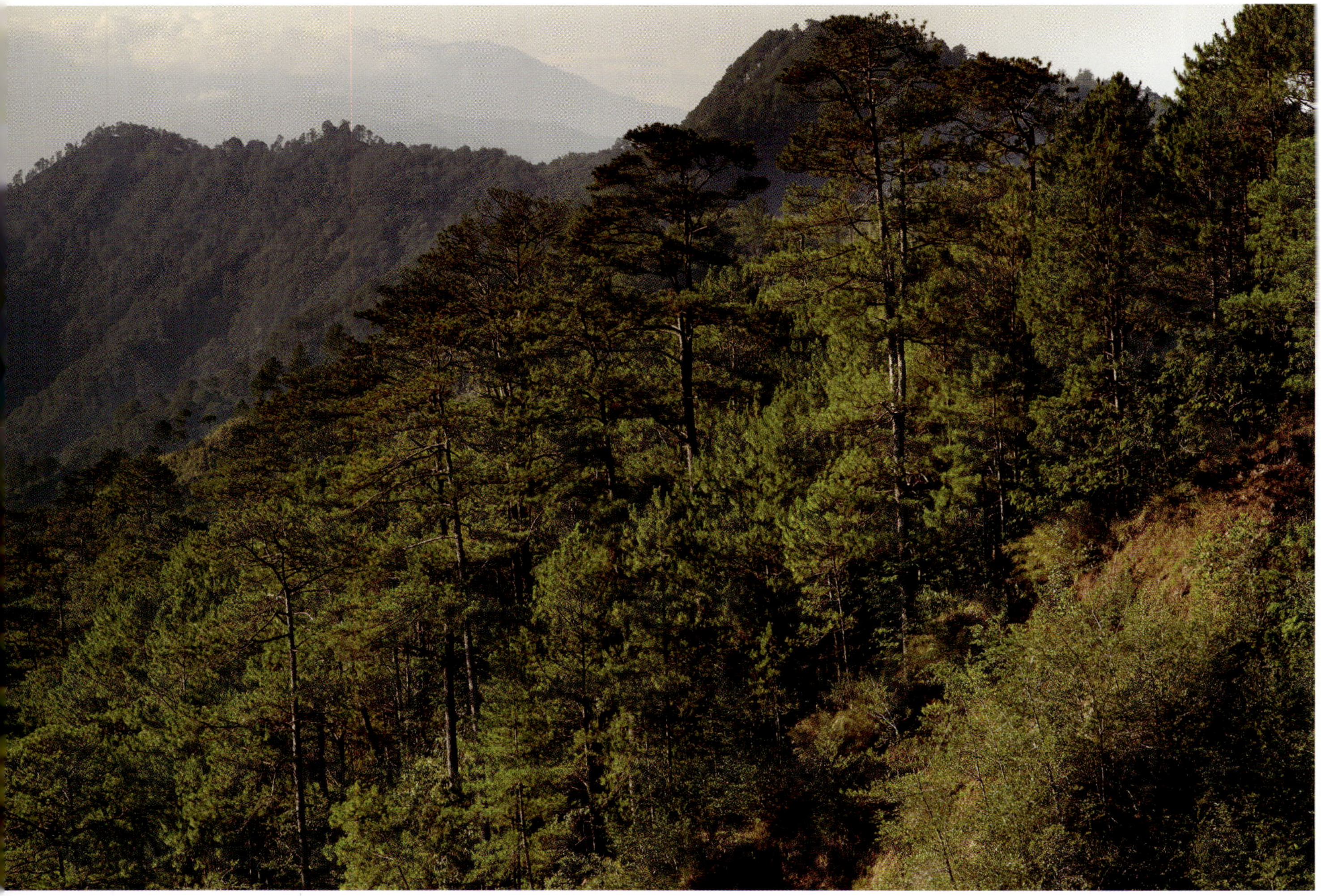

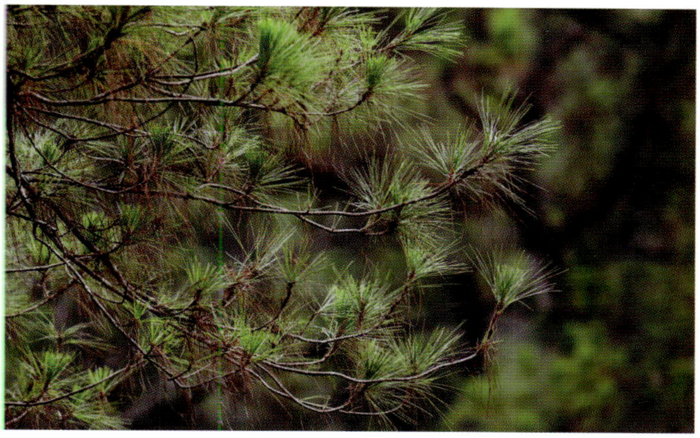

Above: A close-up of the needles of the Benguet Pine in the forest that surrounds the scenic mountain village of Sagada, in a northern part of Luzon's Cordillera Central.

Right : Karst limestone cliffs, near the village of Sagada, high in the mountains of the Cordillera Central, stand surrounded by a forest of Benguet Pine, an unusual scene in an otherwise tropical country like the Philippines.

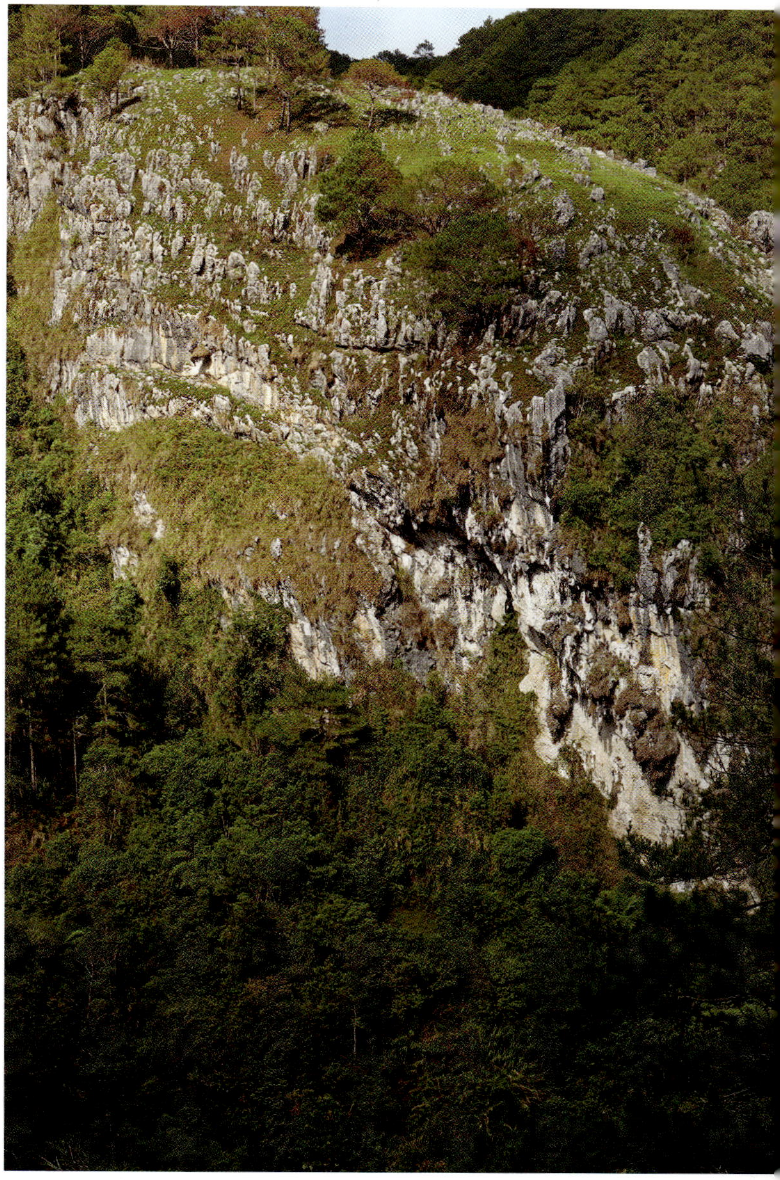

fragmented. According to the IUCN, however, there are extensive plantations in Indonesia for production of its resin as well as for timber.

Benguet Pine

This is the Philippines' best-known pine due to its widespread presence across the mountains of Luzon's Cordillera Central. It occupies many steep drier slopes at about 1,000–2,400 m (3,300–7,900 ft) altitude, usually over a ground vegetation consisting mostly of grasses, one of the results of frequent fires. In this environment, the pine-forest canopy generally remains quite open, allowing plenty of light to reach the ground. Unlike Merkus's Pine, the Benguet Pine is widely found across large parts of upland South and Southeast Asia, so is not in any way threatened.

In many areas of the Cordillera Central, the upper limits of the pine forest blur with montane rainforest, though the balance between the two is usually tipped in favour of the pines by man-made fires, which are set during the dry season to clear ground for agriculture. If the land is being completely cleared for farming then the pines may be cut down too, though often the pines are left and vegetables are grown among the trees. In the few areas where fires are no longer deliberately set, montane rainforest can start to grow in among the pines, if rainfall is high enough, eventually overwhelming them.

Some of the best views of extensive forests of Benguet Pines are along the Halsema Highway, the country's highest road, running through the Cordillera north of the city of Baguio. Some of the most attractive and explorable pine forests are around the mountain village of Sagada, while Mt Pulag – Luzon's highest peak – also has some extensive pine forests, though they are usually mixed in with vegetable farms.

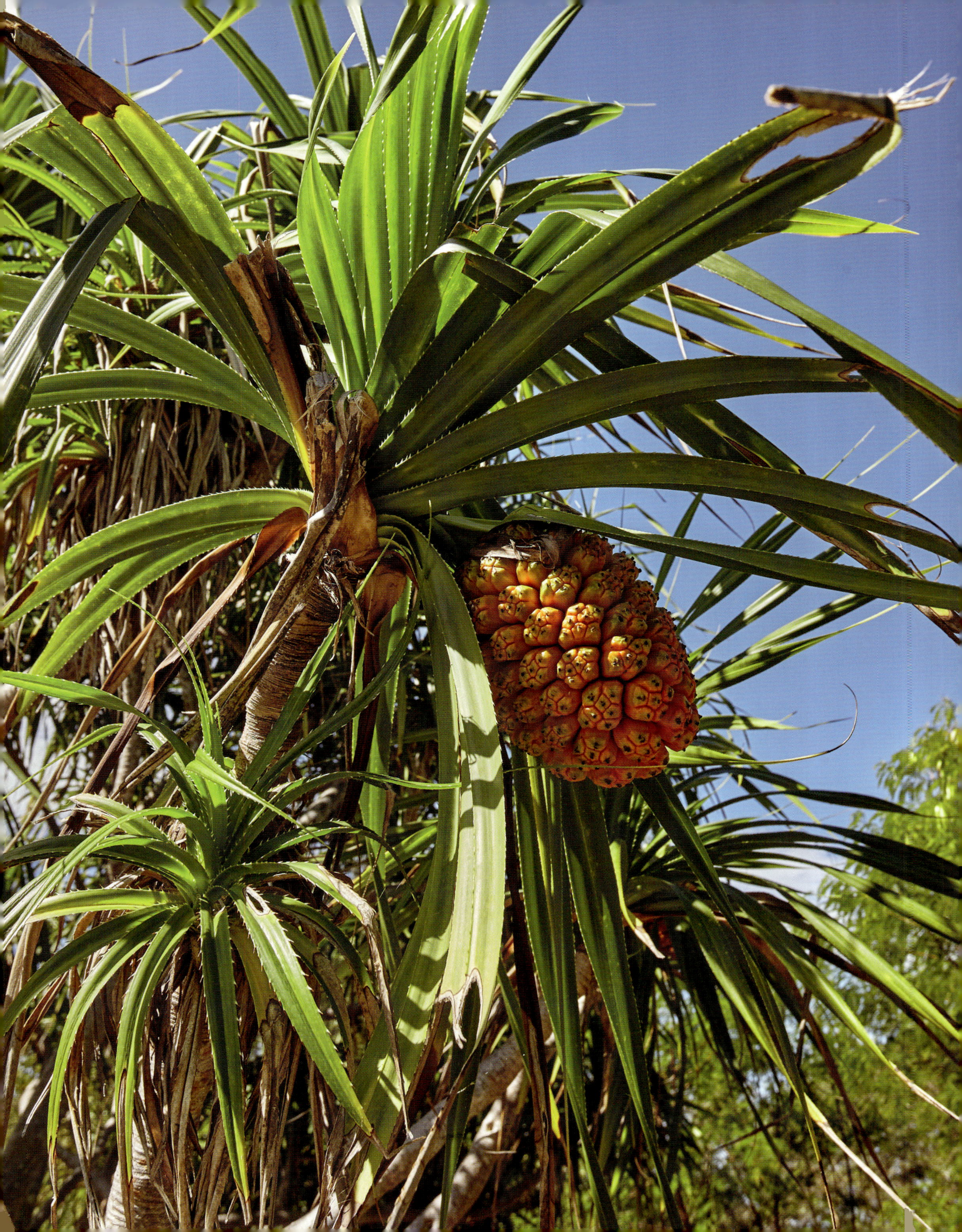

4

Plant and Animal Wildlife

Iconic species of the Philippines

For many years the plant and animal wildlife of the Philippines was largely overlooked in favour of its neighbours. The country undoubtedly lacks the large charismatic mammals of Asia and Australasia, providing a home – or so it seemed – to just a few rodents and bats. But when, in the latter part of the twentieth century, scientists looked more closely, the truth about the Philippines became clearer, such that when it comes to preserving as much of the world's plant and animal wildlife as possible, this nation is one of the most important places on Earth.

Recent estimates indicate the Philippines to be home to a staggering 52,000 species of plant and animal already known to science, 36,000 of them animals and 16,000 plants, the plant count equating to 5 per cent of the world's entire flora.

Among the animals, an estimated 22,000 are insect species, many still barely understood. Of the remaining 14,000 many more, like the insects, are invertebrates with no backbone. The backboned vertebrates still come to several thousand, including rather more than 3,000 species of marine fish, a little over 600 birds, about 350 reptiles and amphibians, and nearly 200 mammal species.

When calculated on a species per hectare basis, all these amazing numbers make the Philippines the most biodiverse place on Earth, surpassing all the other much more globally famous megadiverse regions, including Madagascar, the Amazon and the Galapagos Islands.

Philippine endemism

Even more amazing is the number of plants and animals that are unique, or endemic, to the country. With the Philippine islands rising up from below the sea between 20 and 5 million years ago, most of the islands' plants and animals have evolved down their own, quite separate paths. It is estimated that more than 60 per cent of the Philippines' plant, mammal and reptile species are endemic to the country, and a staggering 85 per cent of amphibians. Birds come in with a rather lower rate, at about 30 per cent.

Introducing Philippine species

The following pages introduce a selection of some of the most interesting and important species from the perspective of the Philippine environment and conservation.

Opposite: *Though resembling palms,* Pandanus *plants are actually trees. Hugely varied and adaptable, they can be found in a host of habitats from beach forest, to montane rainforest, to swamp forest. This image shows the striking red-orange fruit of* Pandanus tectorius, *a common species along sandy beaches and in beach forest.*

Animals of the sea

With its position at the very hub of the Coral Triangle, the global epicentre for oceanic distribution of coral and many other tropical marine species, the Philippines is one of the world's most important places for marine conservation. It is also one of the most exciting places for scuba diving, with the range of wildlife that can be encountered in just a handful of dives being quite staggering.

Each of the sprawling archipelago's 7,000-plus islands makes its contribution to the country's 35,000 km (21,700 miles) of coastline and 25,000 km^2 (9,650 mi^2) of coral reefs. The 3,000-plus species of fish, approximately 600 species of coral and barely counted numbers of molluscs (ranging from giant clams down to tiny sea slugs, or nudibranchs), crustaceans (including shrimps, crabs and lobsters) and echinoderms (such as starfish) that call those reefs home together constitute a marine treasure house that should make a nation wealthy beyond measure.

Forming the very base for all this coastal marine biodiversity are the corals, most especially those that build the reefs, thereby creating a physical habitat that huge numbers of animals can live in, both coral and otherwise. While it would be impossible to describe every species of Philippine coral here (in fact many have barely been described anywhere), the next few pages summarize some of the fundamentals, concentrating on the species that are most important in building the reefs and are most likely to be seen during many reef dives.

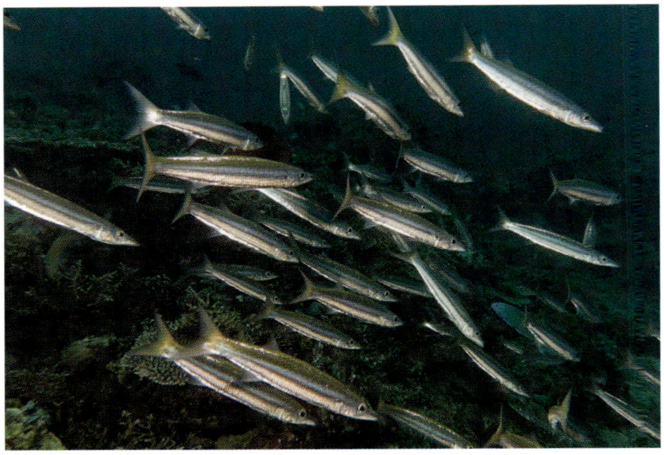

Above: *A shoal of Big-eye Barracuda (*Shyraena forsteri*) in Bacuit Bay, El Nido, Palawan.*

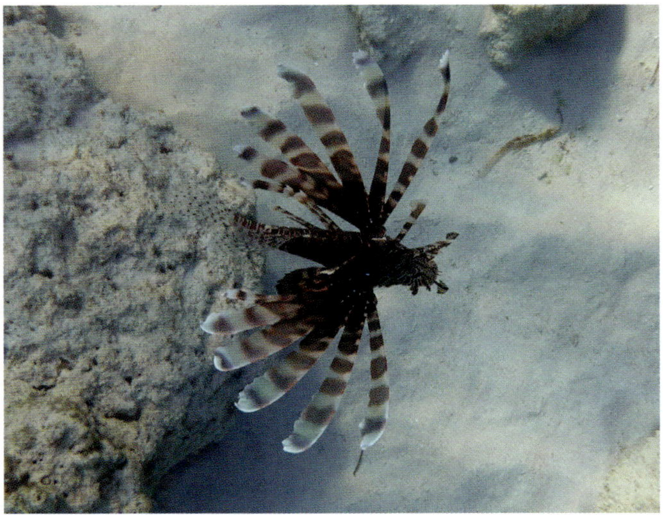

Below: *A Common Lionfish (*Pterois volitans*) swimming across reef shallows off Dimakya Island, in the Calamian Islands, Palawan.*

Among some of the most globally threatened marine animals are the sea turtles (see page 90). Five of the world's seven species can be found feeding on Philippine reefs. Their presence, or otherwise, is taken by some observers as something of a barometer of the state of the marine environment, so especially in the larger Philippine marine protected areas their relative abundance is watched with eager eyes.

Then there are the animals that patrol the open seas, the pelagic fish. Usually just silver and grey to camouflage them against the muted colours of the open sea – such a contrast to the rainbow of hues among fish living on the coral reefs – they tend to travel in shoals, probably mainly for safety in numbers. This is less so for many of the sharks, most especially the huge Whale Shark (see page 92), the world's largest fish, which can travel great distances mostly as a solitary figure, making it generally rather hard to find. Around the Philippine coast, however, there are places where it has been found to congregate, as a result of which in recent years it has become something of a national marine icon.

The corals

As already mentioned, the corals and the huge reefs they build up underpin the vast majority of the marine life of the Philippine coast. Their triumph in being so successful here is what makes these islands the very heart of the Coral Triangle and the global distribution centre for a host of tropical marine species. Without those corals, the Philippines' generally nutrient-poor waters would be home to a very limited range of marine life.

Even just a cursory glance at a healthy coral reef tells a tale of vast complexity and diversity in coral types, structures and interactions. It is a hugely complex environment that then leads on to great complexity in any attempt to describe and classify the corals. Indeed, scientists working on this are still struggling with the task. The descriptions given here are necessarily a simplified summary of what is currently known.

Some coral biology

All corals belong to the group of animals called Cnidaria (also known as coelenterates), simple cylindrical animals consisting of a stomach cavity, with a mouth surrounded by small tentacles at the

Below: Acropora *species staghorn hard corals on a reef crest off Miniloc Island, El Nido, Palawan.*

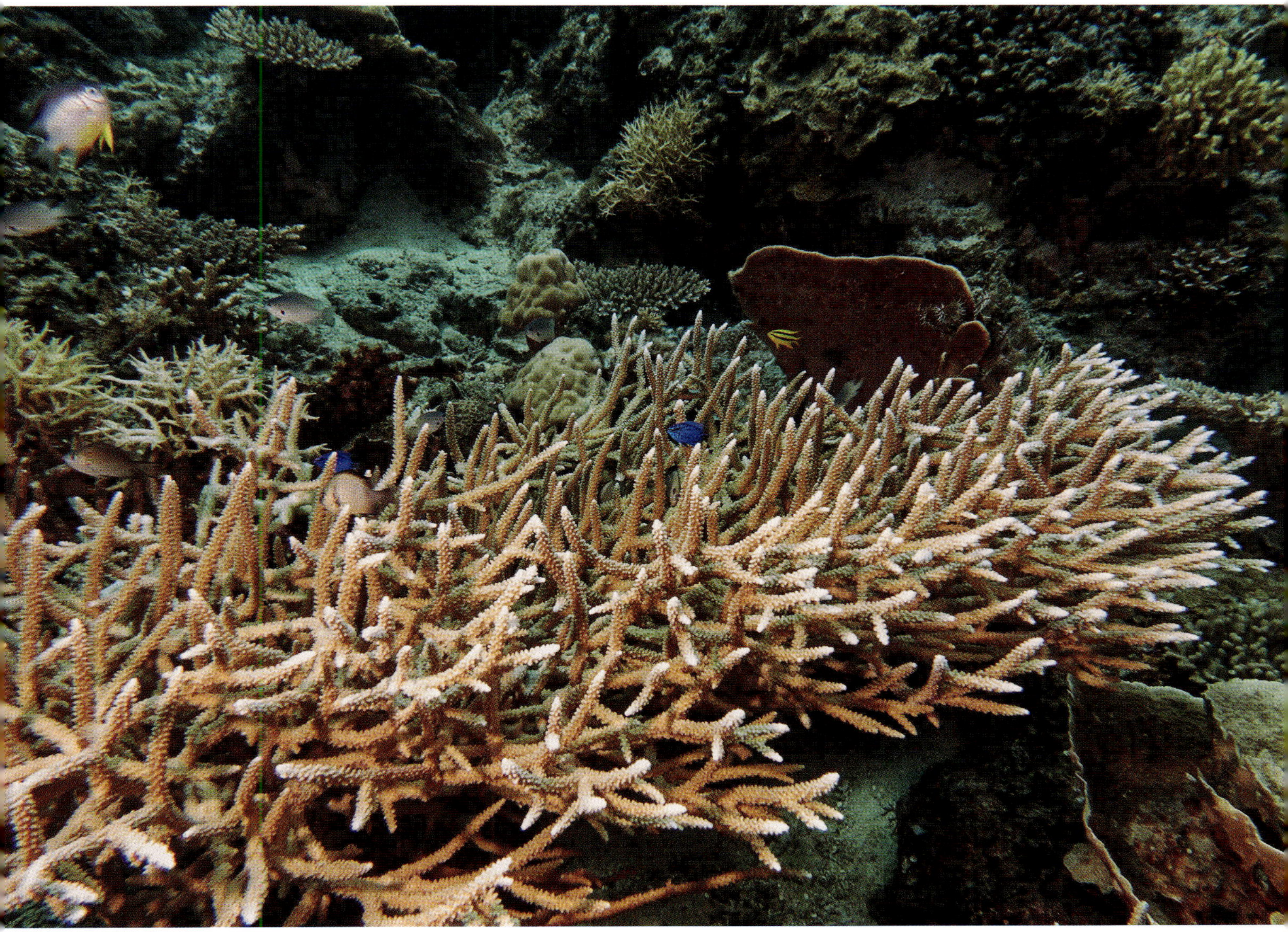

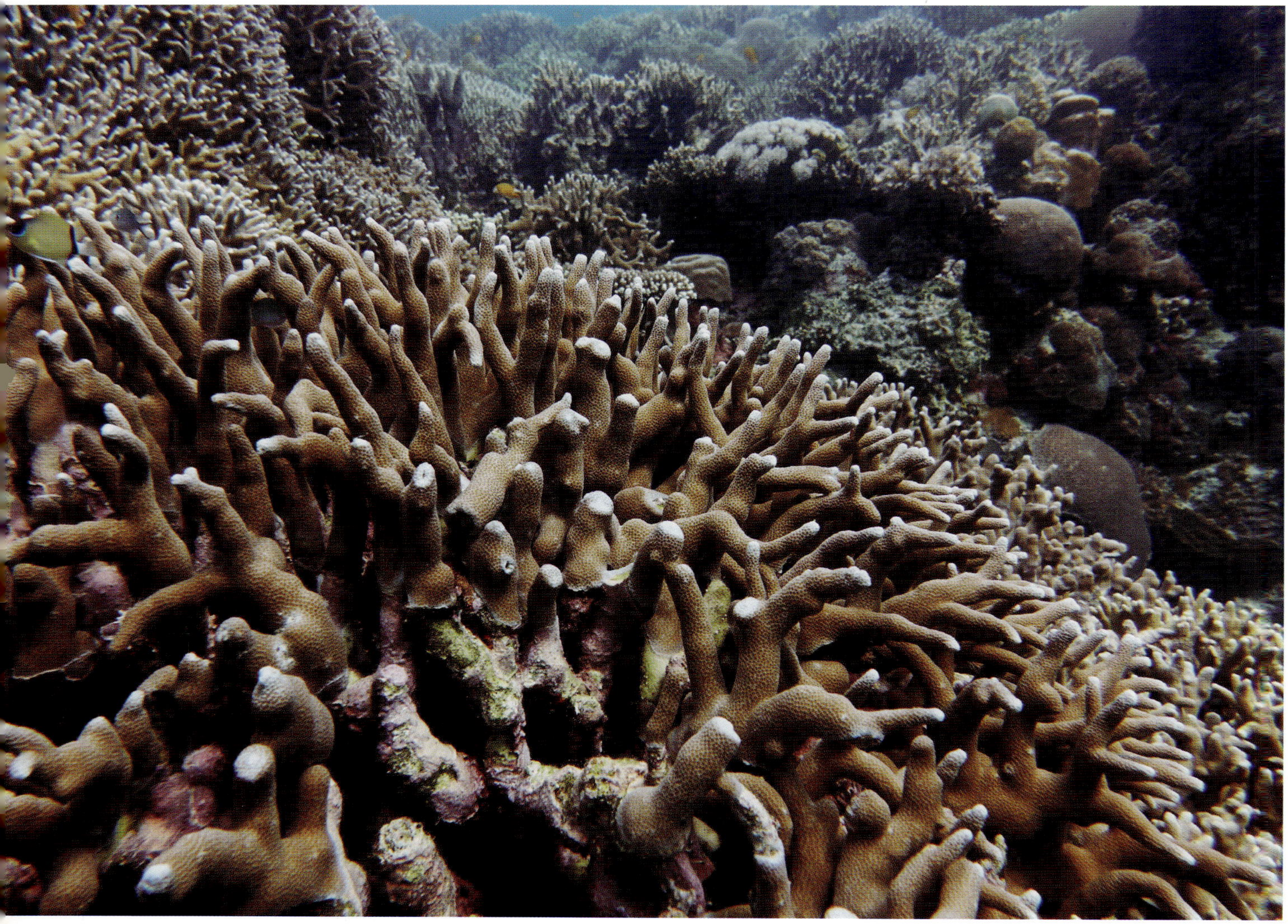

upper end. A single animal is called a polyp, though the great majority of coral species form colonies of polyps, each polyp usually barely a couple of millimetres in diameter, which together build the coral structures that so typify coral reefs.

All the polyps of a colonial coral are attached to, or embedded in, a skeleton (which may or may not be calcium carbonate) produced by the polyps, which in turn is usually coated with the coenosarc, a layer of organic tissue also produced by the polyps. Although the polyps are spaced across the skeleton, they are all connected to one another via the coenosarc that enables them to share digestive systems.

Technically speaking, there are three coral groups: the hydrocorals, hexacorals and octocorals. The first of these are rather simpler, in evolutionary terms, than the other two, and are more closely related to jellyfish than they are to the hexacorals and octocorals. They consist of lace corals (*Stylaster* and *Distochopora* species) and fire corals (*Millepora* species), both of which produce calcium carbonate skeletons. As their name suggests, fire corals are renowned for their painful stings when touched.

Hexacorals consist most importantly of sea anemones, black corals and the scleractinian hard corals. The last of these are the world's major coral reef builders, creating huge and stable calcium carbonate skeletons that build up over the years into the main reef structures. Black corals are so named because of their black skeletons, made of horn, though this is usually hidden beneath a frequently highly colourful coenosarc. The sea anemones, unlike most hard and

Opposite: A huge, crowded mass of hard corals jostling for space on the crest of a reef off Moalboal, Cebu.

Right: Bushy black coral (Antipathes species) is sporadically present across Philippine reefs, seen here in Bacuit Bay, El Nido, Palawan.

Below: The beautiful, swaying tentacles of a Heteractis species sea anemone, a common resident of virtually all Philippine reefs, here seen in Bacuit Bay, El Nido, Palawan.

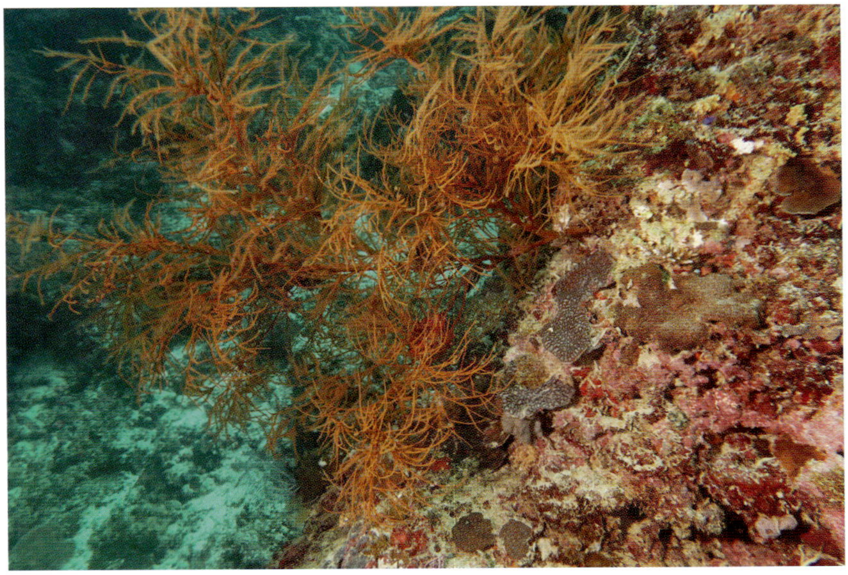

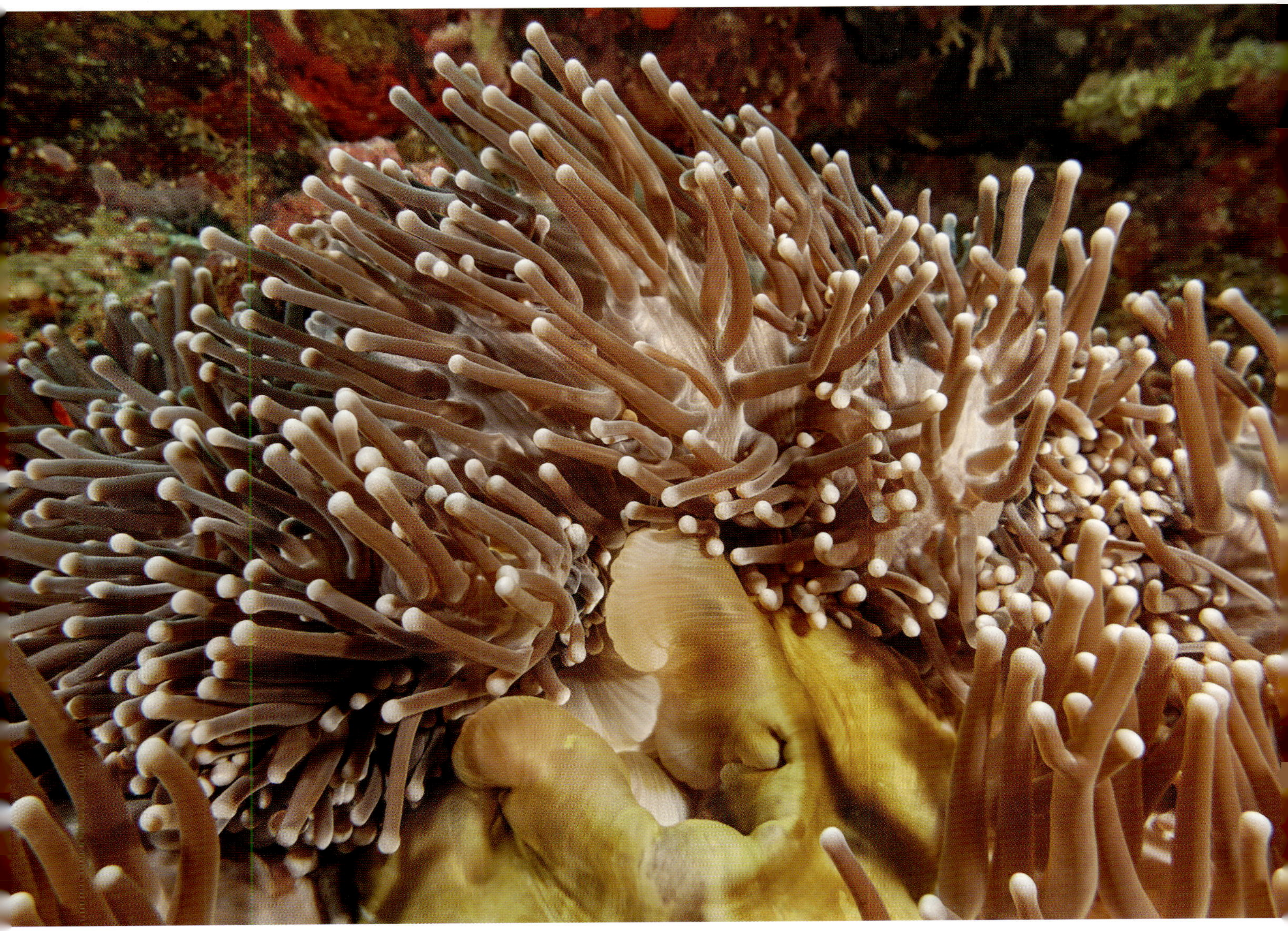

black corals, do not live as colonies but as single individuals, something that is sometimes hard to believe given the great size to which some species grow.

The octocorals consist of the soft corals, blue corals, the gorgonians (which in turn consist of sea fans and sea whips) and sea pens. None of these corals produces large calcium carbonate skeletons, though they may be present in small amounts. Soft corals have a limited skeleton that gives some rigidity but which also allows them to sway in response to currents or wave action. The gorgonians have a rather stiffer skeleton made largely of horn that is nevertheless flexible enough also to allow them to move a little in sea currents. They are among the most delicate and stunningly beautiful structures on a coral reef and are often very colourful. They usually occur in waters about 20 m (66 ft) or more deep, on the drop-off below the reef crest, where wave action is minimal but where currents may be relatively strong.

Feeding and symbiosis

It is sometimes hard to remember that corals are even living structures, let alone predatory animals. Although all the corals can feed by catching passing food with the sting cells (called nematocysts) on their tentacles, they are also able to directly absorb dissolved nutrients from the surrounding water. Many also have a symbiotic relationship with algae called zooxanthellae, which live inside the coral tissues. Being plants, the algae photosynthesize, the resulting

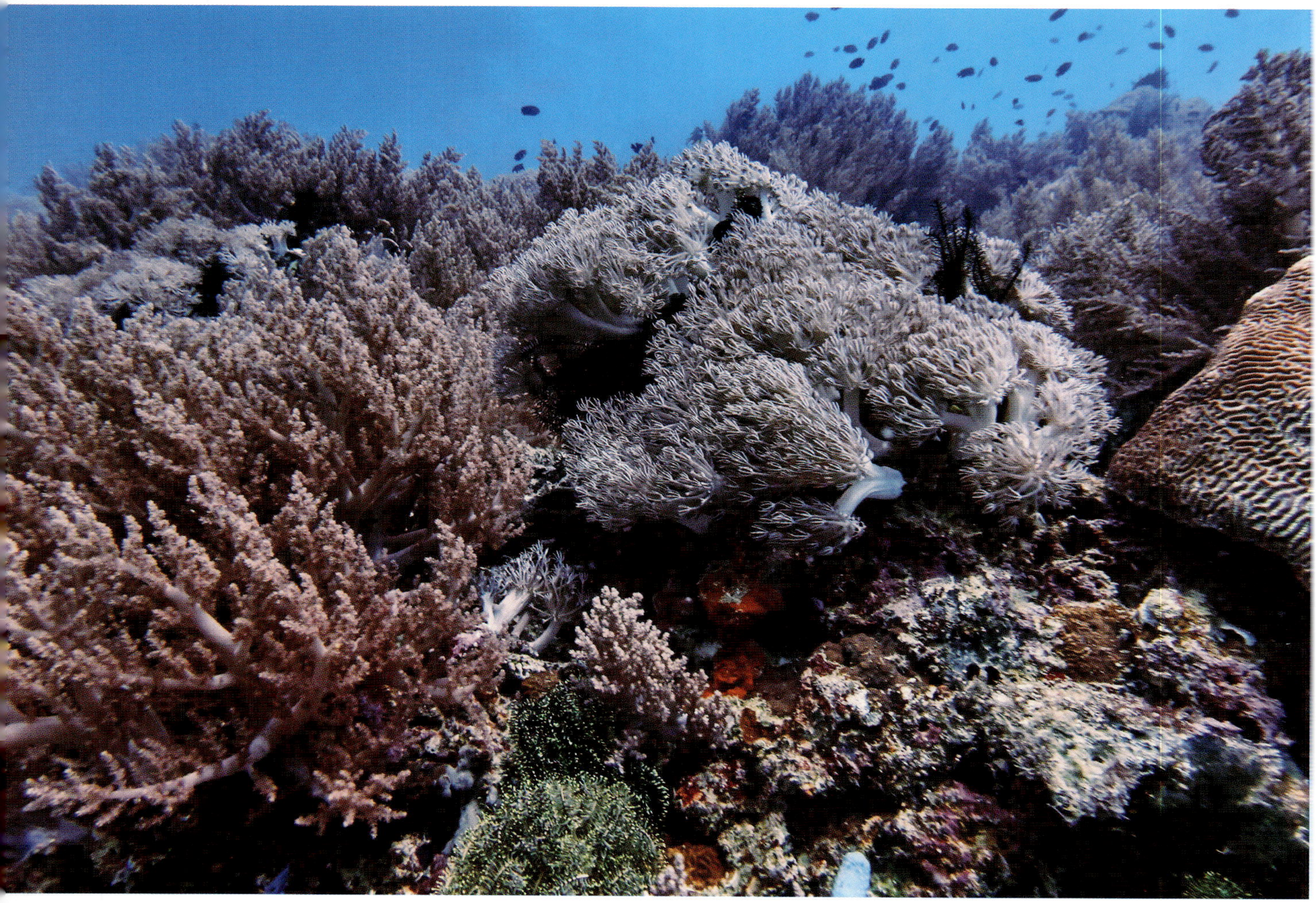

*Below: A forest of soft corals (*Xenia *species on the right,* Nephthea *species on the left) growing in the shallows on a reef flat, swaying with the wave action, off Balicasag Island, Bohol.*

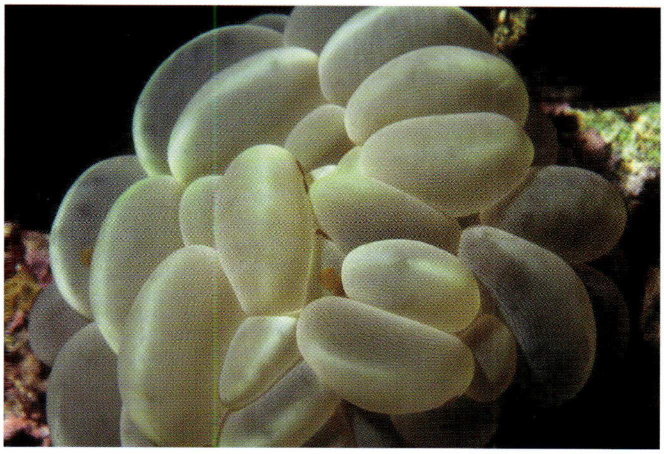

Above: *Bubble Coral (*Plerogyra sinuosa*), an unusal hard coral, though one regularly seen on Philippine reefs.*

nutrients feeding the corals, and the corals in return give the algae a safe place to live. All the reef-building hard corals and many of the others – though not all – include zooxanthellae, and it is for this reason that good light levels are crucial to coral reef survival.

It is often said that it is the zooxanthellae that give corals their colour, and while this is true for many of the scleractinian hard corals it is not so for many of the others. For the sea fans, black corals and lace corals, for example, their bright colours are the result of the polyps' own pigment production.

The corals that do not contain zooxanthellae are, of course, not constrained by light levels, so can grow at much greater depths than is possible for the main coral reefs. Lattice corals, for example, regularly grow at depths down to several hundred metres.

Hard corals

The great calcium carbonate skeleton-producing activities of the scleractinian hard corals are largely responsible for the development of the coral reefs. The structures they generate range from complex sprawling and highly branched 'staghorn' structures, to vast 'tables', to huge boulder shapes, all of which can easily reach 2 m (6.6 ft) across or tall, though shapes can vary within a single species, depending on local conditions and exactly where on the reef they are growing.

The ridges, peaks and ravines of the reef crest – so important in protecting the reef flat and the shore behind – are largely the work of the hard corals, where table, staghorn and boulder corals predominate, the first two generated by a range of *Acropora* species, the boulder corals the product mainly of *Favia*, *Favites* and *Platygyra* species. On the reef crest especially, but also on the reef flat to some extent, different forms of *Acropora* species hard corals in particular can form a total covering across the rocky substrate, crowding each other out and jostling for space. Significantly for the life of the reef as a whole, their intricate shapes and branches create a myriad of holes and shelters that are used by the thousands of other reef species that make a home here.

Below: *One of the most beautiful of the soft corals, a* Dendronephthya *species tree coral, in Bacuit Bay, El Nido, Palawan.*

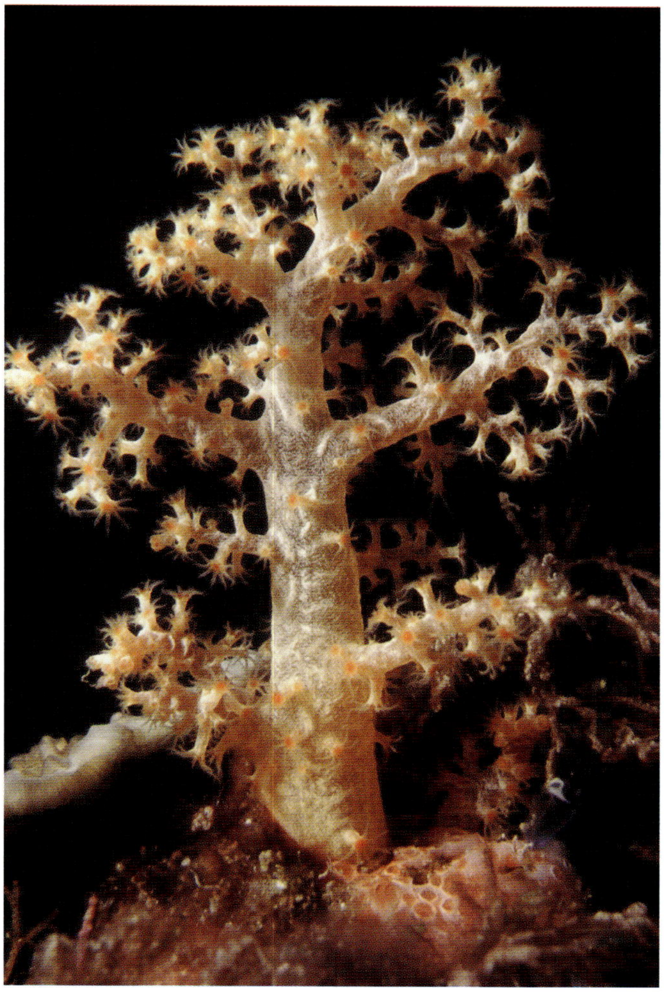

Sea turtles

The Philippines hosts five of the world's seven species of marine turtle, although three are by far the most widely distributed. These are the Green, Hawksbill and Olive Ridley Turtles *(Chelonia mydas, Eretmochelys imbricata* and *Lepidochelys olivacea*, respectively). Less commonly seen are Leatherback and Loggerhead Turtles (*Dermochelys coriacea* and *Caretta caretta*). All five species are threatened with extinction, so are completely protected in the Philippines. However, egg collecting and the hunting of adults still continue, though at a significantly reduced level compared to 20 years ago.

Some turtle biology

Turtles are reptiles, and despite living in the sea they have to breathe air. As a result, they need to come to the surface from time to time, though their highly efficient lungs and blood supply enable them to stay submerged for considerable amounts of time. All five turtle species are, of course, protected by shells. The carapace, which covers the back, consists of a series of plates, the numbers and patterns of which help to distinguish one species from another, along with such features as the number of facial plates, or scutes, and the number of claws on the front flippers.

The turtle species vary quite significantly in size, the Leatherback Turtle being the largest, at up to 600 kg (1,320 lb) in weight and 178 cm (70 in) long. The far more commonly encountered Green Turtle comes in second with a maximum weight of 230 kg (506 lb) and length of 125 cm (49 in). The smallest is the Olive Ridley Turtle, weighing just 50 kg (110 lb) and with a length of 70 cm (27.6 in).

When it comes to food, it is often said that marine turtles are vegetarian, but this is not true. Adult Green Turtles do eat just seagrass and algae, but the juveniles also take jellyfish and crustaceans. Hawksbill Turtles eat mostly sponges, algae and sea anemones, while Olive Ridley and Loggerhead Turtles take quite a

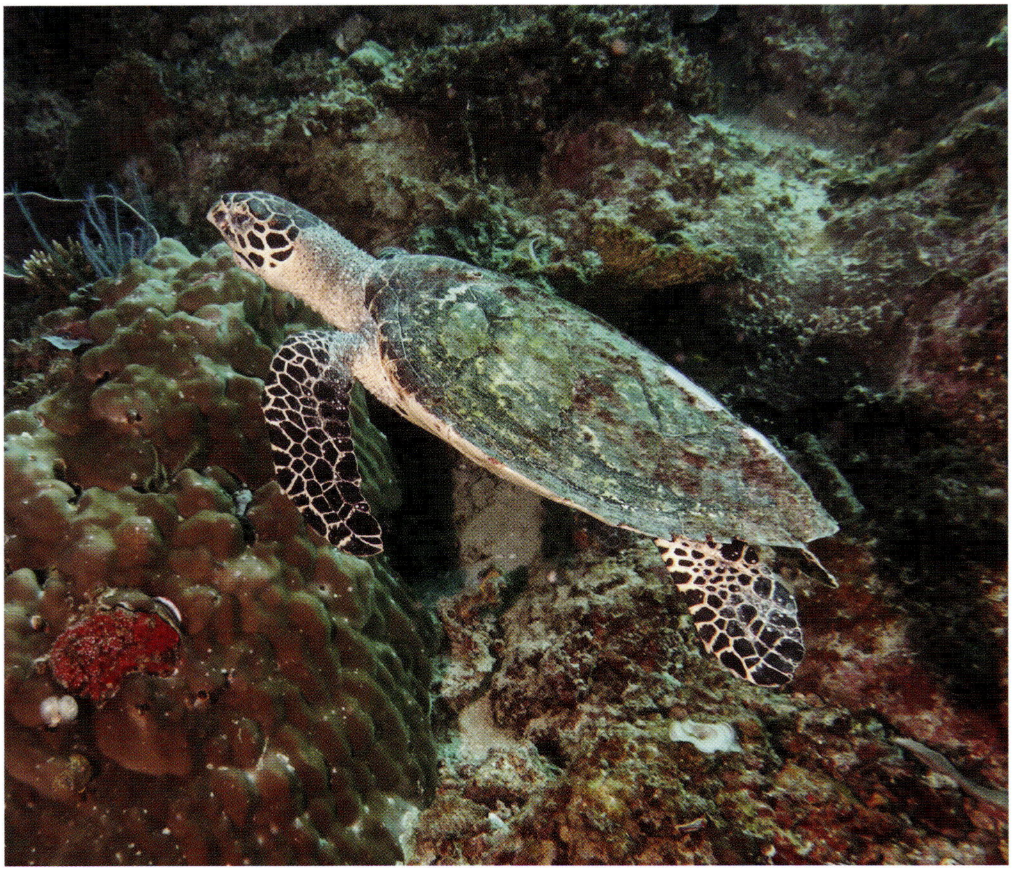

Left: *Hawksbill Turtles are quite common on reefs in some parts of the Philippines, especially - as here - around the islands of Bacuit Bay, El Nido, Palawan.*

Opposite: *Green Turtles are very common around many parts of the Philippines, seen here off Balicasag Island, Bohol. They are the second largest of the country's five turtle species after the Leatherback.*

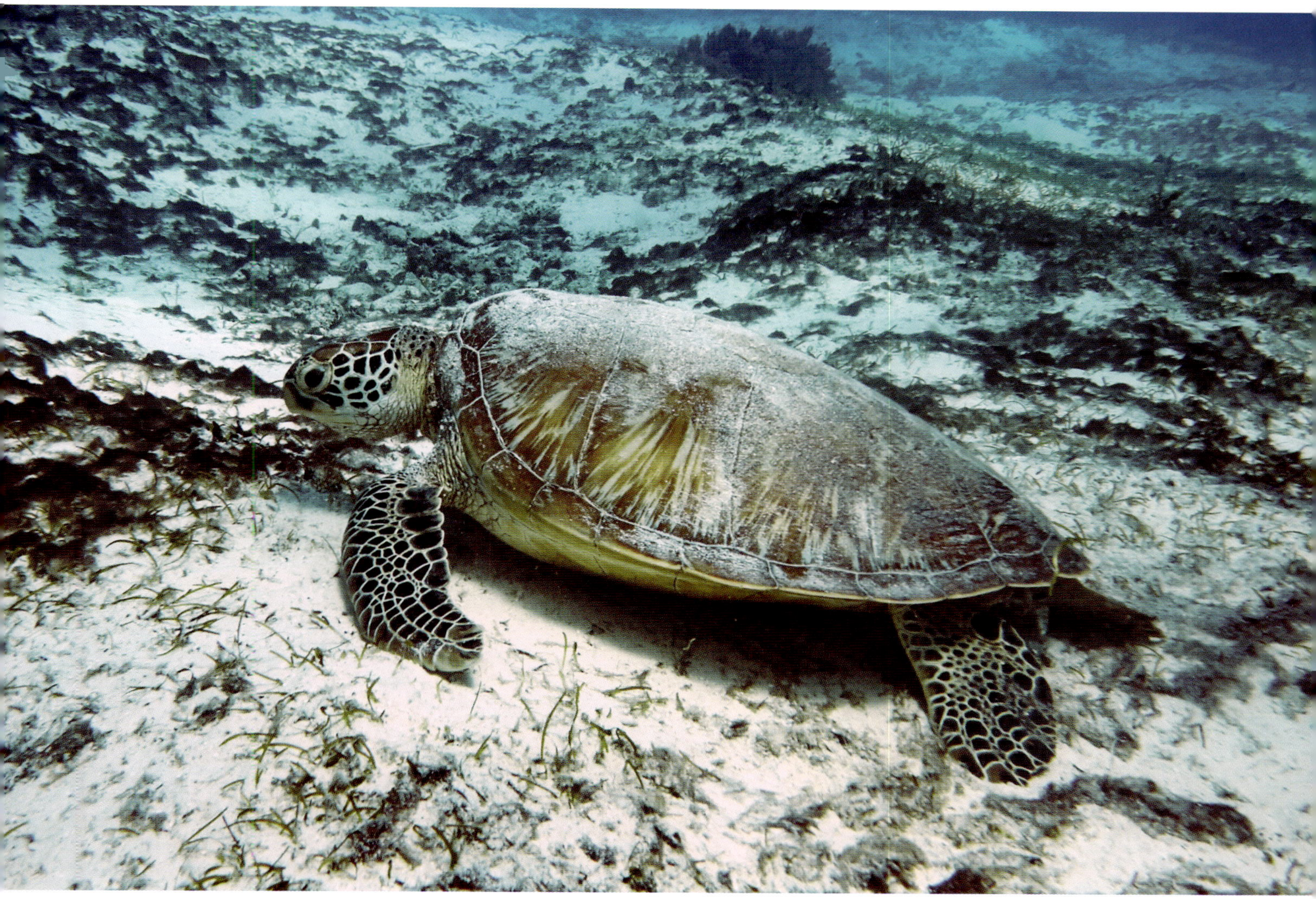

variety of invertebrates and even fish. The huge Leatherback Turtle restricts itself to jellyfish and sea squirts.

Distribution and breeding

Green, Hawksbill and Olive Ridley Turtles are widely spread across the whole country, though the Green Turtle is much more common and more likely to be seen by divers on quite a variety of reefs. By contrast, the Leatherback Turtle has been found only at a relatively small number of sites around the Philippines, while the Loggerhead Turtle is known only at a handful of places in southern Luzon, Basilan Island and Palawan. The vast majority of places where turtles are found in the Philippines are simply feeding sites, with the breeding and nesting areas restricted to just a few places that the turtles return to year after year. Probably the most important of these for Green and Hawksbill Turtles is the Turtle Islands Heritage Protected Area, a cluster of small islands off the coast of Borneo that is administered jointly by the Philippines and Malaysia.

Other breeding sites for the Green Turtle occur in Palawan, south-west Mindanao and Apo Reef, off the west coast of Mindoro. The Hawksbill Turtle is also known to nest along the north and south coasts of Mindanao, on Palawan and on Guimaras. Nesting for the Olive Ridley Turtle is restricted to Luzon's north-west coast. The Leatherback Turtle has been observed to nest only at three sites in the centre of the country, while the Loggerhead Turtle is not known to nest here at all.

Whale Shark

Although Whale Sharks (*Rhincodon typus*), are found throughout the world's tropical and warm temperate oceans, the Philippines is one of the world's leading hotspots, with researchers to date having identified just over 1,000 individuals living in Philippine waters, giving the country the world's third largest population.

Surprisingly little is known about this behemoth, but what is certain is that it is the world's largest fish. It is also a filter feeder, eating by sieving food from the sea. The food consists largely of microscopic zooplankton, though it has also been found to eat small fish. The Whale Shark is a very docile animal and poses no threat to humans.

Whale Sharks are huge. Their typical size has been measured at about 9.8 m (32 ft), with a weight of about 9 tonnes (19,800 lb), though much larger individuals have been reported.

Studying Whale Sharks

Whale Sharks have proved to be extremely difficult to study, largely due to their mostly pelagic and solitary lifestyle, making them very difficult to find in the vastness of the open sea. However, seasonal Whale

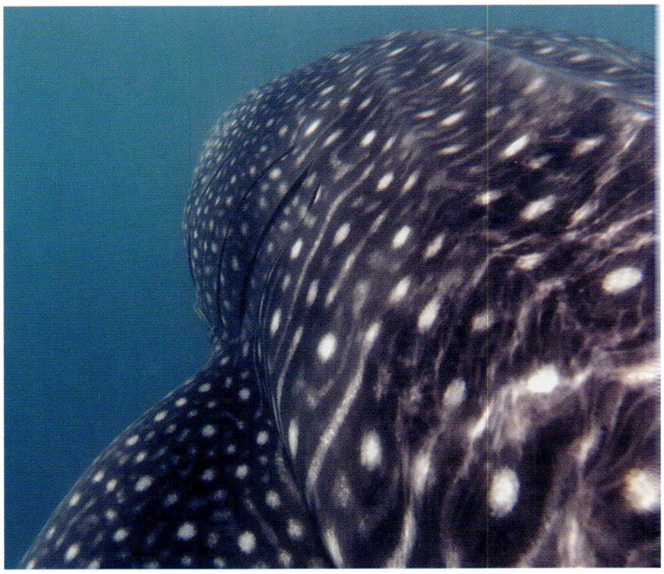

Shark congregations identified at coastal sites all over the world have in recent years made that study a lot easier. In the Philippines, three such congregations have been identified to date; off Donsol in the Bicol region of southern Luzon, around Panaon Island off the southern tip of Leyte, and at Oslob, on the south coast of Cebu. Quite apart from these congregations, Whale Sharks have been found to be relatively common in a number of areas, such as off the north

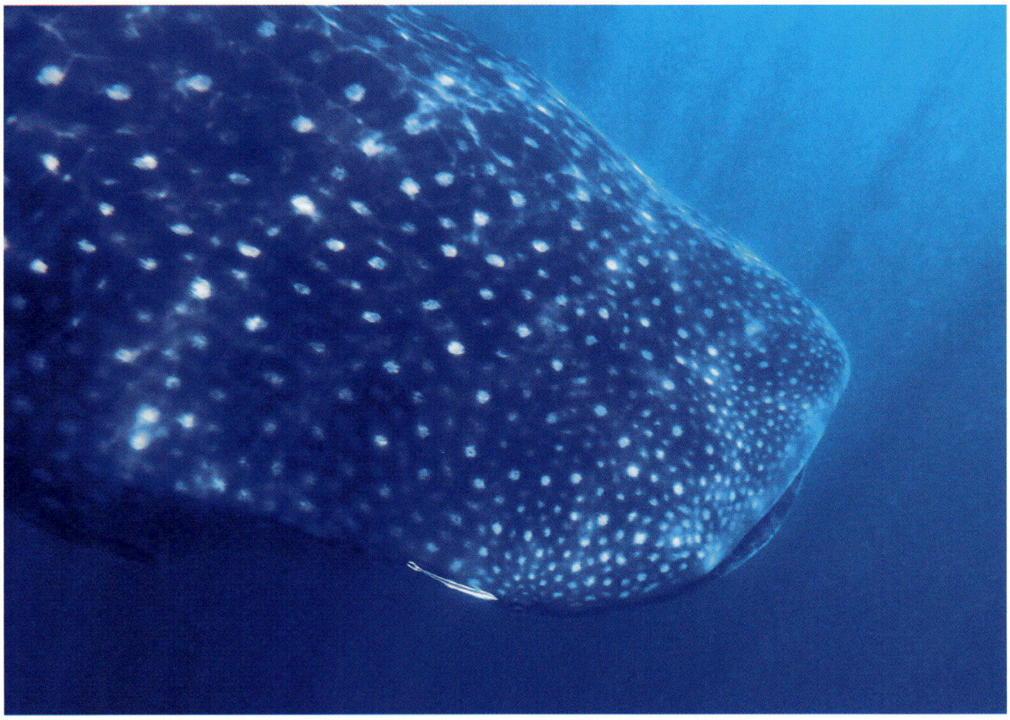

Top: When a Whale Shark glides past at close quarters, you get a very strong sense of just how enormous it is.

Left: The head of a Whale Shark appears out of a 'fog' of plankton in the waters off Donsol, southern Luzon.

Opposite: A Whale Shark's head, illustrating not only its wide, gaping mouth, adapted to sieving plankton, but also its tiny left eye, and the interplay between its pattern of skin spots and the light streaming down from above.

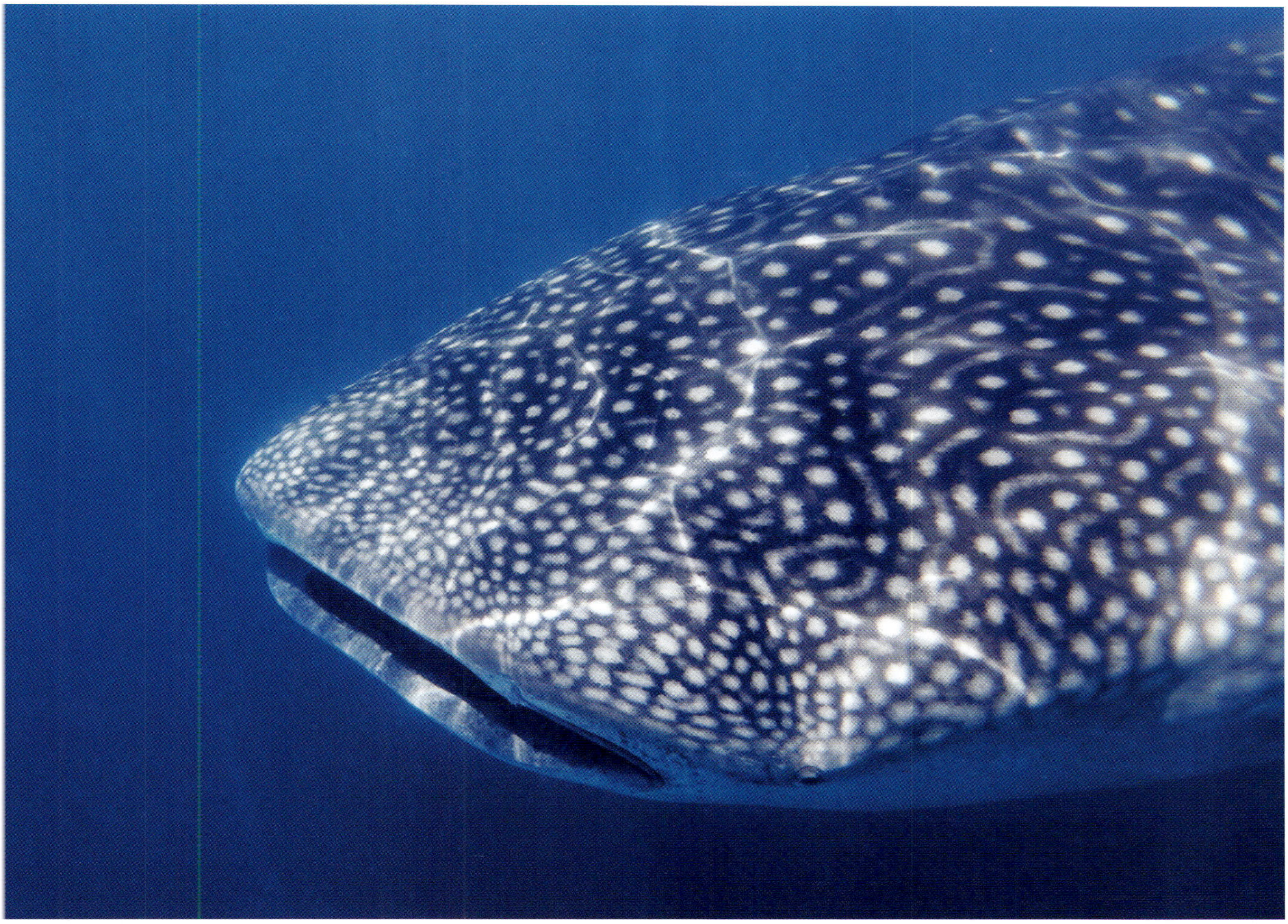

coast of Mindanao and around Tubbataha Reefs, in the Sulu Sea. These populations have been studied by a Philippine research body, the Large Marine Vertebrates Research Institute (Lamave), which has used the Whale Sharks' patterns of body spots to identify individuals. It has also used tagging and satellite tracking to study movements, work that has shown the animals to be resident around the Philippines' coasts, though also highly mobile, migrating large distances within the Philippines' marine area.

Whale Sharks and tourism

Whale Shark-watching tours have been developed at all three congregation sites around the Philippines, with the one at Donsol being the best known and longest established. Set up in 1997, its development was highly instrumental in the 1998 nationwide ban on all Whale Shark hunting and trade. Since then, from November to March each year, well-controlled Whale Shark-watching and snorkelling trips have been a major part of the Donsol economy.

Less well controlled are the activities at Oslob, where fishermen feed the Whale Sharks in order to encourage them to stay close to increasingly large groups of visitors. Both the sustainabillity and ethical nature of this activity have been questioned. Research by Lamave has shown that hand feeding does alter Whale Shark behaviour, encouraging the animals to stay in the area longer than they might otherwise do. Many of the sharks involved (more than 50 were recorded) also have scars from propellor hits.

On a rather smaller scale, and possibly the most environmentally sustainable of all three tourism activities, are the Whale Shark-watching trips organized by the local communities of Panaon Island.

Animals and plants of the land

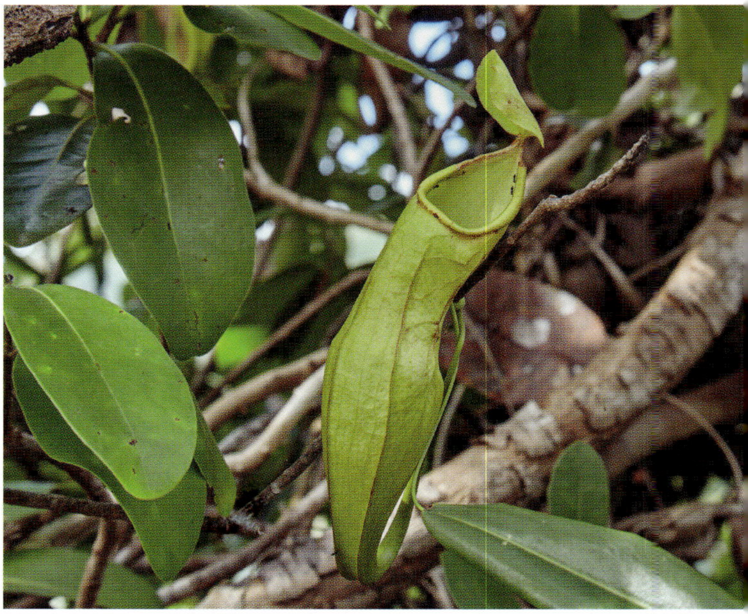

As already mentioned, the Philippines' vast terrestrial plant and animal wildlife makes this country one of the biologically richest on Earth. And yet, new species are still being discovered at a significant rate, at least in part because large areas of the country, along with many environments and habitats, remain barely studied by scientists.

Unfortunately, with the great majority of the Philippines' forests already cleared, we will never know what species have been lost without us ever even seeing them. However, the race is on to understand and protect as much as possible of what remains, work that is revealing ever greater numbers of plants and animals, continually widening the Philippines' already huge biodiversity, and resulting in the establishment of ever more protected areas.

*Below: The Philippine Eagle-owl (*Bubo philippensis*) is the country's largest owl. It is also an endemic species, found in forests across the country.*

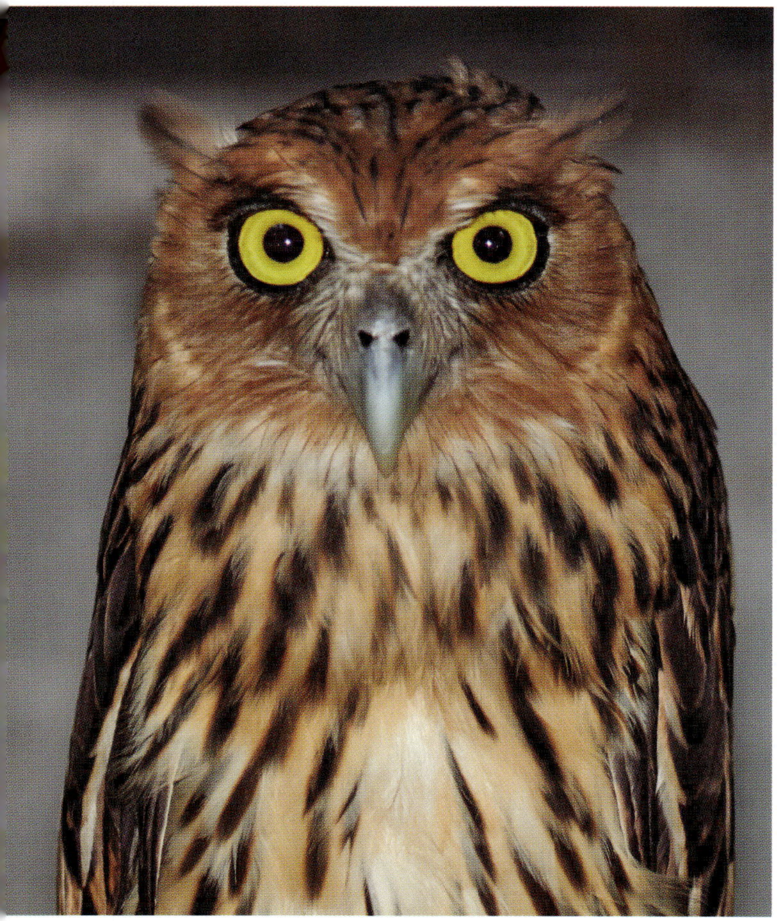

Not surprisingly, the species diversity is far greater than could possibly be described in any single publication. In this chapter just a handful of plants and animals are introduced, those that are particularly relevant in the Philippines context.

In terms of plants, summaries of the country's orchids, pandans, rattans, rafflesias and medinillas are given, all plants that have a special place in the Philippine flora, illustrating well the recent speed of discovery of new species, the rarity of many of these species and the endemism that separates many of them from cousins in mainland Asia. Tree species are not summarized here as some of them, such as the dipterocarps and pines, have been covered in descriptions of the country's habitats in Chapter 3.

When it comes to animals, inevitably such iconic birds as the Philippine Eagle (*Pithecophaga jefferyi*) and the numerous hornbills are summarized, along with the hugely important pigeons. Among the mammals the unique Tamaraw (*Bubalus mindorensis*) and Philippine Tarsier (*Carlito syrichta*) are introduced, along with the remarkable and still-growing collection of largely mountain-living rodents. Equally remarkable are the fruit bats, another vital animal in the ecology

*Opposite: A pitcher plant (*Nepenthes philippinensis*) that is endemic to the country, found only in Palawan, here seen in ultramafic forest on Snake Island, in Bacuit Bay, El Nido.*

Right: The Philippine Tarsier, the world's smallest primate, is a Philippine endemic species, restricted to forests of the Greater Mindanao Faunal Region.

*Below: A female Red Ground Skimmer (*Neurothemis terminata*), a species common across Southeast Asia, seen here in woodland near Davao in the south of Mindanao.*

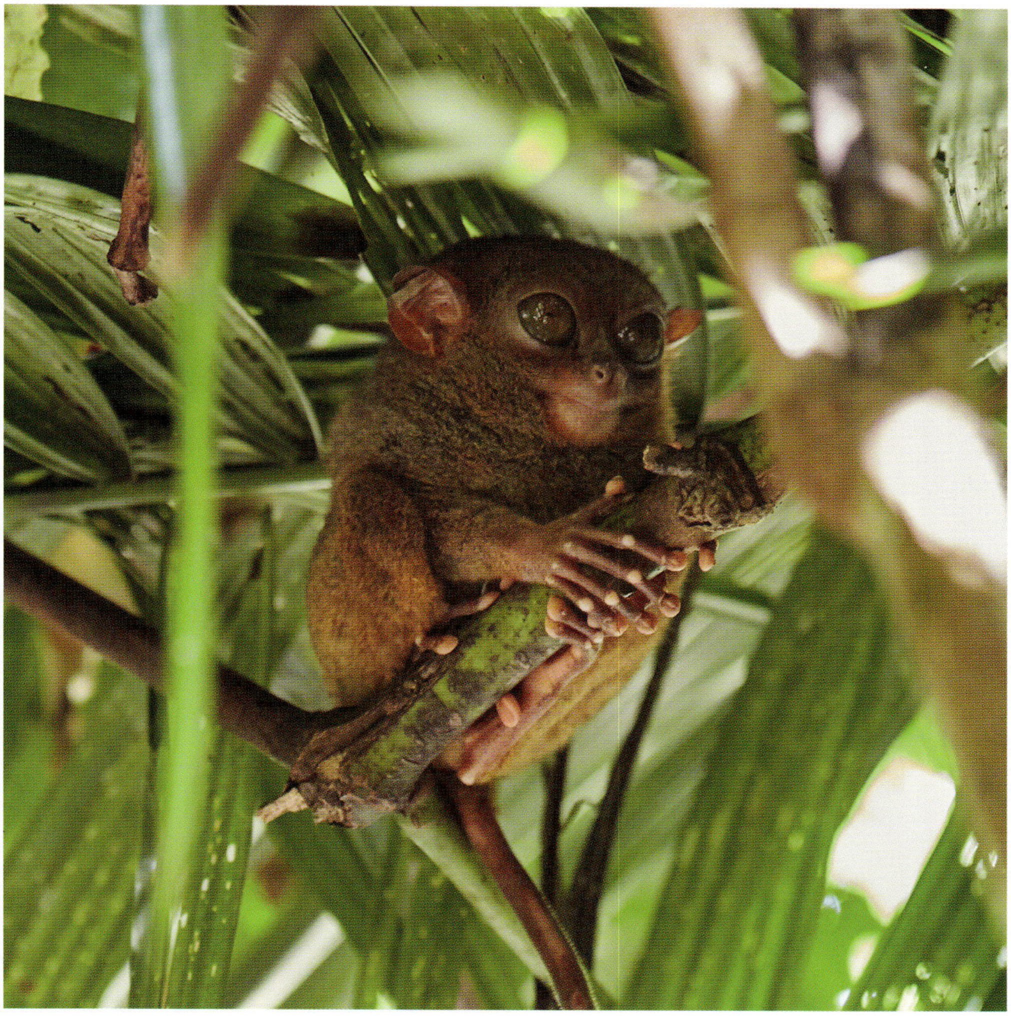

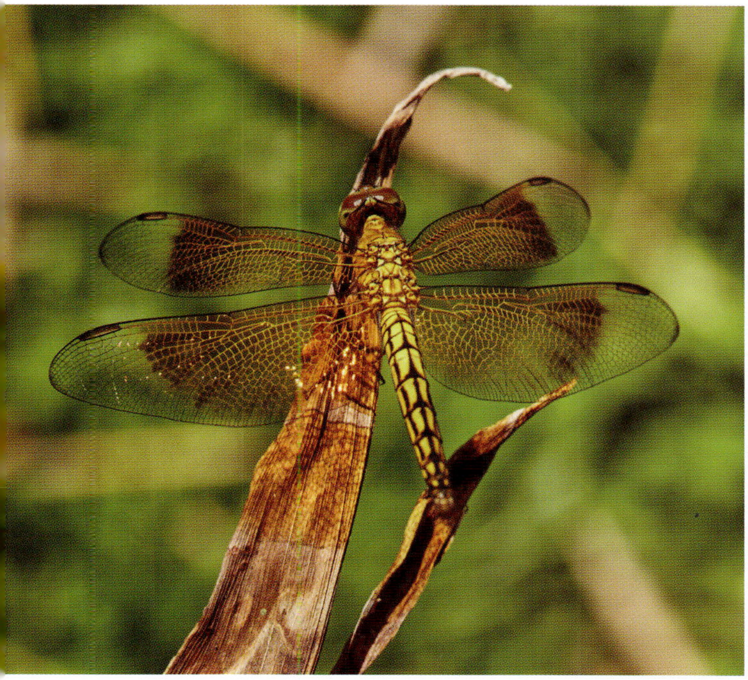

of the Philippines' forests. In addition to the Tamaraw, some of the country's other larger mammals are described, including the several species of deer, pig and civet. As is the case with the plants, these animal species illustrate not only the endemism prevalent in the Philippines, but also the complexity involved in understanding species distinctions and – when it comes to the rodents at least – the continuing flow of new discoveries. More so than the plants, the distributions of many of these animals quite closely illustrate the boundaries of the Philippines' major faunal regions, those divisions set up during the last Pleistocene ice age (see p. 26).

Although this little selection is just a tiny sample of the Philippines' plants and animals, it gives a reasonable cross-section of types of wildlife that call these islands home.

Mammals

As already mentioned, one of the distinguishing features of the Philippines' list of mammals is that it is completely devoid of the iconic large species for which the rest of Southeast Asia is so famous. Indeed, there are very few large mammals at all – they are restricted mainly to just a dwarf buffalo, one species of macaque, a few deer and pigs, and a couple of civets and leopard cats. The vast majority of the 180 or so mammal species are actually really rather small, with the great majority consisting of rodents and bats, two groups that have tended to have a bad press over the years, rather unfairly.

Superficially, this may seem like a somewhat disappointing collection of mammals, but closer inspection reveals them to be ever-more fascinating, their attractions rather more subtle than the in-your-face larger mammals of mainland Asia. Firstly, there is of course the impossibly cute and tiny Philippine Tarsier, whose only relatives are in Sulawesi, Sumatra and Borneo, and whose distribution in the Philippines is closely tied to the Greater Mindanao Faunal Region. At the other end of the size scale, some of the larger mammals, particularly those found in Palawan, clearly illustrate their links to and divergence from their relatives in mainland Southeast Asia. The one big exception, of course, is the Tamaraw, a wholly unique animal restricted simply to the island of Mindoro.

The bats, particularly the fruit bats, once you get used to looking at them upside-down, are remarkably attractive, with rather soft, dog-like faces. The fruit bats consist of quite a mix of endemic species and

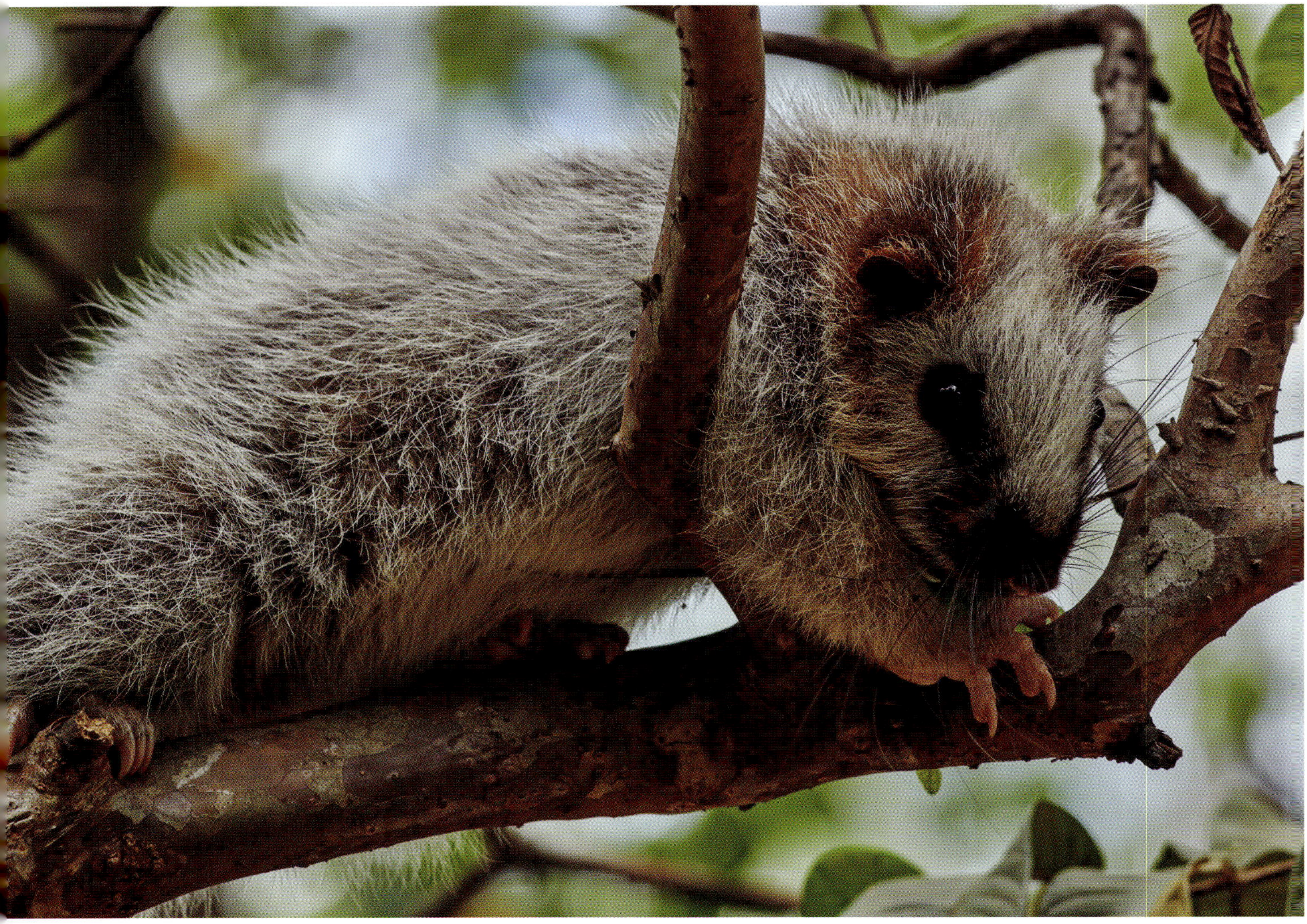

Right: A female Long-tailed, or Crab-eating, Macaque seen in forest near Davao, in Mindanao. These macaques are the Philippines' only monkey species, and are widespread across the country.

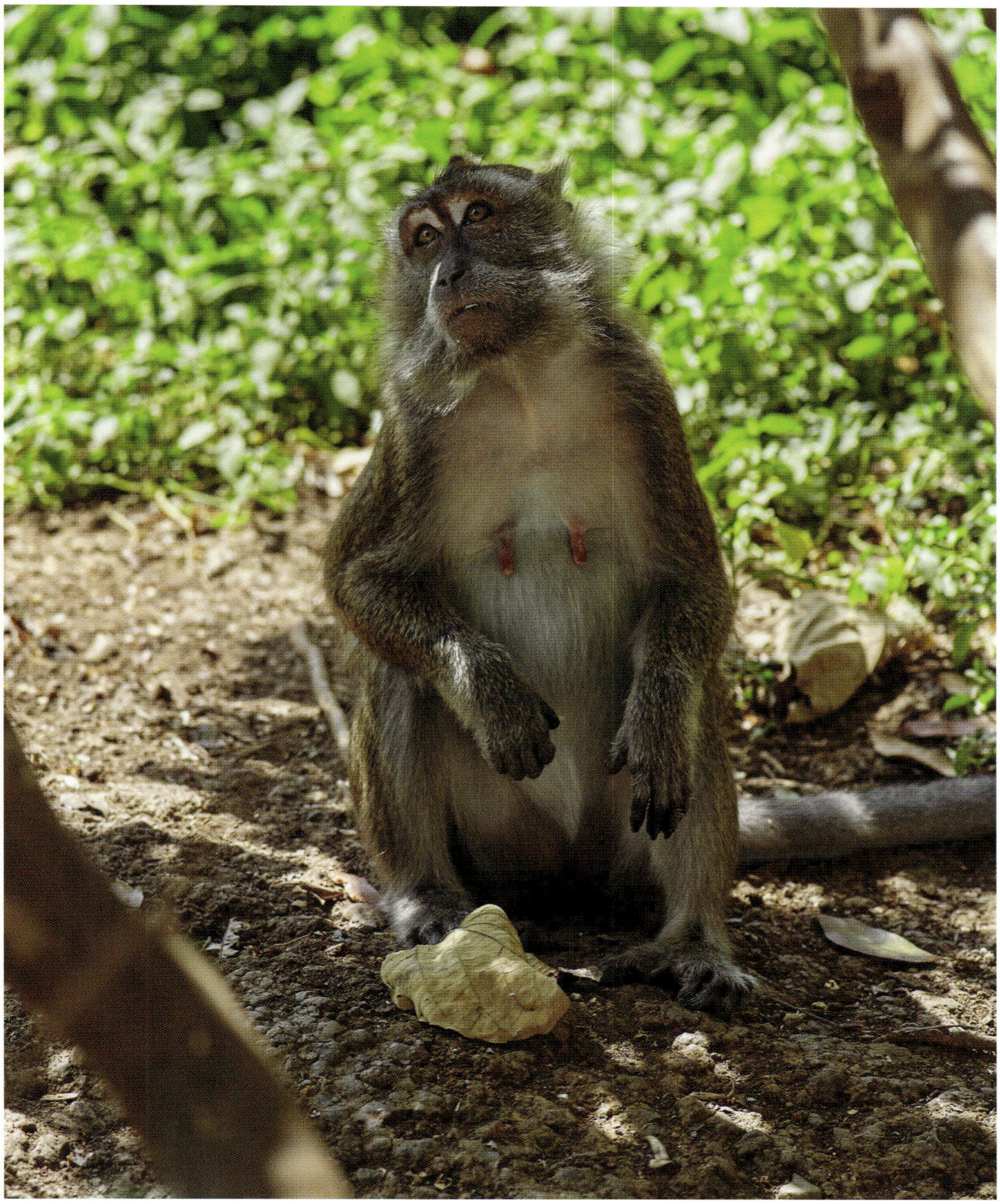

Opposite: The rather cute-looking Northern Luzon Giant Cloud Rat, is an endemic species restricted to the northern half of Luzon, mostly in montane forests, though it can also be found at low altitude.

Below: The Philippine Warty Pig is one of several pig species still found across the country, all of them endemic to the Philippines.

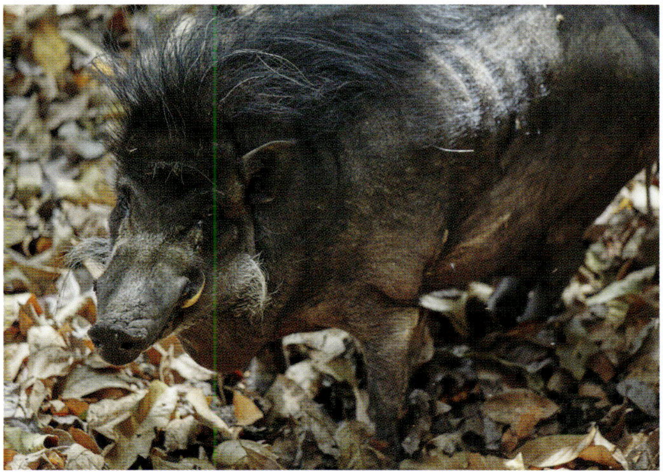

those found across Southeast Asia, but they can be difficult to distinguish in a large group hanging high up in forest trees.

Philippine rodents are proving to be endlessly fascinating. Living mostly high up in remote areas of the country's forested mountains, the number of species is continually growing, ranging from small, worm-eating animals, to large, furry and rather attractive vegetarians. Barely heard of outside the Philippines, within the country they still remain largely overlooked, but are somehow slowly managing to achieve the status of icons of the Philippine fauna.

Tamaraw

The Philippines' largest wild terrestrial animal is the Tamaraw, also known in English as the Mindoro Dwarf Buffalo, a name that accurately describes exactly what it is. It is a relative of the significantly larger domesticated Water Buffalo (*Bubalus bubalis*), which is widespread across Asia, but is clearly a distinct species. The Tamaraw is endemic to the island of Mindoro, and today it teeters on the brink of extinction, clinging on in small groups mostly in the west of the island. The IUCN classifies the Tamaraw as Critically Endangered, meaning that extinction is an imminent risk.

***Below:** A lone male Tamaraw spotted in tall grass (its favourite food) in the heart of Mts Iglit-Baco National Park, on Mindoro.*

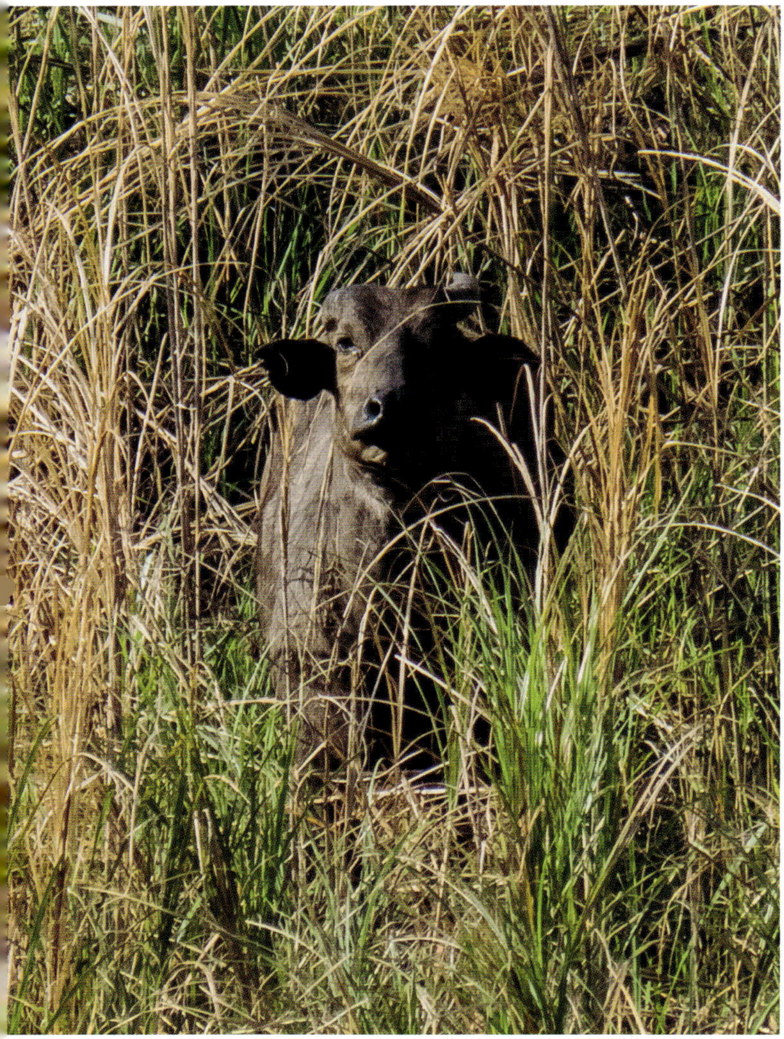

Some Tamaraw biology

The Tamaraw grows to a shoulder height of only 100–105 cm (39–41.4 in), with a weight of up to 300 kg (660 lb), making it a good bit smaller than the Water Buffalo. A characteristic feature is that its horns come together on the skull in a distinct V, quite different from the sweeping U-shape of Water Buffalo horns. Unlike other bovine animals, the Tamaraw does not live in large herds. It is generally said to be a shy, solitary animal, but several recent studies have identified it in small family groups that include a single adult male, with some single males living alone

Habitat and distribution

Tamaraw seem to prefer dense forest or scrub, though with open areas of grassland in which they can feed, eating mostly grasses such as Cogon (*Imperata cylindrica*), and ranging in altitude from 200 m (660 ft) to 1,800 m (5,900 ft). According to the IUCN, only three sub-populations are known about, in Mts Iglit-Baco National Park and on Mt Aruyan, both in the west, and Mt Calavite in the north of Mindoro. There have been reports of sightings in other areas, such as on Mt Halcon, in the north-east, though these have not been confirmed.

Saving the Tamaraw

The Tamaraw is under huge pressure on Mindoro, as a result of massive deforestation, slash-and-burn farming, cattle ranching, a growing human population and direct hunting. It is estimated that in the early 1900s there were about 10,000 Tamaraw across the island, but by the late 1940s this had crashed to no more than about 1,000, at least in part due to an outbreak of rinderpest, a disease spread from newly introduced cattle. By 1969, according to the IUCN, there were barely more than 100 left in the wild, though by 2007 they had recovered to about 200 individuals.

In the early 1980s an attempt at captive breeding starting with the capture of 20 animals ended in failure, with only one Tamaraw successfully produced. He lives on as the sole survivor of that group.

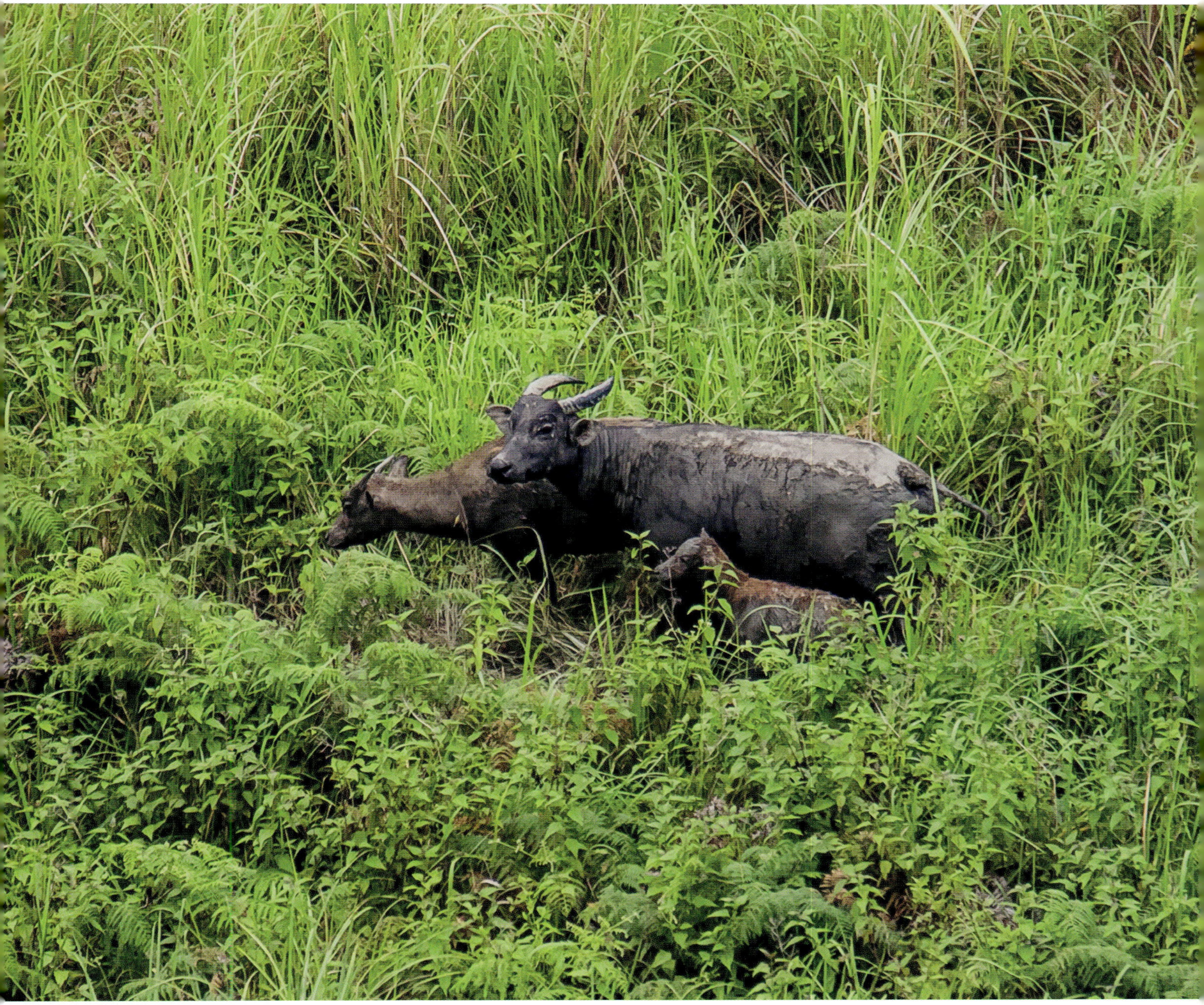

Above: A Tamaraw family group comprising an adult female and two juveniles, a rare but hugely encouraging sight, in Mts Iglit-Baco National Park, the species' main surviving stronghold.

Since the early 2000s, conservation efforts have been focused on the main surviving wild group, in the south-west of Mts Iglit-Baco National Park, with close patrols and enforcement of anti-hunting laws, along with support from local communities of Mindoro's indigenous people, the Mangyan. As a result, over recent years annual counts have shown a steadily rising population, with the latest (in April 2018), revealing 523 Tamaraw in the park's survey area. The rising numbers are, of course, hugely encouraging. However, they may also reflect improving survey methods, or the migration of animals from other areas where they are under threat. If the latter is happening then this would actually suggest a worsening situation, not an improving one, with the Tamaraw's safe home range shrinking. This remains to be clarified.

Deer, pigs and cats

Apart from the Tamaraw, the Philippines' 'large' mammals consist of a collection of deer, pigs and cats or cat-related animals, as well as a single macaque, none of which would be classed as large in just about any other country. However, in the Philippines' context they are about as big as mammals get. Typically for the Philippines, most of these mammals are either endemic species, or at least subspecies of relatives found in mainland Southeast Asia. Most are threatened, with limited natural ranges combined with hunting and massive habitat loss working together to bring about rapid falls in their populations. Some have already been wiped out from parts of their natural range, and it is likely that only highly innovative conservation measures will save them from extinction.

Deer

The Philippines is home to three species of deer, plus one mouse-deer. All are endemic to the Philippines and have ranges restricted to certain areas.

The most widespread species is the Philippine Deer (also called the Philippine Brown Deer), known by two scientific names, *Rusa marianna* or *Cervus mariannus*. It is small by the standards of deer internationally, reaching only about 70 cm (28 in) at shoulder height. It is found in forested areas across much of the Philippines, with the exception of the western Visayas, Palawan and some of the country's outer islands. It is divided into four subspecies, one living on Luzon, one on Mindoro and two on Mindanao.

The islands of the western Visayas are home to the Visayan Spotted Deer (*Rusa alfredi*, or *Cervus alfredi*), a region to which it is endemic. Originally spread across Panay, Negros, Guimaras, Cebu, Masbate and Ticao, it

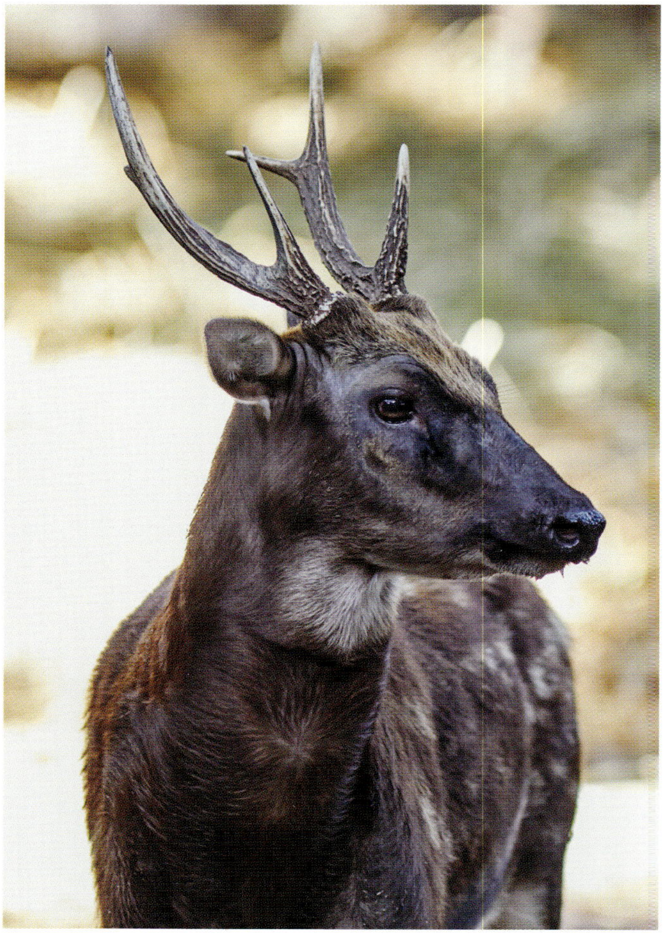

Above: Male Visayan Spotted Deer. A highly endangered species, this one was seen in Bacolod, on Negros, at one of the captive-breeding centres that is working to save it from extinction.

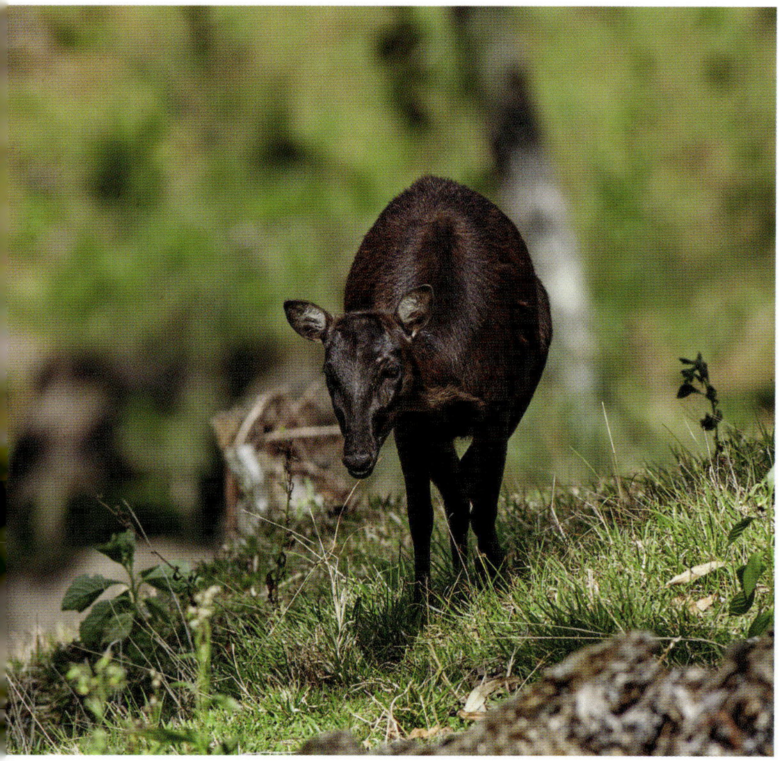

Left: Female Philippine Brown Deer wandering free in woodland at the Eden Nature Park, near Davao, in Mindanao.

has been wiped out from all but Panay and Negros, and even here it is restricted to some of the few remaining forested areas.

The Calamian Deer (*Axis calamianensis*) is endemic to the Calamian Islands of northern Palawan, and in particular the islands of Calauit, Busuanga and Culion. It is not found on Coron Island. The Philippine Mouse-Deer (*Tragulus nigricans*), by contrast, is restricted to Balabac and neighbouring islands, at the opposite, southwestern end of Palawan.

All four deer species are losing out rapidly to loss of forest habitat and hunting. Even the widely spread Philippine Deer is increasingly fragmented and scattered, with sub-populations retreating into ever more remote forested mountains. It is believed to have already been wiped out from Bohol and Catanduanes.

It is the Visayan Spotted Deer that is in the most serious trouble. Eliminated from more than 90 per cent of its natural range, it is now quite rare, with the IUCN estimating its adult population to be as low as 700, fragmented into the few remaining forests of Negros and Panay. In 1990 a captive-breeding programme was started for this species, with herds at two centres on Negros, one on Panay, and several more in zoos scattered around the world. The programme has been quite successful, to the point that there have been some trial releases of captive-bred deer.

Below: *The Philippine Mouse-deer (*Tragulus nigricans*), a small species that is restricted to Balabac Island.*

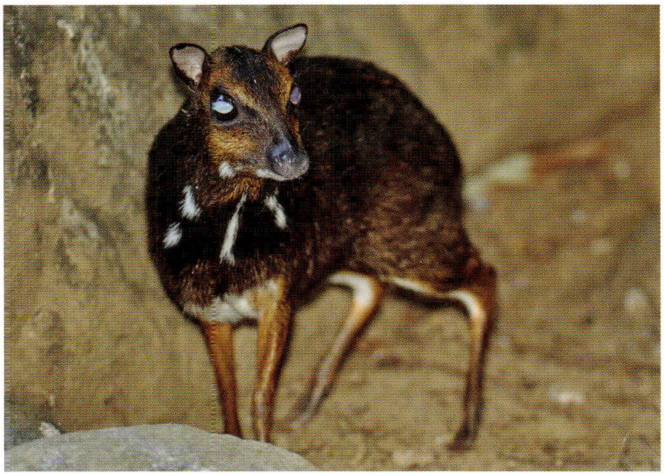

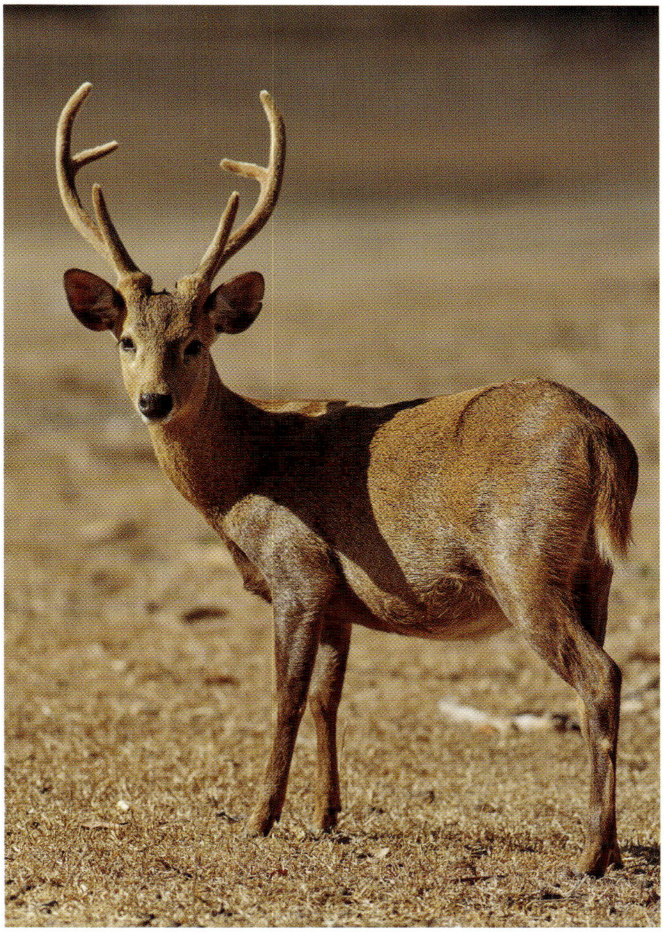

Above: *Male Calamian Deer, seen on Calauit Island, one of northern Palawan's Calamian Islands, and one of the very few places where these deer can be seen in the wild.*

Pigs

Four species of wild pig live in the Philippines, like the deer each showing regional distributions that coincide well with the country's major faunal regions. The most widespread is the Philippine Warty Pig (*Sus philippensis*), which is spread across Luzon and Mindanao. Though currently divided into two subspecies, one for each of the main islands, it is likely that sooner or later they will each be given full separate species status.

Oliver's Warty Pig (*S. oliveri*), named after the late William Oliver, a prominent and very passionate advocate for Philippine wildlife conservation, is restricted to the mountains of Mindoro. Palawan is home to the Palawan Bearded Pig (*S. ahoenobarbus*), a

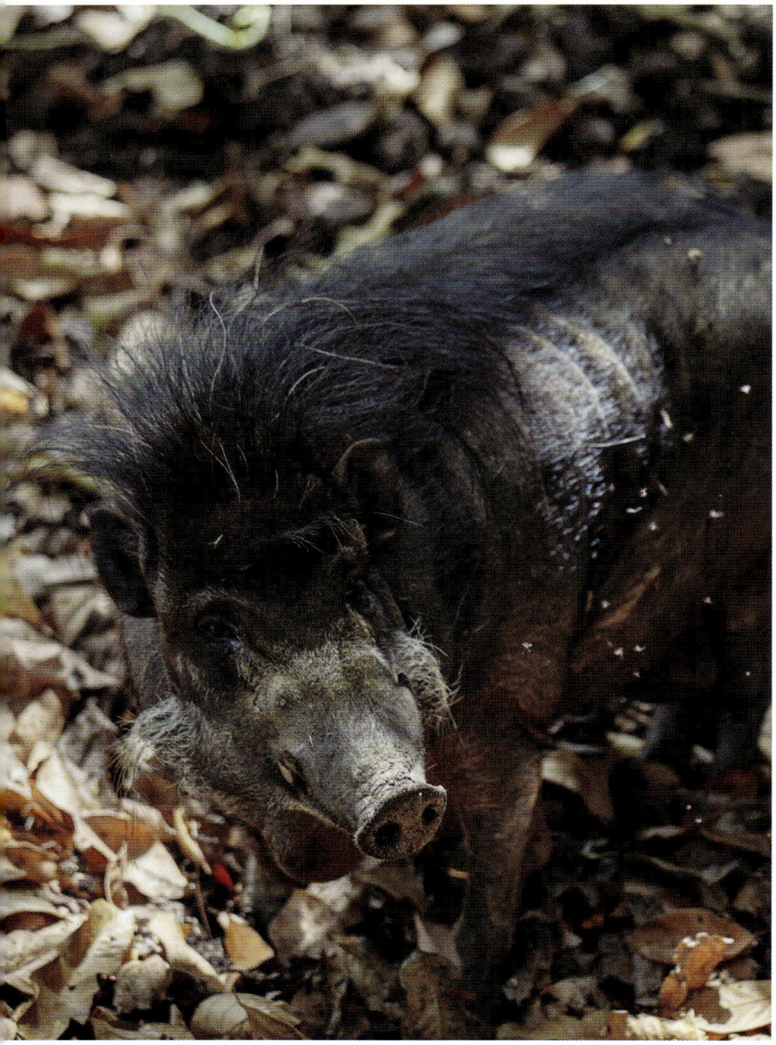

There still remains, however, an urgent need to halt deforestation and hunting, if there is to be any chance of saving this animal from extinction.

Cats

The Philippines is home to just one true cat, the Asian Leopard Cat (*Prionailurus bengalensis*), which in this country is restricted to Palawan and the western Visayas. Widely spread across South and Southeast Asia, this is not a Philippine endemic species, and the leopard cats found in Palawan appear to be very similar to those in the rest of Asia. Those in the western Visayas, however, are rather different, forming a separate subspecies called the Visayan Leopard Cat (*P. bengalensis rabori*). Due to the Asian Leopard Cat's ability to adapt to a host of different habitats, including those with heavy human influence, it does not seem to be seriously threatened, and even the Visayan subspecies is classed by the IUCN as no worse than Vulnerable.

Left: *The Philippine Warty Pig is still found in forests across much of the country, though its numbers are in decline due to extensive hunting and loss of its forest habitat.*

Palawan endemic species that is closely related to the Bornean Bearded Pig (*S. barbatus*).

Again as is the case with the deer, the Western Visayas has its own wild pig species, the Visayan Warty Pig (*S. cebifrons*), which has also been wiped out from much of its natural range.

All four species are suffering from habitat loss and extensive hunting, but with the exception of the Visayan Warty Pig they remain reasonably widespread across their ranges. The IUCN classifies these three as either Vulnerable or Near Threatened, indicating that they are at risk. The Visayan Warty Pig, however, is ranked as Critically Endangered, having been wiped out from all the Visayan islands with the exception of Negros, Panay and possibly Masbate. As for the Visayan Spotted Deer, a captive-breeding programme was established in the 1990s, using the same Negros and Panay centres, plus a number of overseas zoos.

Below: *Adult male Visayan Warty Pig. Now Critically Endangered, it is very rare to see this animal in the wild. This individual was seen in a captive-breeding facility at Bristol Zoo, in the UK.*

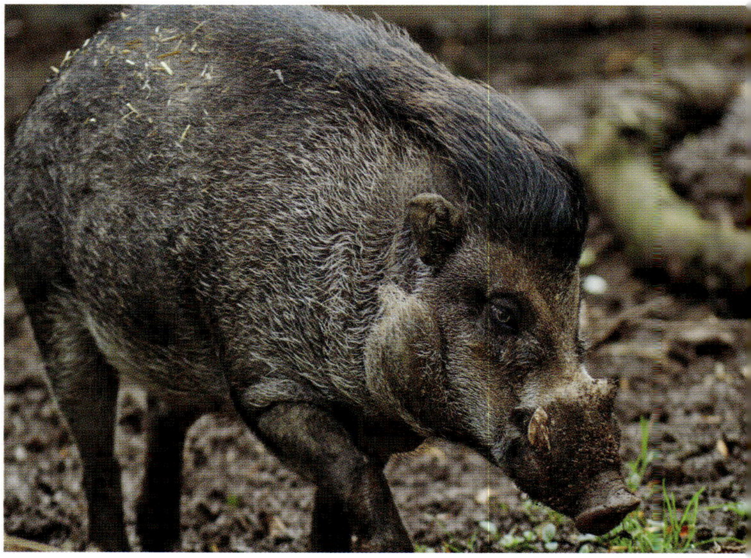

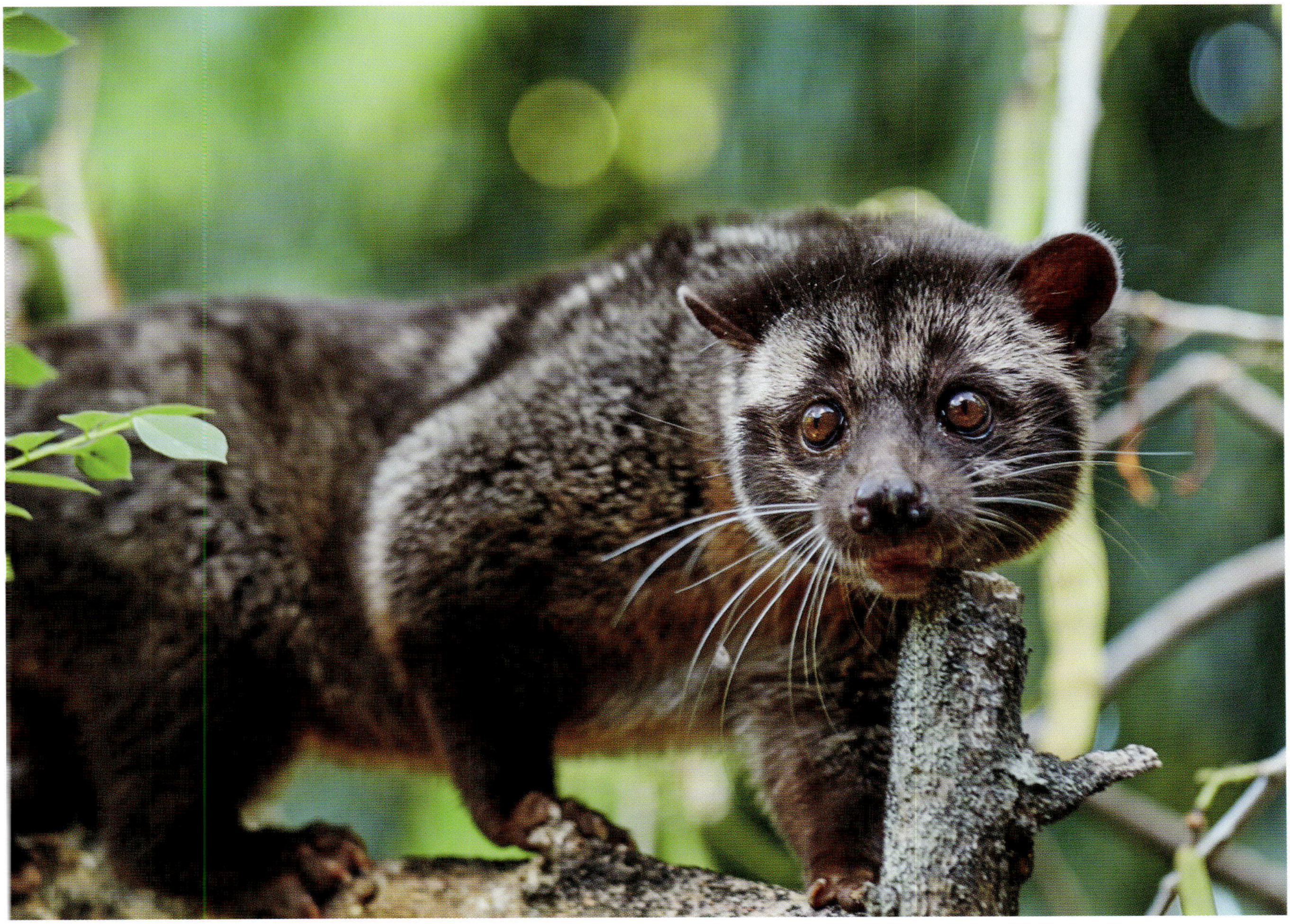

Above: The Common Palm Civet is widespread across the Philippines. Although introduced, it has evolved from its relatives in Southeast Asia to become a separate species.

There are also three cat-related species, one of which is native to the Philippines, while the other two were almost certainly introduced by humans. The former is the Palawan Binturong, or Bearcat (*Arctictis binturong whitei*), a subspecies of the Binturong found across much of Southeast Asia, which in the Philippines is restricted to Palawan. This is one of those species (along with the Asian Leopard Cat, Palawan Bearded Pig and Palawan Mouse-deer, among others) that highlights Palawan's past links with mainland Southeast Asia, and distinctiveness from 'oceanic' Philippines.

The two introduced species are the Malay and Common Palm Civets (*Viverra tangalunga* and *Paradoxurus philippinensis*, respectively), both of which are highly successful in a range of habitats across the Philippines. Introduced about 4,000 years ago, the Common Palm Civet has evolved into a distinct species, though it is not a Philippine endemic as it is also found on Borneo. Both civets are widespread across the country and still quite common, though they are suffering from hunting for both food and the pet trade. Palm civets are also under increasing pressure from the new industry of civet coffee. These civets love to eat coffee-tree fruits, the beans of which are then passed out in their droppings. Beans passed in this way have a unique flavour, and have become a hugely fashionable and expensive drink. As a result, growing numbers of wild palm civets are being captured and caged in cruel conditions, simply to increase the production of civet bean coffee, a faddish (and probably short-lived) consumer product.

Philippine Tarsier

With a body that is barely 10–15 cm (3.9–5.9 in) long, though with a tail that is twice that, the Philippine Tarsier (*Carlito syrichta*) is one of the world's smallest primates. It is also one of the cutest, largely due to its enormous, imploring eyes, which dominate its head. The result of an almost entirely nocturnal lifestyle, the large eyes allow the tarsier to see even in the dark. The eyes cannot, however, be swivelled, though this apparent limitation is overcome by the tarsier being able to rotate its head by 180 degrees.

Habitat and distribution

The only wholly carnivorous primate, feeding on insects as well as small lizards, snakes and birds, tarsiers live in primary and secondary forests, preferring dense vegetation. They spend their entire time in trees, though often only 2 m (6.6 ft) or so off the ground, especially during the daytime. They are rather shy and sensitive, and if disturbed are capable of leaping 2–3 m (6.6–10 ft), an ability gained from their elongated ankle bone, the tarsus, from which their name is derived.

In the Philippines, tarsiers can be found across Mindanao and its associated islands, including Basilan, Bohol, Samar and Leyte, conforming closely to the Pleistocene Epoch Greater Mindanao Faunal Region. There is just one species of tarsier in the Philippines, though it appears to be divided into three subspecies, each with a separate range, one covering much of Mindanao, another present in north-east Mindanao, Dinagat and Siargao Islands, and a third in Bohol, Samar and Leyte.

Although the Philippine Tarsier is endemic to the south of the country, related species are also found in Borneo, Sulawesi and eastern Sumatra. Until recently, every species was classified in a single genus, *Tarsius*, but the Philippine species has been separated into its own genus, and as a result its scientific name has been changed from *Tarsius syrichta* to *Carlito syrichta*. The new genus name is in honour of Carlito Pizarras, a Bohol-based hunter-turned-conservationist who, since the 1990s, has done more to promote conservation of the Philippine Tarsier than anyone else, through his own Philippine Tarsier Foundation (PTFI).

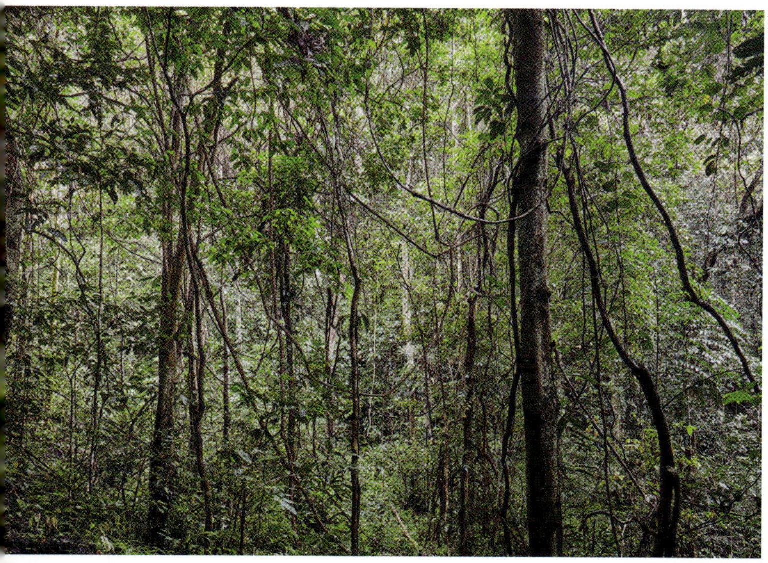

Conservation of the Philippine Tarsier

The IUCN classifies the Philippine Tarsier as Near Threatened, meaning that its population is in decline to the extent that it could soon become officially threatened. This may seem a little at odds with findings suggesting that in Bohol at least, the population is down to just a few hundred, making it already threatened on that island. However, overall the fact that Philippine Tarsiers are quite happy with secondary forest is likely to help with their survival. That said, they are suffering from considerable habitat loss, putting them under pressure. They are also a victim of their own cuteness, a large pet trade causing significant damage to wild populations. Both Filipinos and overseas tourists are buying them as pets, despite it being well known that they survive very poorly in captivity. This is a major threat to the species that could be stopped by the public very quickly simply by refusing to be a part of the trade.

Left: *Typical dense limestone rainforest that is home to the Philippine Tarsier on Bohol.*

Opposite: *The incredibly cute Philippine Tarsier is endemic to forests of Mindanao, Basilan, Bohol, Samar and Leyte.*

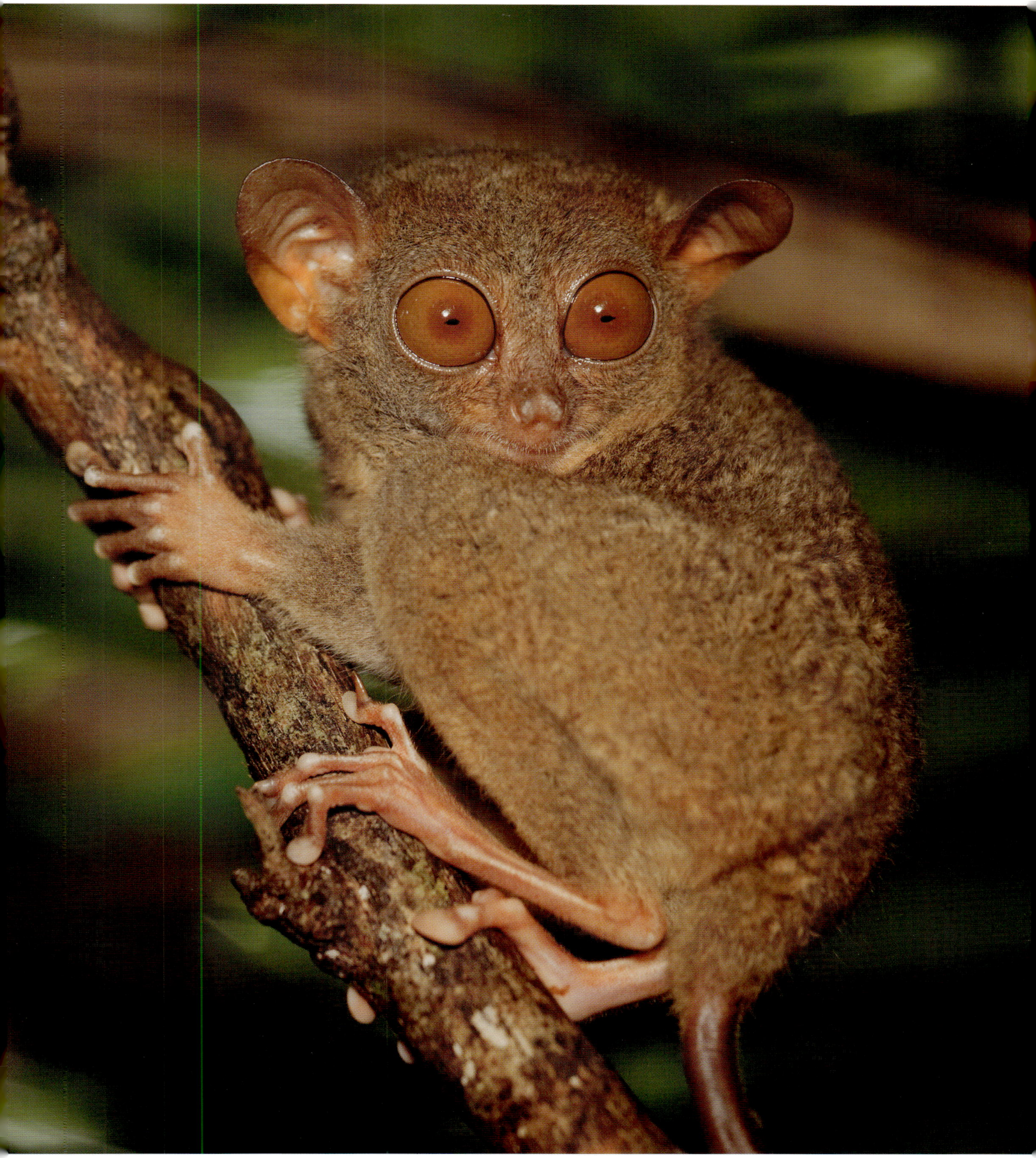

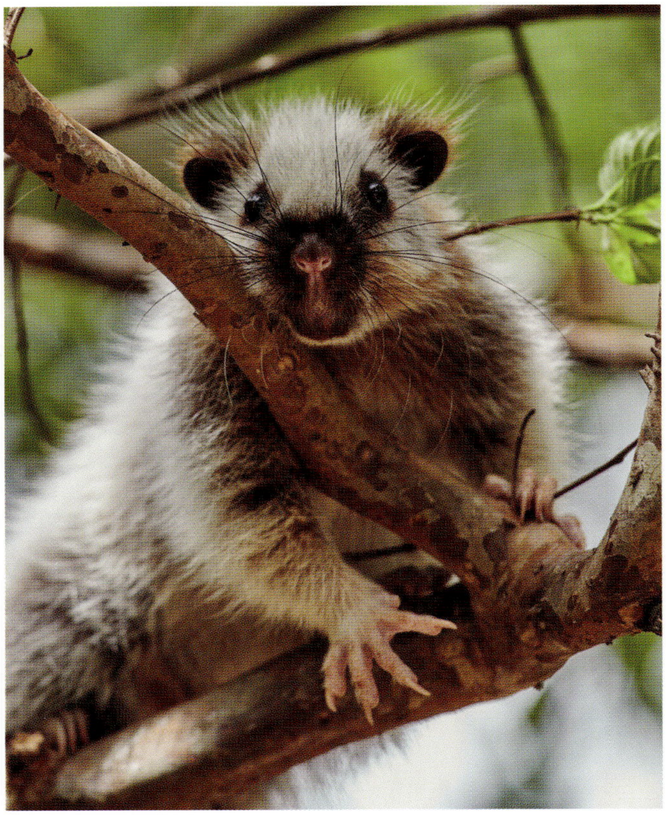 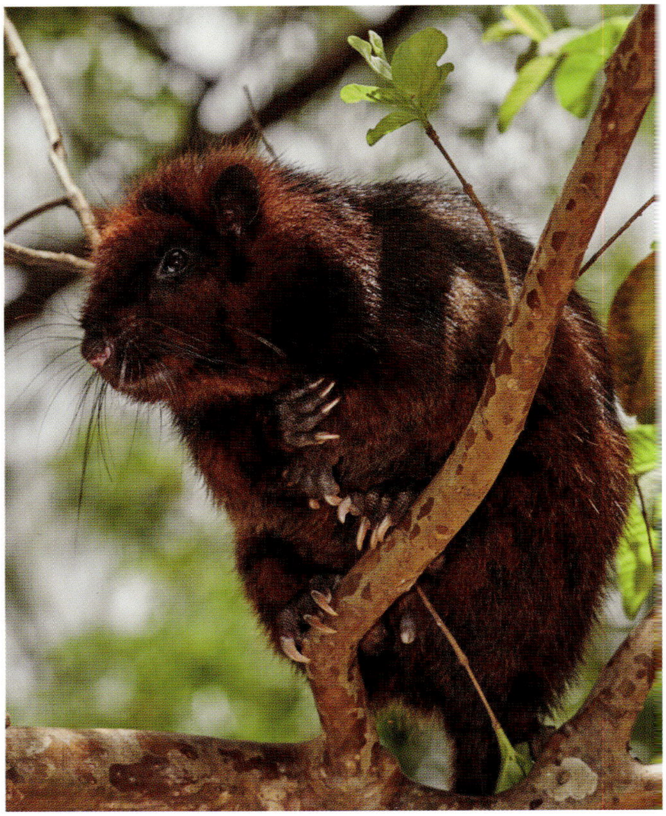

Endemic mountain rodents

In most countries rodents do not generally get the limelight, but in the Philippines things are rather different. Given that they make up nearly one-third of the country's terrestrial mammalian species and that many of them are really quite remarkable, in the Philippines they constitute a significant and fascinating group of animals. Nationwide, probably up to 70 rodent species are known, the vast majority of them endemic to the country. Fifty-six are found on Luzon, though this heavy bias almost certainly largely reflects where most scientific studies into rodents have been carried out. As studies expand to other parts of the Philippines, it is likely that the nationwide rodent species counts will grow.

Cloud rats and earthworm mice

The most abundant rodents so far identified are quite different from those found around the world. They fall mainly into two groups, cloud rats and earthworm mice, the former so named because most of them live high in the mountains, the latter because they feed mainly on earthworms. Each is derived from a single species that is believed to have arrived in the newly forming Philippines 14 million and eight million years ago, respectively, before then developing into a whole series of Philippine endemic species.

Among the cloud rats there are currently known to be 18 species, 12 of them found on Luzon. They are variable in size, ranging from 15 g (0.58 oz) for the smallest *Musseromys* species, up to 2.7 kg (5.9 lb) for the largest of them all, the Northern Luzon Giant Cloud Rat (*Phloeomys pallidus*). The latter, along with its cousin, the Southern Luzon Giant Cloud Rat (*P. cumingi*) and the Luzon Bushy-tailed Cloud Rat (*Crateromys schadenbergi*), are perhaps among the best known of the cloud rats, if only because they are the largest, the most beautiful and the most widespread.

The earthworm mice, by contrast, are mostly small Thirty species have so far been identified across Luzon 14 of them belonging to the genus *Apomys*. However, there are almost certainly many more still to be discovered. They are abundant across the forested mountains, and it is common to find several species

Opposite, far left: The Northern Luzon Giant Cloud Rat is commonly found in the montane forests of northern Luzon.

Opposite left: Closely related to the northern species, the Southern Luzon Giant Cloud Rat is found in forests across southern Luzon.

Right: Upper montane rainforest on Luzon's Mt Pulag: classic habitat for most of the country's endemic mountain rodents.

Below: The Palawan Treeshrew (Tupaia palawanensis) is endemic to Palawan, found in forests from sea level to 1,400 m (4,600 ft).

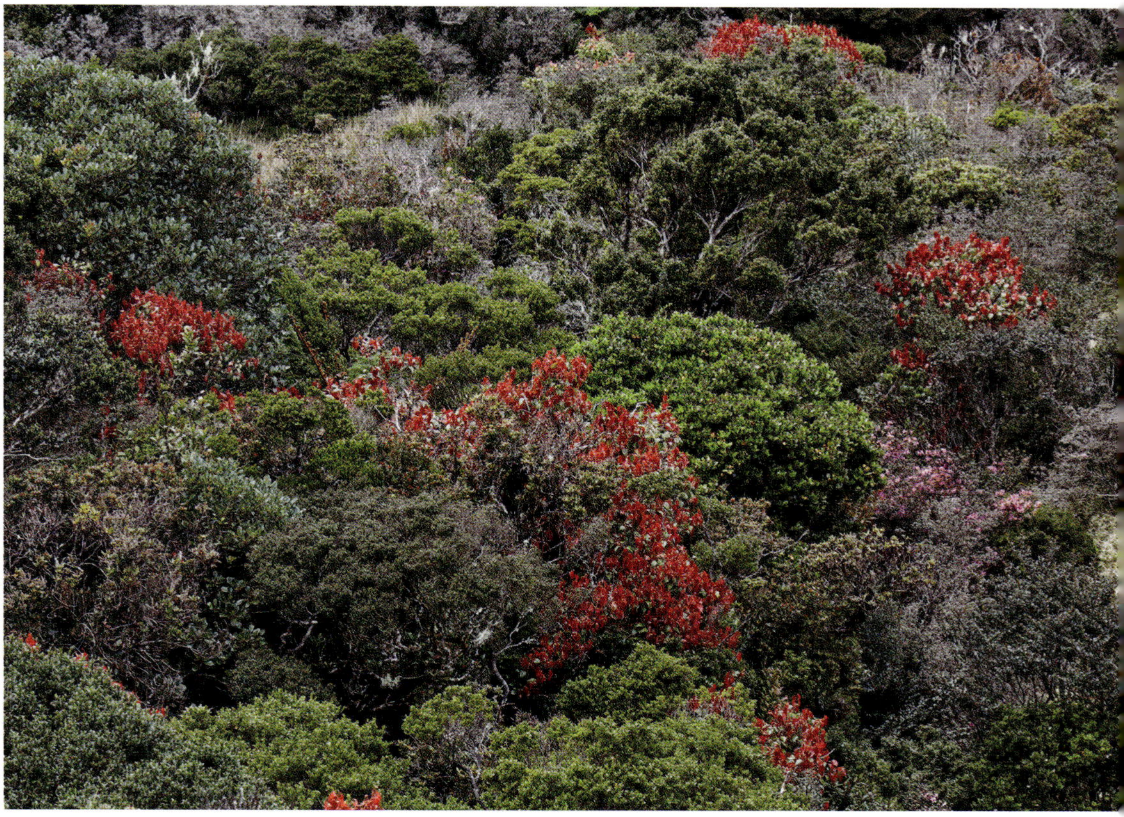

living in the same area but occupying slightly different ecological niches. Two, for example, are arboreal, so find their food in the trees, others scratch around among the leaf litter and still more dig into the topsoil.

One of the striking features about both rodent groups is that their abundance and species diversity increase with altitude, quite the opposite of what one normally expects. Many of them are very happy in the montane forests, and are rarely found in the lowlands. An exception to this are the two giant cloud rats, which range from sea level up into the high mountains.

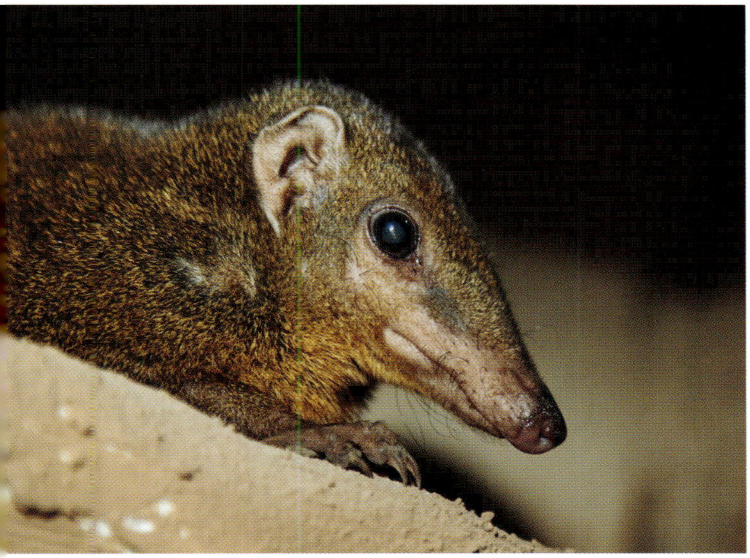

One of the consequences of this preference for mountain forests is that separate populations are quite isolated from one another, something that has helped with the generation of new species. As a result, many of the cloud rats and earthworm mice are restricted not just to the Philippines, but actually to a single mountain or mountain range. The Luzon Bushy-tailed Cloud Rat, for example, is restricted to some of the highest montane forests in the Central Cordillera, while the Banahao Tree-mouse (*Musseromys gulantang*) has been found only on Mt Banahaw, and the Mt Isarog Shrew-mouse (*Archboldomys luzonensis*) is restricted to Mt Isarog, both mountains in southern Luzon.

Again, the two giant cloud rats are exceptions, the northern species being widespread across northern Luzon, and the southern species found across the south. The two overlap in the Southern Sierra Madre mountains and the Calabarzon region south of Manila.

Fruit bats

With almost 80 species, bats make up nearly half of all the Philippines' mammals. Most of them are small insectivorous bats that roost primarily in caves throughout the Philippines and which are hugely important in controlling the numbers of agricultural pest insects. However, 25 species are fruit bats, feeding largely on fruits, though also sometimes on flowers, leaves and nectar. They are almost all much larger than the insect-eating bats, and can frequently be seen roosting in trees, though some species also use caves. They are hugely important in the ecology of the rainforests, their fruit diet making them responsible for distributing seeds over large areas, helping not only to maintain primary forests, but also with regeneration of degraded forest areas.

Biology and distribution

Fruit bats do not have the echolocation (that is, sonar)

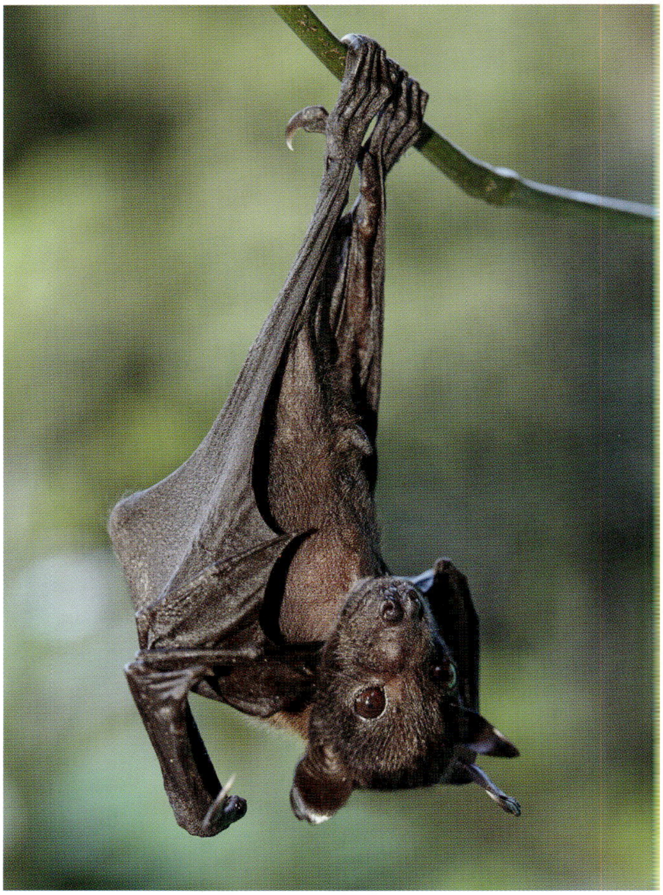

Right: *A juvenile male Island Flying-fox in forest in Mt Kanlaon Natural Park, Negros.*

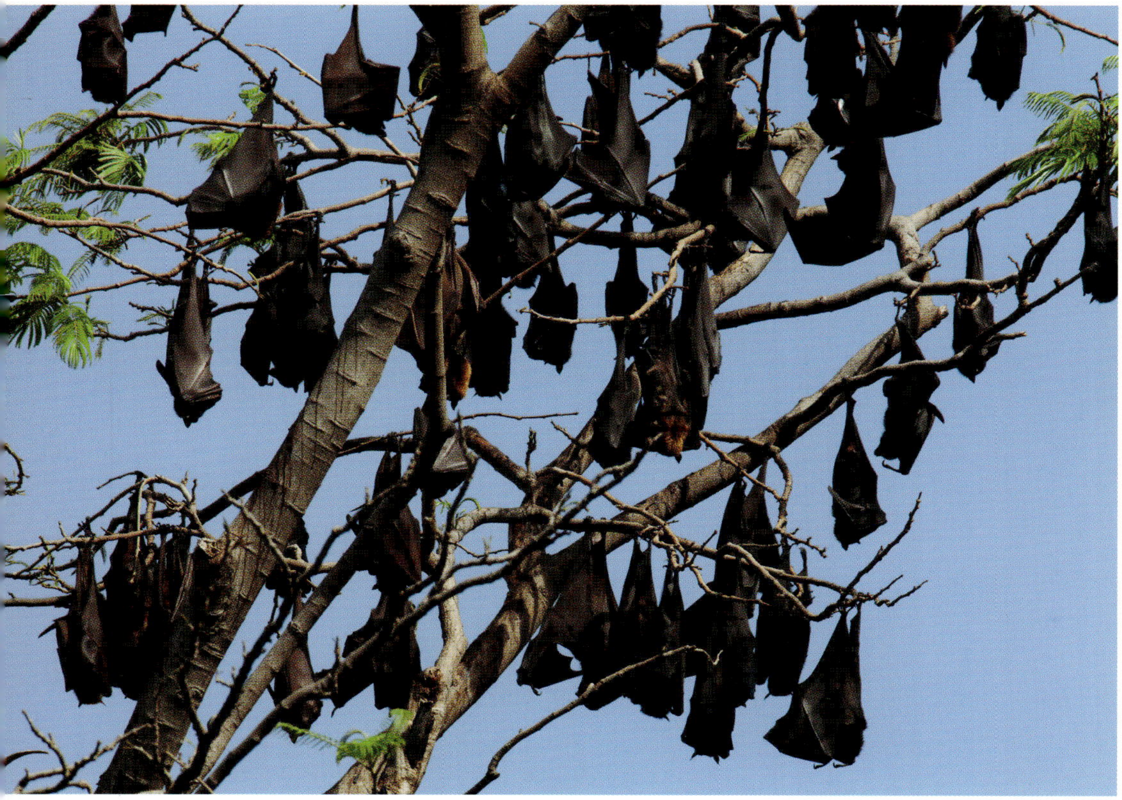

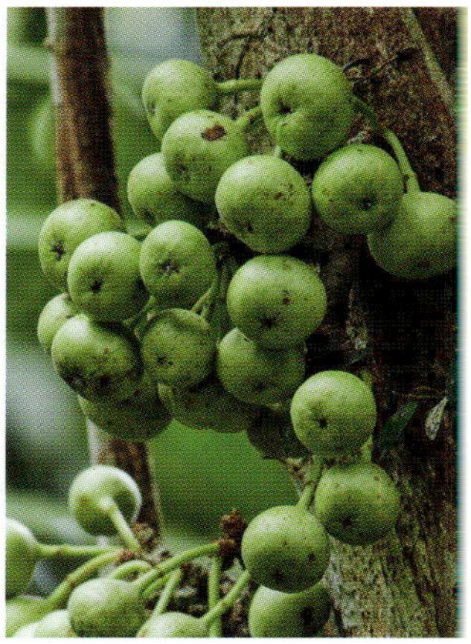

Above: *Fruit of the many widespread fig (*Ficus *species) trees, a major food for fruit bats.*

Left: *A colony of roosting fruit bats, in lowland rainforest in Subic Watershed Forest Reserve, Luzon.*

navigational abilities for which insectivorous bats are so famous. Instead they rely on sight, and as a result have relatively larger eyes and smaller ears, as well as simpler noses. Overall, fruit bats have quite recognizably attractive dog- or fox-like facial features, hence the frequently used name for them, 'flying fox'.

Fruit bats are spread throughout the Philippines, often in seemingly quite large numbers, though in fact they are in serious decline due to deforestation. In complete contrast to the rodents, they are mammals primarily of the lowlands, with species diversity rapidly declining with altitude. Very few forage in the upper montane forests, though one that does is the Luzon Pygmy Fruit Bat (*Otopteropus cartilagonodus*).

The Philippine fruit bat population includes the world's largest bats. The endemic Golden-crowned Flying Fox (*Acerodon jubatus*) is the largest by weight, reaching almost 1.5 kg (3.3 lb) in Mindanao, while the Giant Flying Fox (*Pteropus vampyrus*), a species found right across Southeast Asia, has the largest wingspan at almost 2 m (6.6 ft). These two fruit bats often roost together in colonies that number several thousand. The most accessible of these, in the forest in Subic Bay, has become an important tourist attraction.

Endemism

Fifteen of the country's 25 fruit bats are endemic to the Philippines, including the Golden-crowned Flying Fox and Luzon Pygmy Fruit Bat. Those endemic bats that are spread across the country, such as the Golden-crowned Flying Fox, are nevertheless absent from Palawan, this latter area being occupied only by fruit bats that are common across Southeast Asia, such as the Giant Flying Fox and Common Island Flying Fox (*Pteropus hypomelanus*).

A few of the endemic fruit bats have highly restricted ranges. The Luzon Pygmy Fruit Bat is limited to just Luzon, though it is quite common here. Two Critically Endangered fruit bats, however, are restricted to the western Visayas, namely the Philippine Tube-Nosed and Negros Naked-backed Fruit Bats (*Nyctimene rabori* and *Dobsonia chapmani*, respectively). The former is now limited to a few small populations in remaining forests on Negros and Cebu. The Negros Naked-backed Fruit Bat, the country's largest cave-dwelling bat, was declared extinct in the 1970s but was rediscovered in the early 2000s. Recent surveys have found it to be doing better than expected in the forests of Negros, and to some extent on Cebu too, though its numbers are still dangerously low.

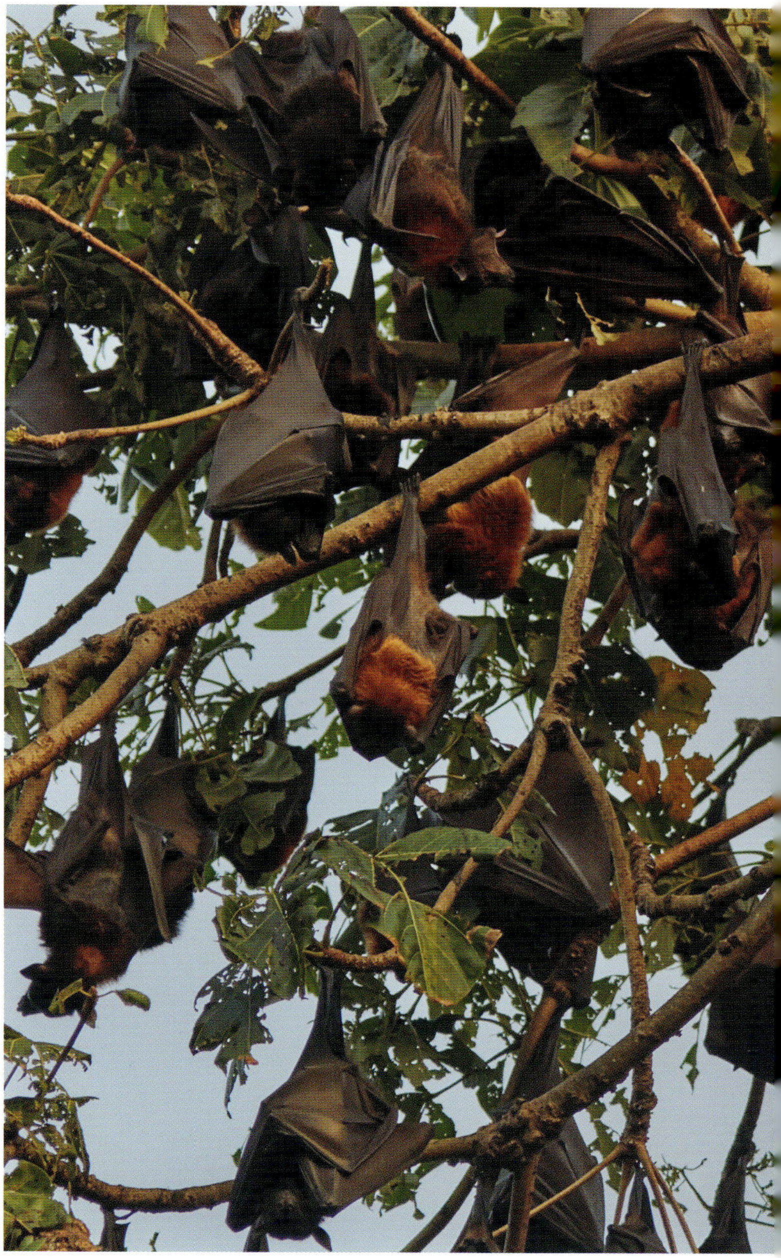

Above: *A colony of Giant Flying-foxes, one of the world's largest bats, roosting in a tree in lowland rainforest at Mambucal, on the edge of Mt Kanlaon Natural Park, Negros.*

Birds

Birds are almost certainly the best-studied component of Philippine wildlife, as a result of which the species lists, distributions, habitats and conservation status are more complete and more regularly updated than is the case for just about any other sector of the country's natural environment. They range vastly in size from the mighty Philippine Eagle, one of the world's largest birds, down to the tiniest but wonderfully colourful sunbirds and flowerpeckers, all perfectly adapted to different habitats, roles and foods. Most are, of course, adapted to rainforest environments, though a considerable number also make a home in the wetlands and along the coasts. A large proportion of these are migratory, simply passing through the Philippines for a few months en route between summer and winter grounds in northern Asia and Australia.

Just over 600 bird species have been identified in the Philippines, representing a modest increase over numbers known at the start of this century. This rise is not so much due to new discoveries but to reclassification of known, closely related groups of birds, slightly different varieties formerly classed as subspecies being split into distinct species. A good example of this is the Philippines' hornbills, whose subspecies have recently been split into separate species, changing five species into 11 (see page 114).

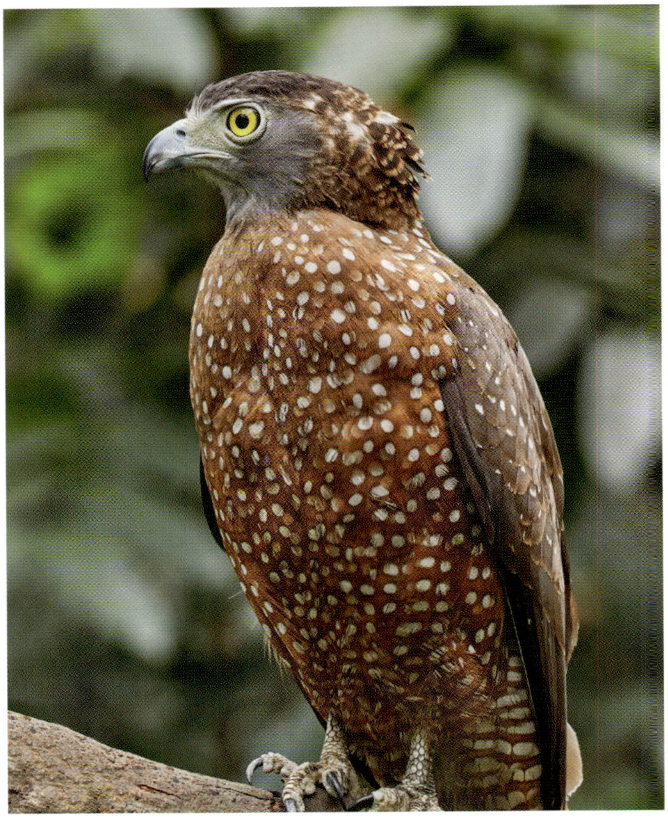

About a third of the Philippines' bird species are endemic, a rather lower proportion than is typical among other categories of the country's plant and animal wildlife, something that is not surprising given their great mobility, particularly in the migratory species. Most of the endemic species are forest birds,

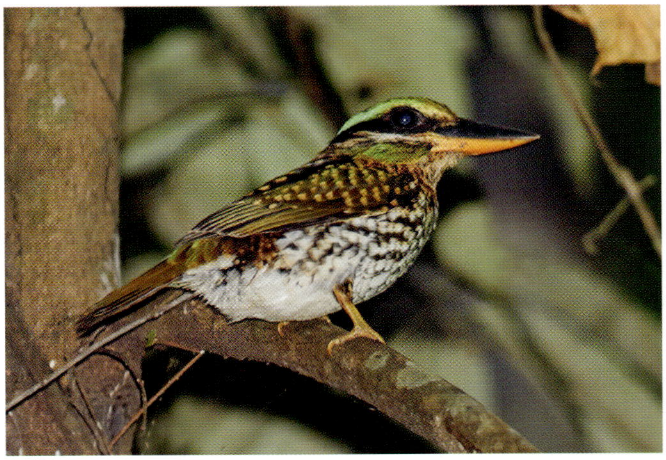

***Below:** The Spotted Wood Kingfisher (*Actenoides lindsayi*) is a rare bird, endemic to forests in Luzon, Panay and Negros.*

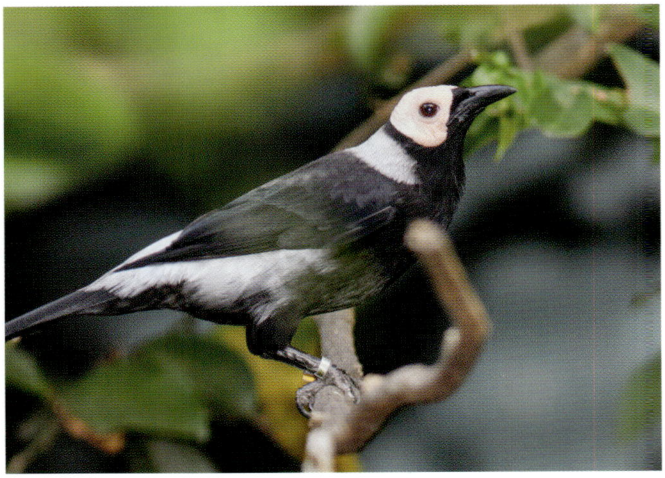

***Below:** The Coleto (*Sarcops calvus*) is endemic to the Philippines and quite common in much of the country, though not found in Palawan.*

Opposite: *The Philippine Serpent-eagle (*Spilornis holospilus*) is endemic to the country but closely related to the Crested Serpent-eagle found in South and Southeast Asia.*

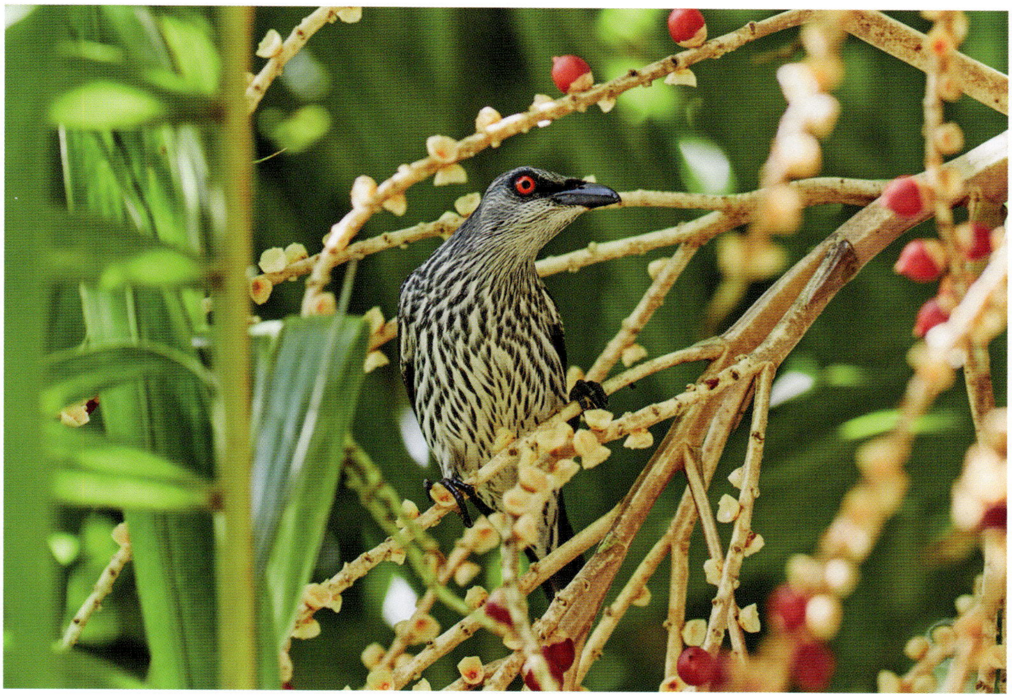

Right: *Juvenile Asian Glossy Starling (*Aplonis panayensis*), a species widespread across Southeast Asia and also found throughout the Philippines, often in noisy gregarious flocks.*

adapted to life specifically within Philippine lowland or montane rainforest, so often subject to quite strict habitat restrictions. While some of these endemics – such as the Philippine Eagle (see page 112) and Philippine Bulbul (*Hypsipetes philippinus*) are spread around the country, others are restricted to just certain areas, often closely linked to the country's Pleistocene major faunal regions. Examples include the bleeding-heart pigeons (*Gallicolumba* genus), of which there are five species, all endemic to the Philippines. They all have their own specific island range, each one restricted to the Greater Luzon, Mindoro, the Western Visayas, Greater Mindanao and Greater Sulu faunal regions (see page 118).

Not surprisingly, with so many birds being forest specialists and having restricted ranges, deforestation across the Philippines threatens many with extinction. Indeed, one of the world's most endangered birds is the tiny Cebu Flowerpecker (*Dicaeum quadricolor*), once believed to be extinct but rediscovered in some of the few remaining pockets of forest left in its Cebu habitat, and today hanging on with just a handful of individuals remaining.

To help promote bird conservation, Birdlife International and the Haribon Foundation have identified many Important Bird Areas (IBAs) across the country, sites that are especially vital to a wide range of birds, particularly threatened and/or endemic species. This information is being used to map out where new national protected areas are needed to make up for gaps in the existing network, something that – provided protection is actually real and not just on paper – should help to conserve as many bird species as possible.

Below: *Female Olive-backed Sunbird, a species that is often seen in woodland and gardens, feeding on flower nectar.*

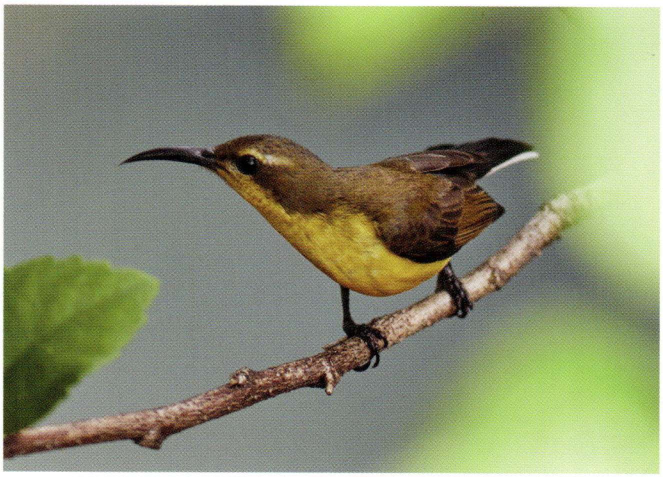

Philippine Eagle

The Philippine Eagle (*Pithecophaga jefferyi*) is the mightiest bird in the Philippine rainforest, standing up to 1 m (3.3 ft) tall. Equipped with a huge bill and powerful talons it is clearly the ultimate predator, the pinnacle of the Philippines' natural food chain. Inhabiting mainly primary lowland rainforest across the country, it is known to feed on a wide range of animals, from macaques, bats, cloud rats and flying lemurs, to a variety of snakes and birds. As the top predator, it would always be quite a thinly spread animal, but this has been greatly exacerbated by the rapid loss of its lowland forest habitat, coupled with intense hunting. As a result, it is now teetering on the brink of extinction, the IUCN estimating there to be only 250–750 Philippine Eagles left in the wild.

Most of the surviving eagles are concentrated on Mindanao, with literally just a handful of birds so far known to live in the forests of Samar, Leyte and the Sierra Madre and northern Cordillera mountains of Luzon, both within and outside existing protected areas.

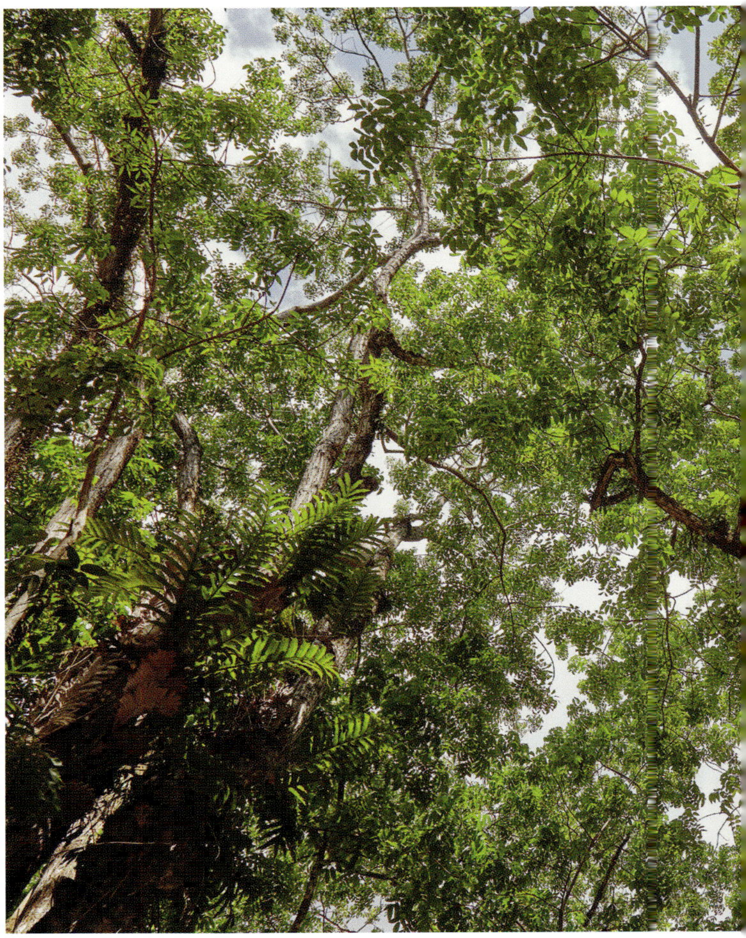

Below: *Portrait of a Philippine Eagle, showing its magnificent crest and huge, powerful bill.*

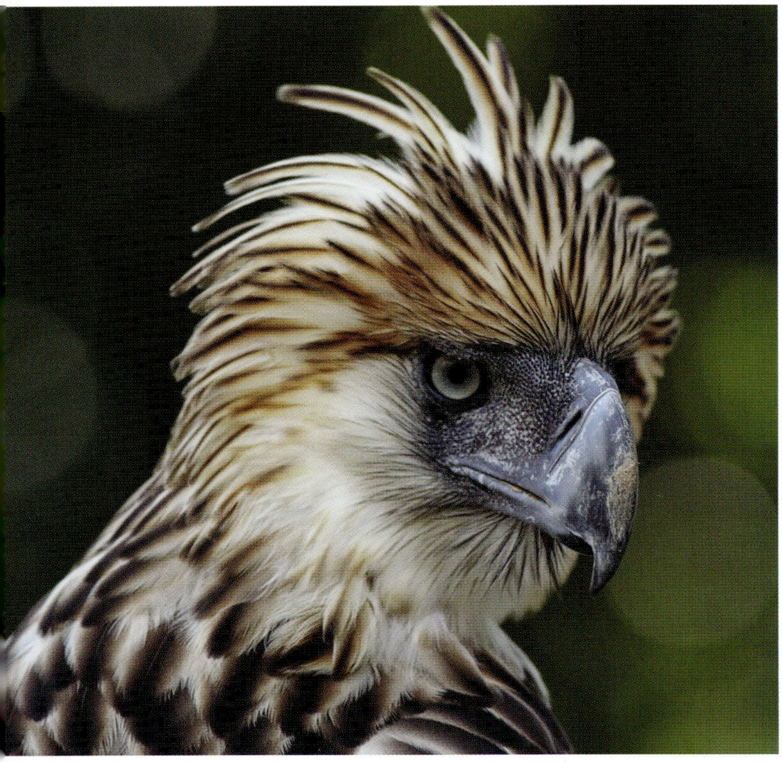

Saving the Philippine Eagle

There are some strenuous efforts going into saving the Philippine Eagle from extinction. It is wholly protected, with both hunting and the cutting of its habitat being illegal. Furthermore, an NGO, the Philippine Eagle Foundation (PEFI) has been active for many years. It has had great success in turning the Philippine Eagle into a *cause célèbre*, making it an icon of Philippine conservation. It also conducts conservation and education work among rural communities close to known eagle home ranges and nest sites, and has been successful in setting up a network of nest guards to protect known nests from poachers.

Scientists at the PEFI have also established a captive-breeding programme at their headquarters, the

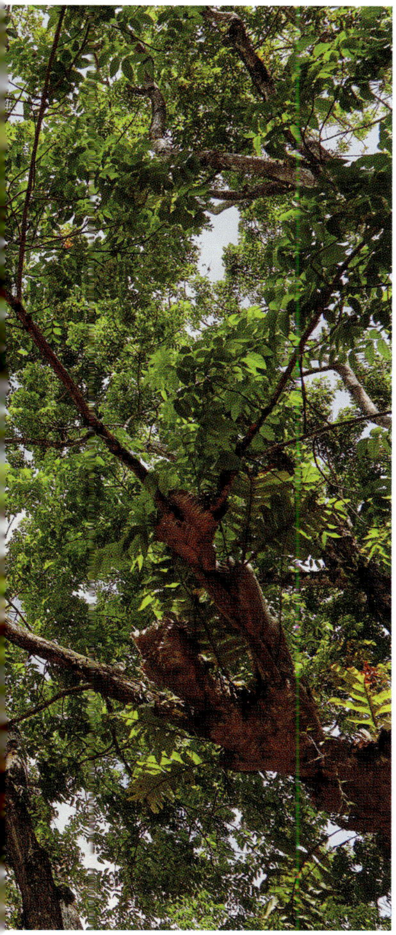

Left: Classic Philippine Eagle habitat: healthy lowland rainforest, here seen near Davao in southern Mindanao. Rapid loss of this habitat is one of the main reasons for the decline in this great eagle.

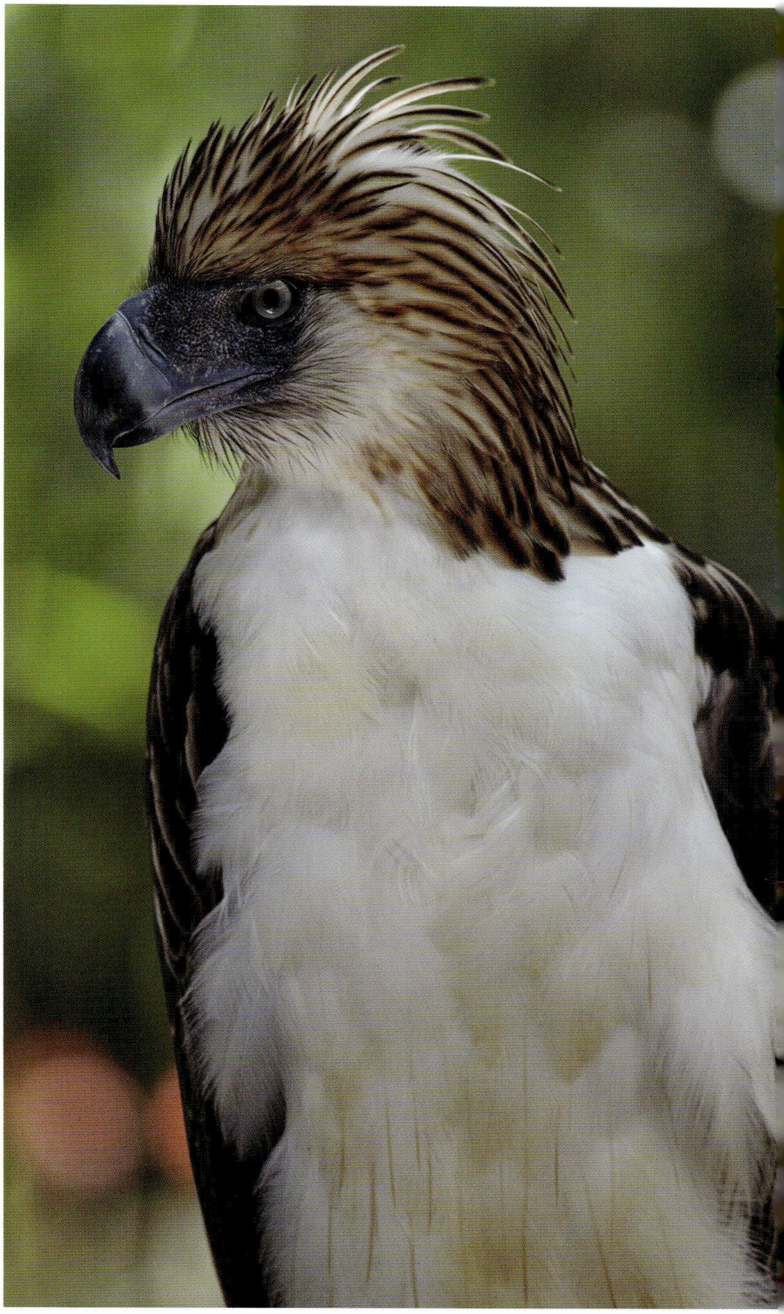

Right: A magnificent Philippine Eagle, photographed at the Philippine Eagle Foundation's headquarters near Davao, in southern Mindanao. This is the centre for a captive-breeding programme aimed at rescuing the eagle from extinction.

Philippine Eagle Center, near Davao, in the south of Mindanao. Despite initial problems, over the past 25 years it has been possible to breed nearly 30 eagles in captivity.

There have been initial trials at releasing captive-bred eagles, the first of these in 2004, with a release into Mt Apo Natural Park. This went well initially, until the eagle was fatally electrocuted on electricity cables. Since then, a number of rehabilitated wild eagles – those brought to the Center after being found injured – have been rereleased quite successfully. Unfortunately, all have been subsequently killed, shot by hunters, which very neatly illustrates the massive hurdle successful eagle conservation still has to climb.

While it is quite clear that the PEFI is doing fantastic work in studying the eagle and promoting its conservation, its status as a conservation icon is somewhat double edged. Not only is it a wonderful symbol of the beauty of Philippine wildlife and the importance of conserving it for the future, but it is also emblematic of the state of Philippine conservation: staffed by fantastic, committed people and supported by fine words from politicians and business people, but hugely under-resourced and faced with a serious disconnect between the apparent support and what is actually happening on the ground. The eagle is still being hunted and its forest habitat is still being cut down despite the protection it supposedly has. Its future survival still very much hangs in the balance.

Hornbills

Hornbills are among the most charismatic and perhaps most iconic of the Philippines' animal wildlife. They are also among the rarest and most endangered animals on Earth, making them the focus of much concern and work among conservationists.

The classification and naming of Philippine hornbills has proved to be the subject of much debate due to the close similarities among a number of clearly different forms, but there is currently general agreement that there are 11 species in the Philippines, divided into four genera, all of them endemic to the country. Altogether, hornbills are spread around much of the country, but with restricted distributions that closely correlate with the major faunal regions of the last ice age.

Like all hornbills across the tropical world, Philippine hornbills are quite large birds, mostly in the 45–70 cm (18–28 in) height range, and are characterized by their large, often colourful bills, topped by a hollow extension known as a casque. They frequently spend their days in sociable feeding groups, making quite a

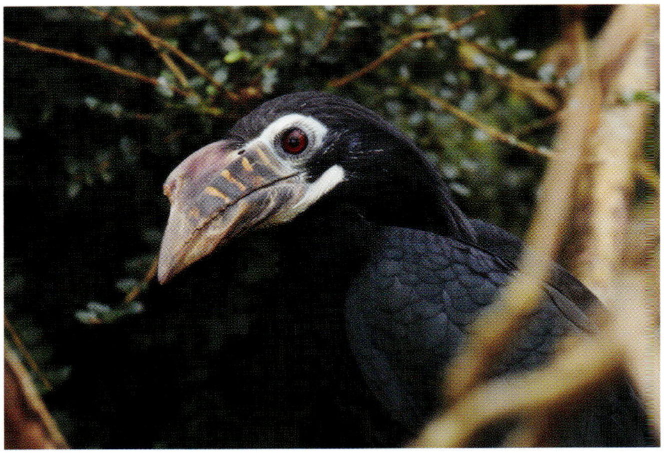

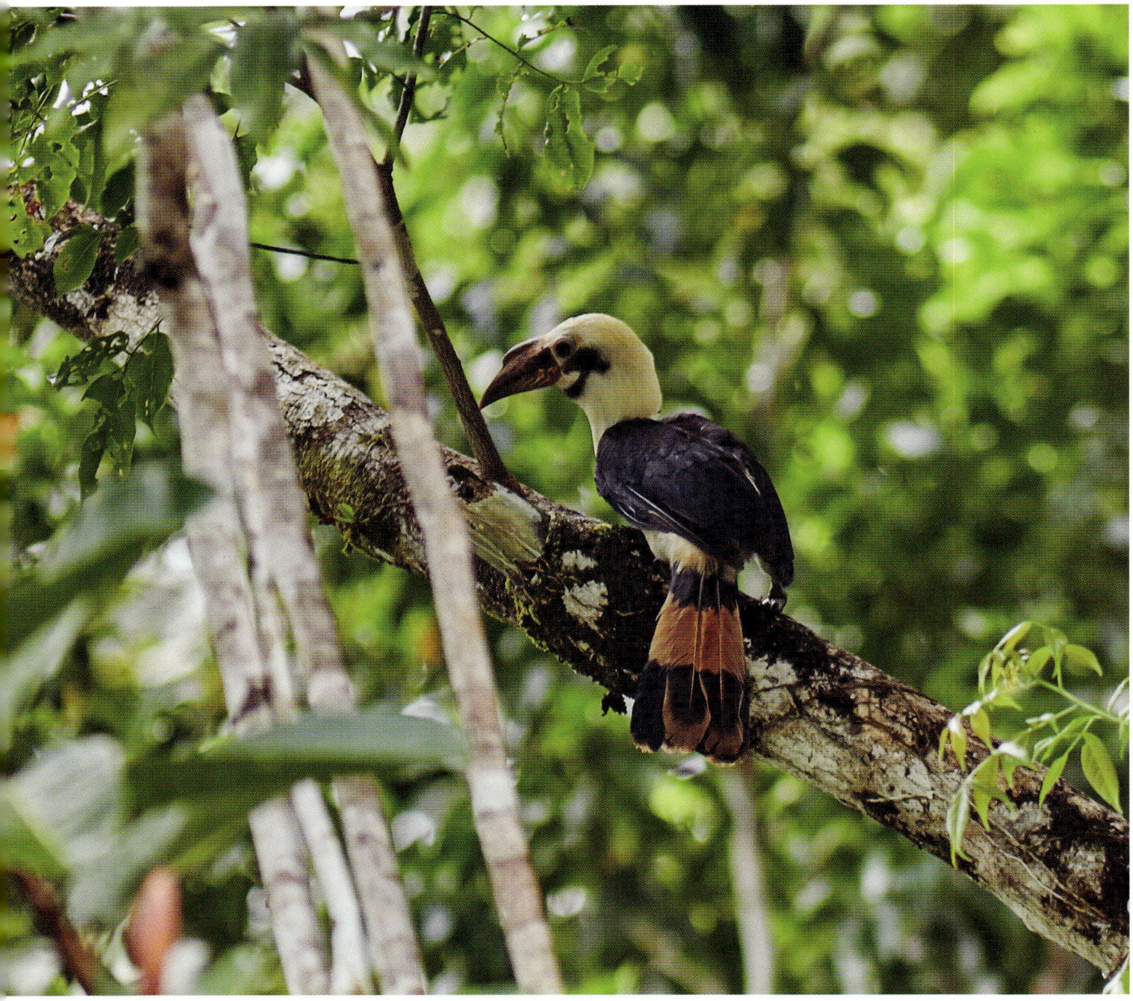

Above: *Female Visayan Hornbill, easily distinguished from the male by its wholly black head and breast, which are white in the male.*

Left: *Male Samar Hornbill, seen in lowland rainforest in Samar Island Natural Park, virtually its only wild home.*

Opposite, top and bottom: *The magnificent rufous hornbills are the largest and most colourful of the Philippines' hornbills, characterized by their scarlet bill and casque. The two species, Northern and Southern Rufous Hornbills, are distinguished by the former having a wholly red bill (opposite), while the latter's bill has a paler yellow or orange tip (opposite top).*

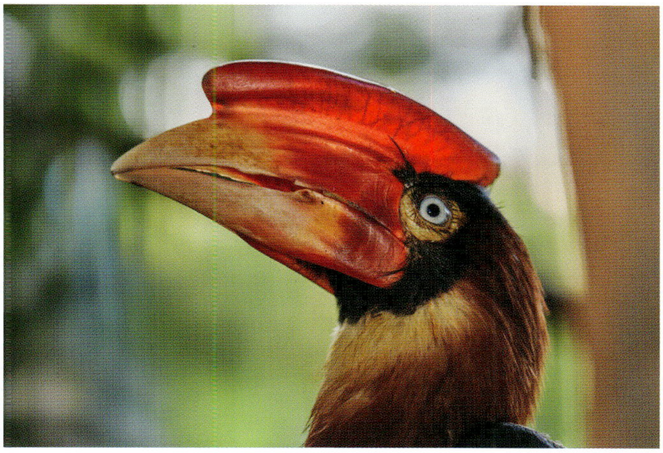

lot of noise, usually a mix of honks and cackling sounds, which can make them easy to find, though not necessarily easy to see in the dense forest foliage.

Philippine hornbill species

The largest single group of Philippine hornbills is the tarictics, relatively small hornbills and so named for their chattering sound. Originally grouped as a single species with several subspecies, they have recently been split into five separate species.

The best known of the tarictics is the Visayan Hornbill (*Penelopides panini*), which is endemic to the islands of the Western Visayas. Over the past 30 years it has been wiped out from most of its range, including the islands of Guimaras and Masbate, and survives only in the few remaining forests of Panay and Negros, with a total population now well below 2,000.

The other tarictics are the Luzon, Mindoro, Mindanao and Samar Hornbills (*P. manillae*, *P. mindorensis*, *P. affinis* and *P. samarensis*, respectively). Each of their names accurately describes their endemic distributions. In four of the five tarictics the males and females are easily distinguished, the latter being mostly black all over, the males having a white head and breast. In the Mindoro Hornbill the sexes look quite similar, both male and female having a white head and underparts.

There are two *Buceros* species hornbills, the Northern and Southern Rufous Hornbills. Like the tarictics, these were until recently classed as a single species, but they have now been split into two, the Northern Rufous Hornbill (*Buceros hydrocorax*) covering most of the northern part of the country, and the Southern Rufous Hornbill (*B. mindanensis*) living across Mindanao, Bohol, Samar and Leyte. Both rufous hornbills are quite large, at 60–65 cm (24–26 in) tall, and typically have a rufous nape and breast, the rest of the body being dark brown or almost black, set against a bright red bill and casque. The two species are easily distinguished by the bill, which in the Northern Rufous Hornbill is entirely red, but which is partially pale yellow in the Southern Rufous Hornbill.

There are two *Anthracoceros* species hornbills, the Palawan Hornbill (*Anthracoceros marchei*), which is endemic to Palawan and the Calamian Islands, and the Sulu Hornbill (*A. montani*), which is endemic to just three islands in the Sulu archipelago. Both have black bodies with white tails, but differ in that the Palawan

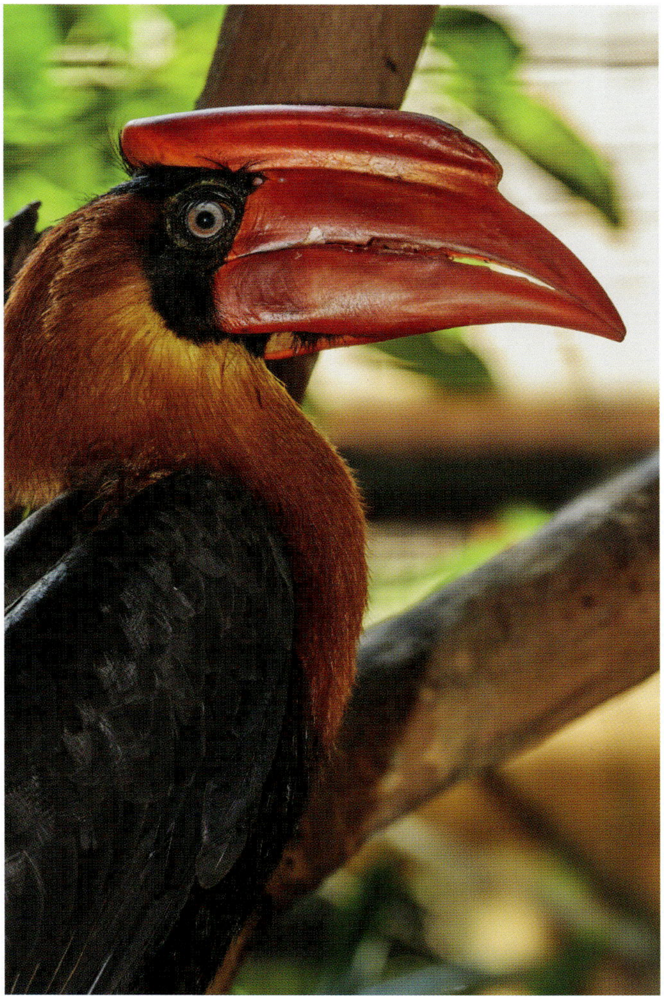

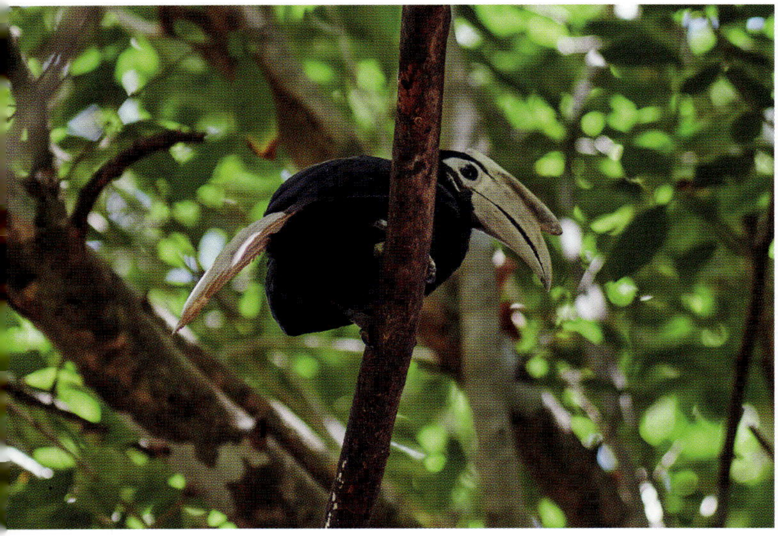

Above: *The Palawan Hornbill, a species endemic to Palawan, in forest on Entalula Island, in Bacuit Bay, El Nido, Palawan.*

Hornbill has a white bill and casque, while those of the Sulu Hornbill are largely black. The Sulu Hornbill is Critically Endangered and may soon become extinct (see opposite).

Finally, there are two *Rhabdotorrhinus* species hornbills, the Mindanao Wrinkled, or Writhed, Hornbill (*Rhabdotorrhinus leucocephalus*) and the Visayan Wrinkled Hornbill (*R. waldeni*). Until recently, the latter was classed as a subspecies of the former, but they are now defined as separate species. They also used to be included in the *Anthracoceros* genus (with scientific names *Anthracoceros leucocephalus* and *A. waldeni*, respectively). To add yet more confusion, the Visayan Wrinkled Hornbill also has several other English names, including Walden's, Rufous-headed (not to be confused with the rufous hornbills) and Writhed-billed Hornbill (not to be confused with the Writhed Hornbill). The Mindanao Wrinkled Hornbill is restricted to Mindanao and the islands of the Greater Mindanao Faunal Region, while the Visayan Wrinkled Hornbill is endemic to the Western Visayas. Not a great deal is known about either hornbill, but both are endangered. The Visayan Wrinkled Hornbill was thought to be almost extinct on Negros, but recent surveys have found a previously unknown population in Northern Negros Natural Park.

Philippine hornbills under pressure

All Philippine hornbills need reasonable quality lowland rainforest, preferably primary, though some of them at least seem to manage quite well in secondary forest. The fact that over the past 50 years lowland rainforest has been decimated across most of the Philippines immediately sets alarm bells ringing about the hornbills' ability to survive. A few species are able to range up to altitudes within the montane forest areas, but these may not provide sufficient substitute food sources. Their diet consists mainly of fruits and seeds, though they do also eat small animals, including insects and frogs.

Over the past 40 years hornbill numbers in the Philippines have been declining rapidly, so that today even the least threatened species probably number no more than 3,000–4,000 individuals, spread over large areas of fragmented forest. The most endangered of all, the Sulu Hornbill, is down to an estimated 40 individuals, and even this may be overly optimistic. Restricted now to forest on just Tawi-Tawi Island, extinction for this bird seems imminent.

In addition, one tarictic subspecies, the Ticao Hornbill (*Penelopides panini ticaensis*), is believed to be extinct already. Endemic just to the Visayan island of Ticao, the last recorded sighting was in 1971, and with all forest long since cleared from the island it seems very unlikely that this bird still survives. If so, then the Philippines has the dubious global distinction of allowing the first ever hornbill extinction.

A captive-breeding programme is in place on Negros and Panay for the Visayan Tarictic Hornbill, and confiscated Visayan Wrinkled Hornbills have successfully been rereleased into the wild. There is also a successful guard network to protect nests of these two hornbills from poachers. Although the Luzon and Palawan Hornbills do not seem to be too badly off at present, that still leaves most of the others struggling on their own, particularly the Sulu Hornbill. Much more needs to be done to safeguard these iconic birds.

Right: *Male Mindanao Wrinkled Hornbill, a Critically Endangered species restricted to the forests of Mindanao.*

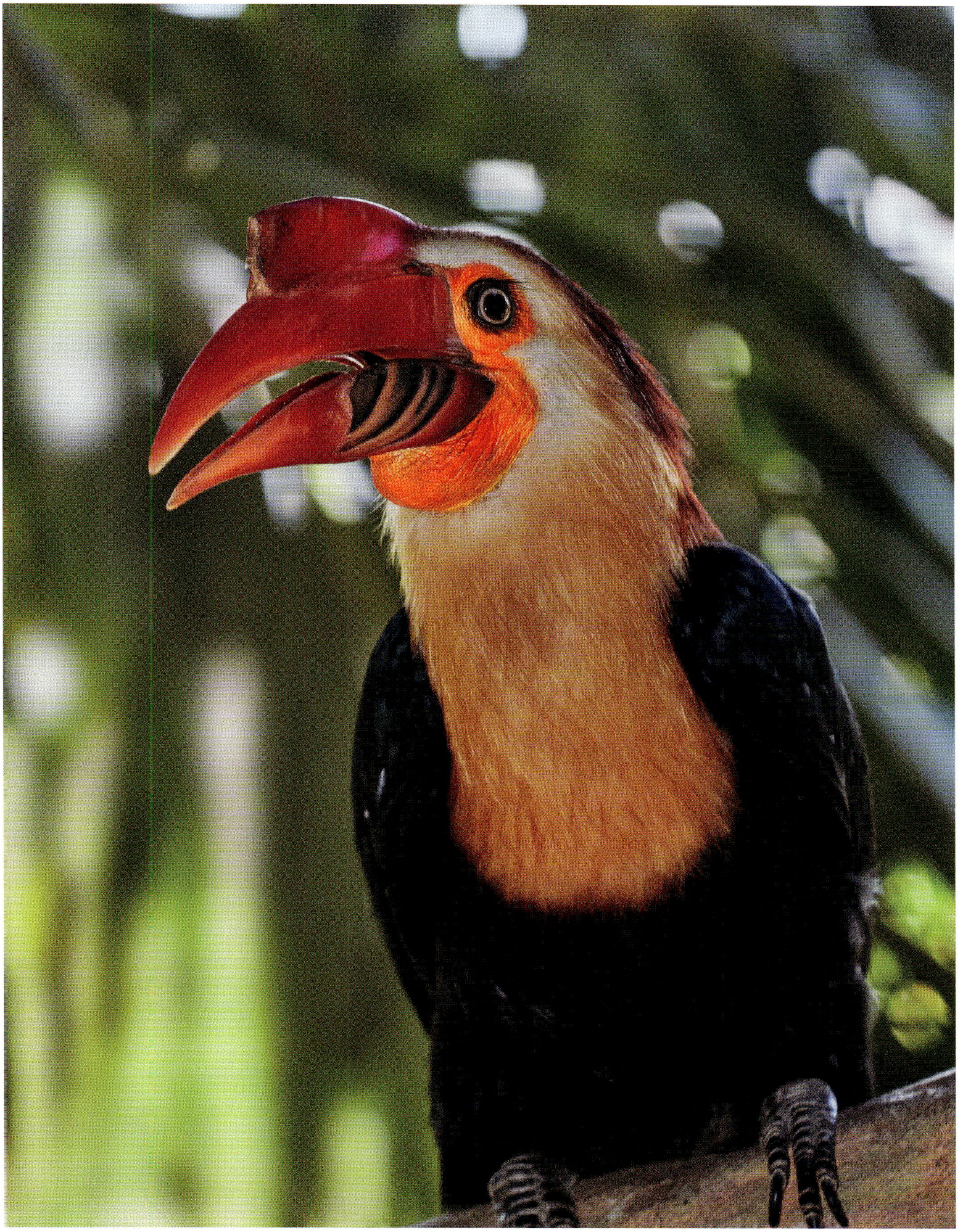

Pigeons and doves

With more than 30 species in the Philippines, the pigeons and doves constitute one of the largest groups of birds in the country. They are also one of the most important, like the fruit bats living mainly on forest fruits, and thus serving as major seed dispersers. As such, they help to maintain healthy rainforest and restore damaged areas, with the seeds they drop germinating into new trees. They are found throughout just about every terrestrial habitat, from mangroves, to upper montane forest, to gardens and farmland, with some species overlapping in both their choice of habitat and favoured fruiting trees. Furthermore, they are multi-coloured and highly beautiful birds, most (with a few exceptions) rather shy of people, and living well out into forested or agricultural countryside.

Many species found in the Philippines are widely spread around South and Southeast Asia, and even

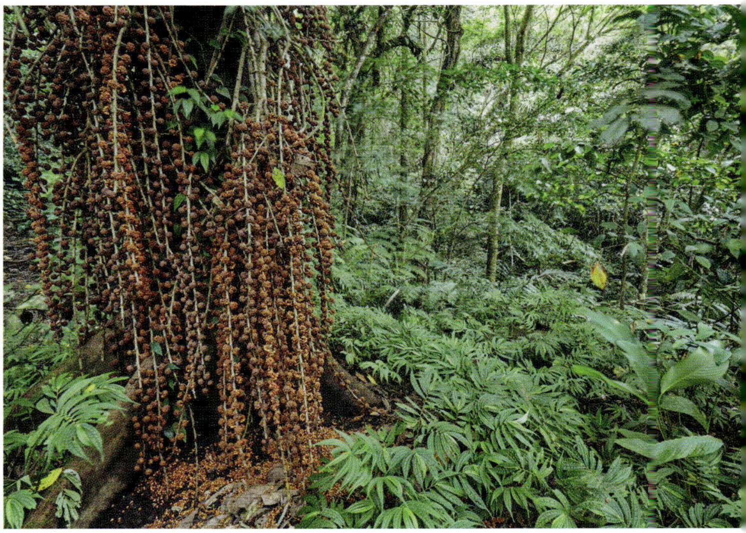

Above: The fruits of forest palm trees are a major food source for the Philippines' largely forest-dwelling pigeons and doves.

Below: The beautiful Nicobar Pigeon is restricted mostly to small, forested off-islands.

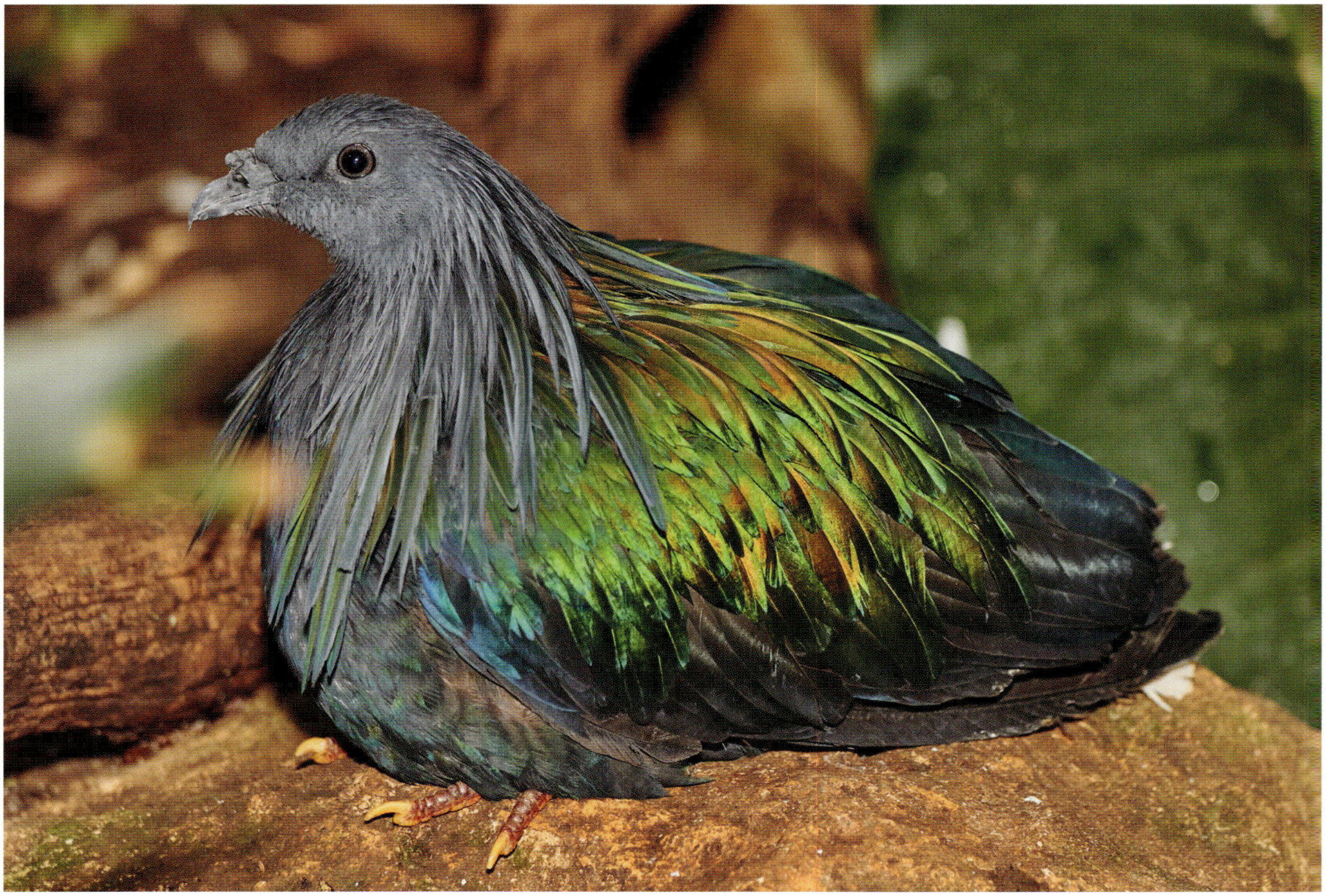

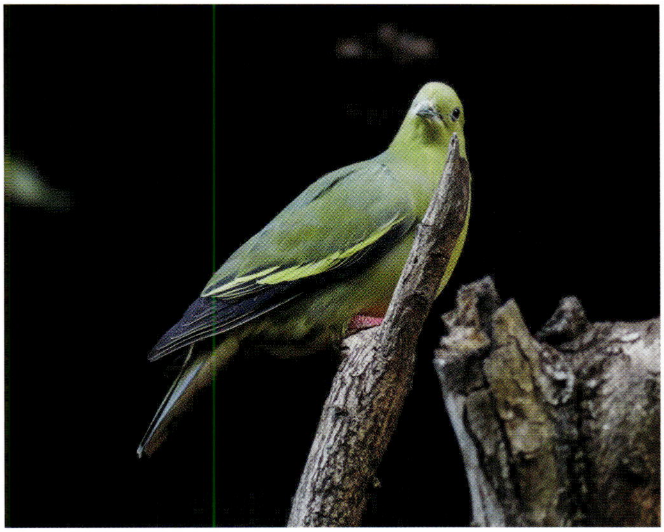

Above: *The Philippine Green-pigeon (*Treron axillaris*) has only recently been established as a Philippine endemic species.*

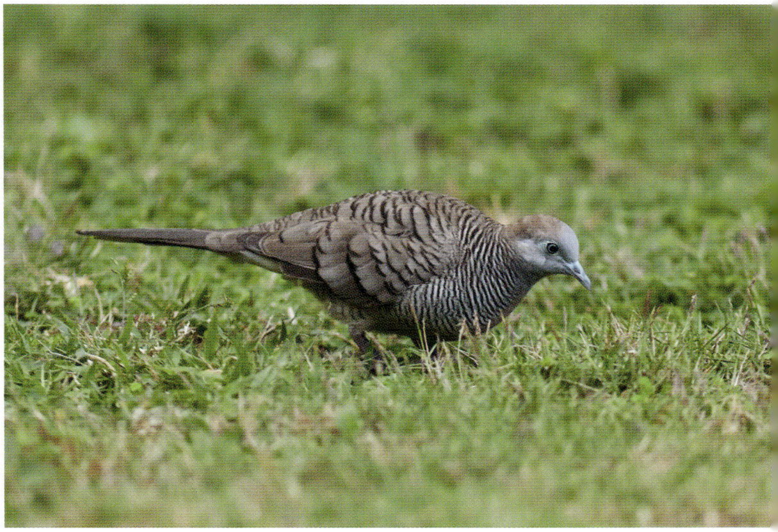

Above: *The Zebra Dove is a small pigeon of parks, gardens and forests.*

Australia, and are really quite common birds, at least in their chosen habitat. Examples include the Pink-necked Green-pigeon (*Treron vernans*), which in the Philippines is common in mangroves and coastal scrubby forest, and the Pied Imperial Pigeon (*Ducula bicolor*), which is something of a small island specialist. Another pigeon that likes the remote islands is the Nicobar Pigeon (*Caloenas nicobarica*). Though widely spread from the Andaman and Nicobar Islands to the Solomon Islands, it is quite rare in the Philippines, being restricted mainly to remote, predator-free forested islands. Less widespread is the Black-naped Fruit-dove (*Ptilinopus melanospilus*), which in the Philippines is found in forested areas of Mindanao and Palawan, and elsewhere only in the islands of eastern Indonesia.

Other widespread pigeons or doves that can regularly be seen, particularly in more remote forested areas, such as the islands of Bacuit Bay (El Nido), the mangroves of Olango Island (Cebu) and the lowland forest of Samar Island, include the Green Imperial Pigeon (*Ducula aenea*) and the Emerald Dove (*Chalcophas indica*). Suburban and urban gardens and parks are regularly the homes of Zebra Doves (*Geopelia striata*) and Spotted Doves (*Streptopelia chinensis*), as a result of which these are probably the most commonly seen of all the pigeons and doves.

Endemic species

Sixteen species are endemic to the Philippines. One example is the White-eared Brown-dove (*Phapitreron leucotis*), a remarkably common bird across the country that is not especially shy, and is frequently seen in wooded rural areas close to villages.

Much harder to see is the stunning Flame-breasted Fruit-dove (*Ptilinopus marchei*), a rather shy bird of upper lowland and montane forests, found only on Luzon. As its English name suggests, it is made quite

Below: *The Black-naped Fruit-dove, a bird found in the forests of Mindanao and Palawan.*

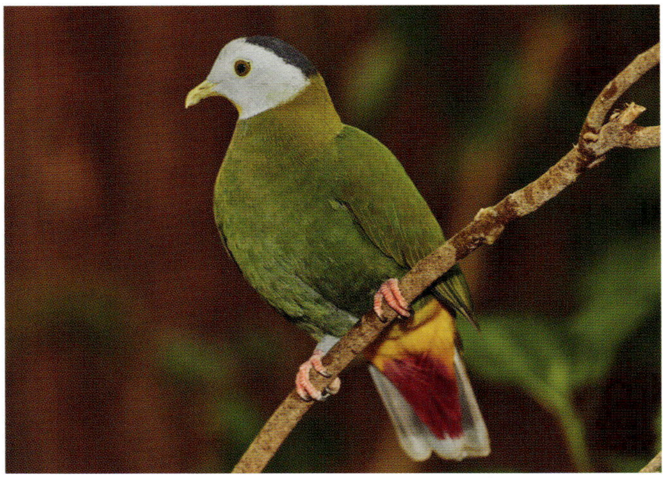

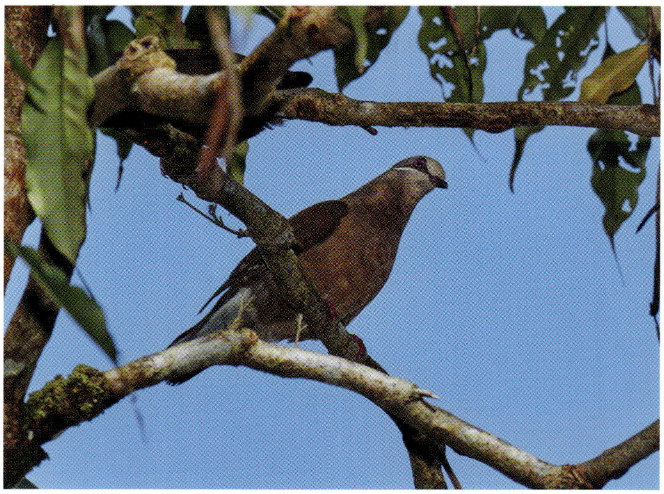

Above: *The White-eared Brown Dove is a Philippine endemic species, widespread in forests and suburban woodland.*

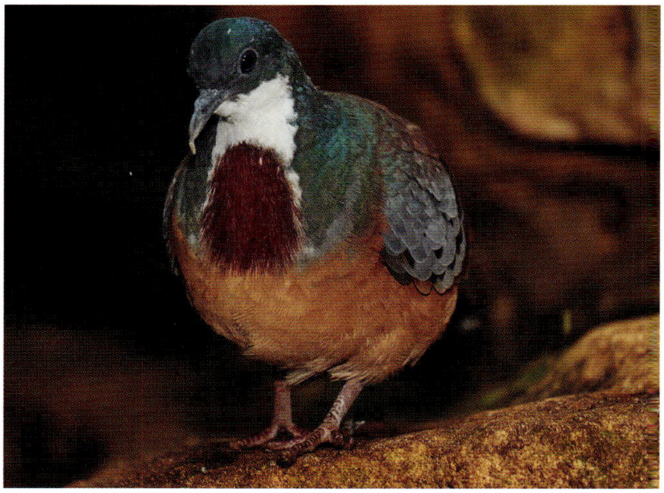

Above: *The Mindanao Bleeding-heart Pigeon can still be seen occasionally on Bohol.*

Below: *The Luzon Bleeding-heart Pigeon is probably the least rare of the bleeding-heart pigeons.*

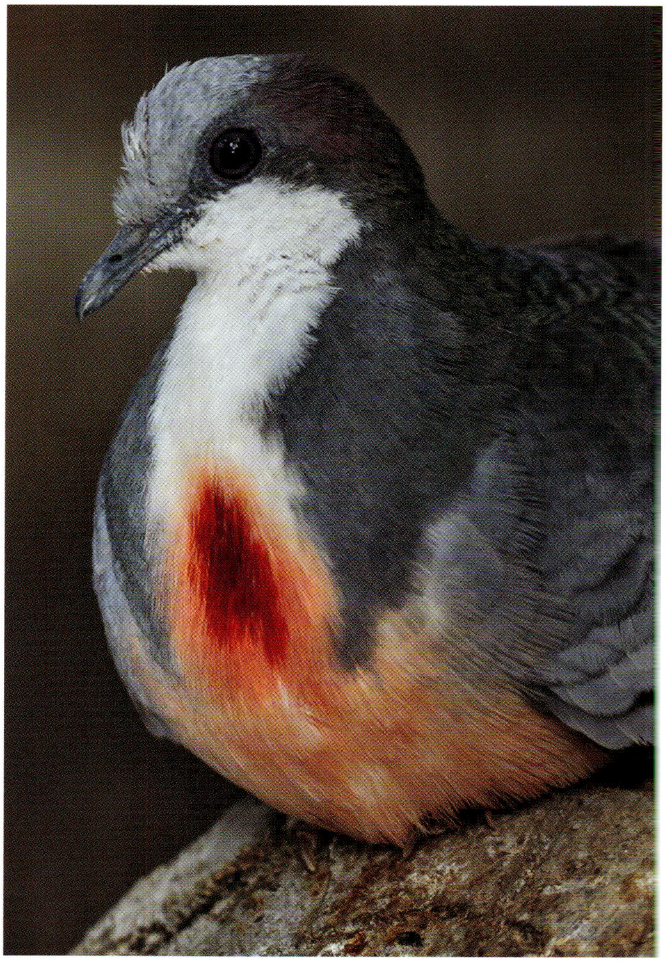

distinctive by its flamboyant orange breast.

Another uniquely coloured pigeon that is endemic to Luzon is the Luzon Bleeding-heart Pigeon (*Gallicolumba luzonica*), one of five bleeding-heart pigeon species, all endemic to the Philippines, and each restricted to one of the Pleistocene major faunal regions (excluding Palawan). All five are unmistakable due to the splash of red – very much like a stab wound – in the chest, which of course gives them their English name. They are all forest-dwelling birds that stay largely on the ground, and all are threatened with extinction.

Classified as Near Threatened, the Luzon Bleeding-heart Pigeon can still sometimes be seen in the wild, typically in Mt Makiling Forest Reserve and the Northern Sierra Madre Natural Park. This species has been subdivided into three subspecies, restricted to northern Luzon, southern Luzon and Catanduanes Island, respectively.

The Mindanao Bleeding-heart Pigeon (*Gallicolumba crinigera*) is restricted to the islands of Greater Mindanao, so should be found on Bohol, Samar, Leyte, Siargao and Basilan Islands, in addition to Mindanao itself. Although classified merely as Vulnerable, it is believed that its only remaining 'stronghold' is Rajah Sikatuna Protected Landscape on Bohol. Like the Luzon Bleeding-heart Pigeon, it has been split into

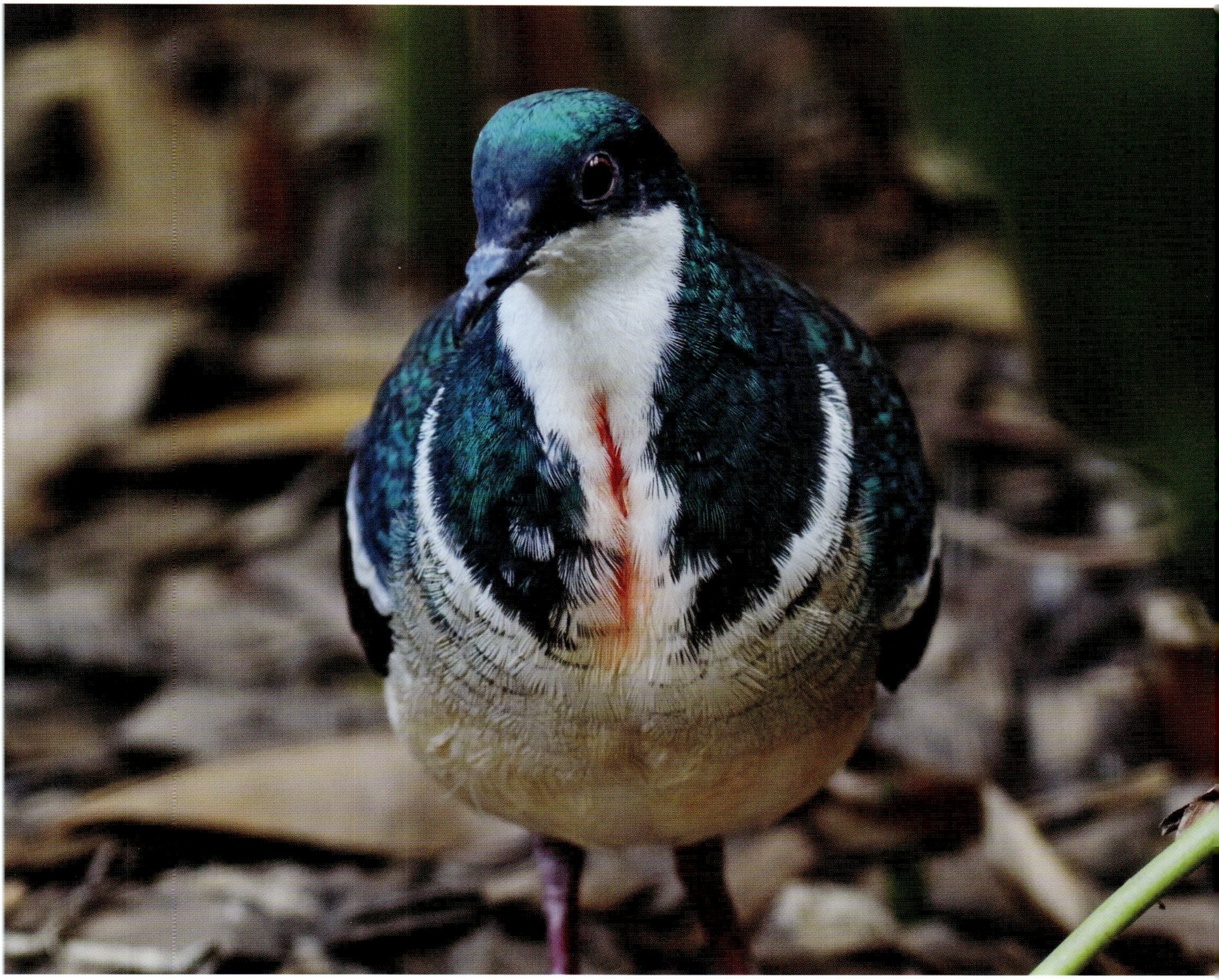

Above: *The Negros Bleeding-heart Pigeon is now Critically Endangered and is the subject of a captive-breeding programme.*

three subspecies, restricted to Mindanao and Dinagat, the Eastern Visayas (that is, Bohol, Samar and Leyte) and Basilan Island.

The remaining three bleeding-heart pigeons – the Mindoro, Negros and Sulu Bleeding-heart Pigeons (*Gallicolumba platenae*, *G. keayi* and *G. menagei*, respectively) – are all classified as Critically Endangered. The first two are both very rare in the wild, though in the case of the Negros species previously unknown populations have been discovered on Panay, and possibly also on Cebu, giving hope for their survival. The Sulu Bleeding-heart Pigeon, however, has not been seen for many years in its, now largely deforested, Tawi-Tawi habitat, and may well be extinct.

To try to head off extinction of the others, in the late 1990s the Bleeding Heart Conservation Programme was launched, combining *in situ* conservation with captive-breeding programmes. Aimed mainly at the Luzon, Mindanao and Negros Bleeding-hearts, successful captive breeding is taking place at the Talarak Foundation, Inc. heaquarters in Bacolod, on Negros, as well in as several overseas zoos.

Amphibians and reptiles

While much is known about the Philippines' birds, studies into its reptiles and amphibians have lagged well behind, though today they are starting to catch up. At present, more than 90 species of amphibian and nearly 250 reptiles are recorded, a very high percentage of them endemic to the Philippines: about 85 per cent for the amphibians and about two-thirds for the reptiles. Not surprisingly, most species are forest dwellers, many of them – particularly the smallest forest frogs – restricted to relatively small ranges within particular islands or island groups.

Frogs

Most of the amphibians can be split into tree frogs and forest frogs, the principal difference between them being that the former are largely arboreal, the latter generally ground living. One of the best known of the tree frogs is the Harlequin Tree Frog (*Rhacophorus pardalis*), which in the Philippines is common in lowland and montane rainforests on Mindanao, Luzon, Bohol and Negros. Many of the forest frogs are *Platymantis* species, a genus that consists of about 60 species, the great majority of them endemic to the lowland and montane rainforests of the Philippines. Any night-time trip along a forest stream is likely to reveal quite a diversity of these.

Lizards

Rather more is known about the reptiles, though even here much is still being learned. The lizards consist of such groups as agamids, flying lizards, skinks and water monitors. The last of these are by some margin the largest, often growing up to 2 m (6.6 ft) long. Originally, they were considered to be subspecies of the Asian Water Monitor Lizard (*Varanus salvator*), but recently they were split into nine separate species, all endemic to the Philippines and each with a distinct range.

Examples include the Palawan Water Monitor Lizard (*V. palawanensis*), which is found mainly in Palawan, Gray's Water Monitor Lizard (*V. olivaceous*), restricted

Below: *Gray's Water Monitor Lizard, known locally as Butaan, is endemic purely to southern Luzon and the adjacent Polillo Islands.*

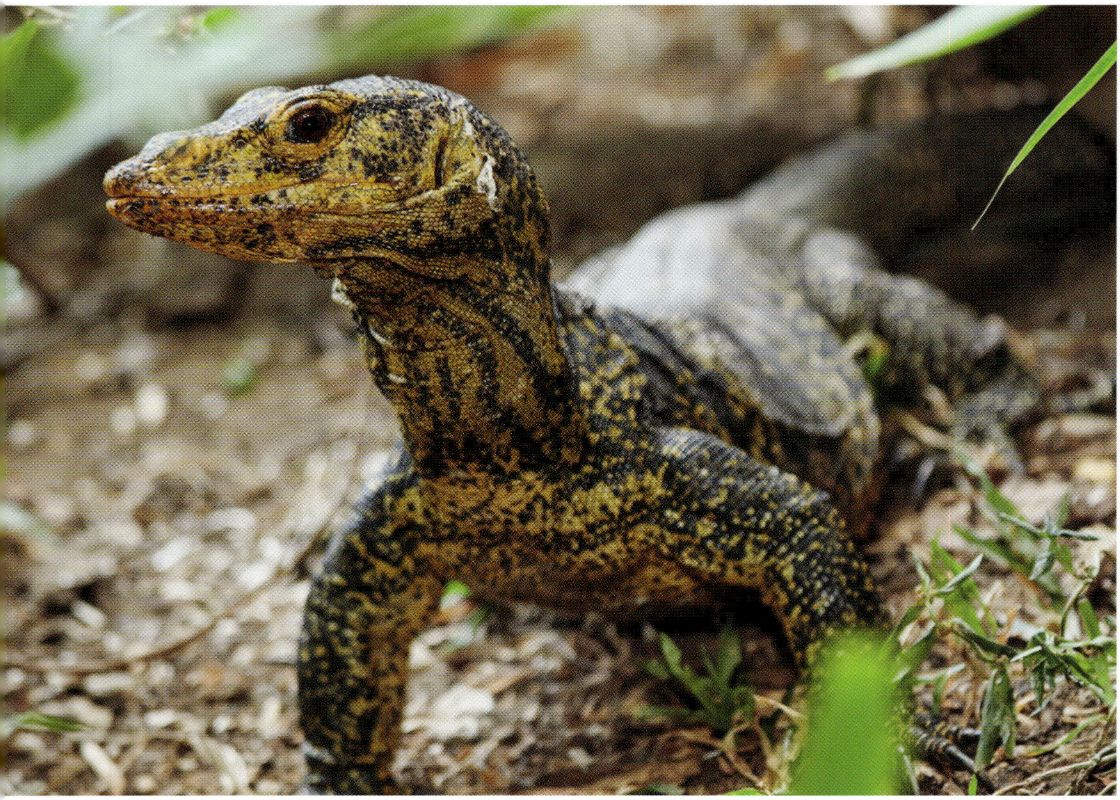

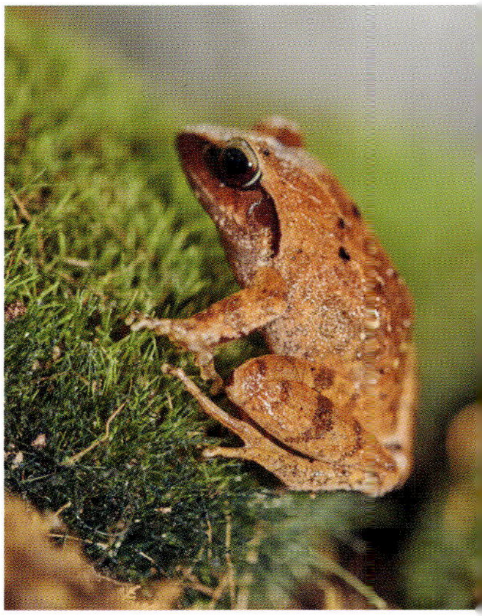

Above: *The Corrugated Forest Frog is endemic to the Philippines and is found right across the country in forests from sea level up to about 1,300 m (4,300 ft).*

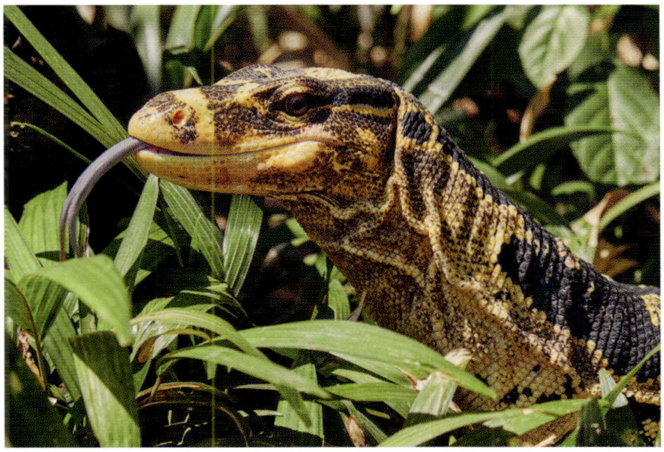

Above: The Mindanao Yellow-headed Water Monitor Lizard, found only on the islands of the Greater Mindanao Faunal Region.

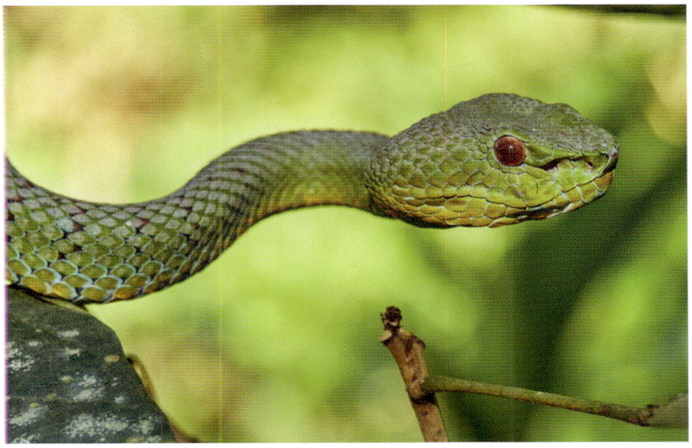

Above: The Philippine Pit Viper is found across much of the Philippines, except Palawan.

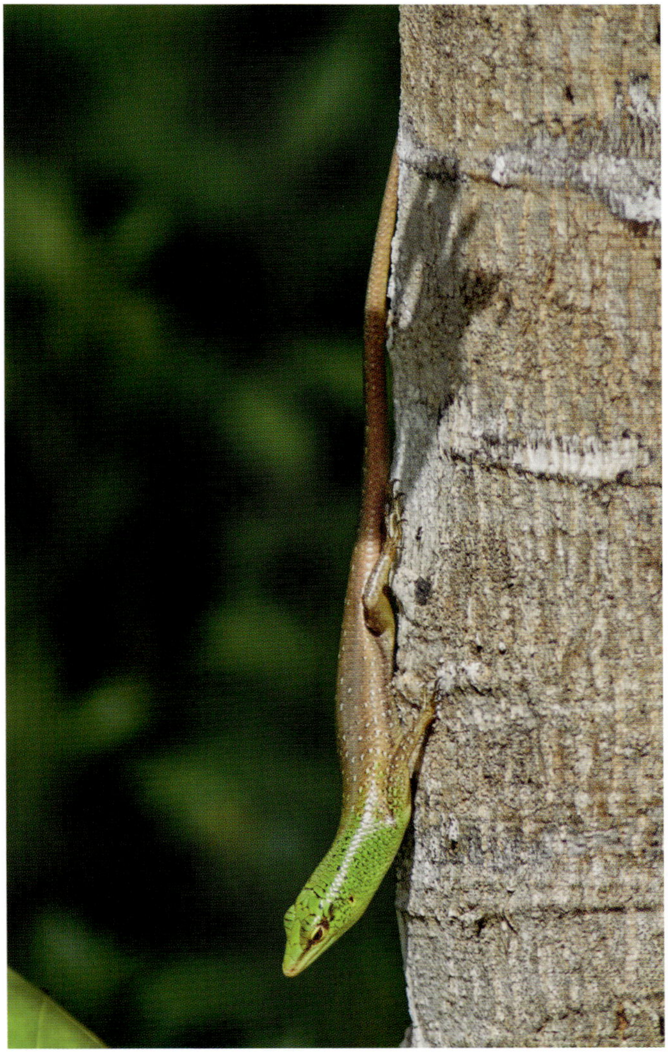

Above: The Emerald Green Tree Skink occurs from Taiwan to New Guinea and the Solomon Islands.

to the Bicol region of southern Luzon, and the Yellow-headed Water Monitor Lizard (*V. cumingi*) which occurs across Mindanao. Fortunately, due to their highly adaptable natures and varied omnivorous diets, they all seem to be quite common, even in agricultural lands.

There is a vast number of skinks in the Philippines' forests. One of the most attractive is the Emerald Green Tree Skink (*Lamprolepsis smaragdina*), which is half bright green and half brown. One of the strangest looking is the Spiny Waterside Skink (*Tropidophorus grayi*), a small and very dark grey or black lizard, covered in spiny lumps, which is usually found lurking under rocks beside streams and pools. Another Philippine endemic, it is restricted to southern Luzon and the Visayas.

Snakes

Snakes are widespread, though quite infrequently seen. The best known is probably the Reticulated Python (*Python reticulatus*), but two other common species are the often vibrantly green Philippine Pit Viper (*Trimeresurus flavomaculatus*) and Northern Philippine Temple Pit Viper (*Tropidolaemus subannulatus*), both venomous, though not dangerously so. The former is endemic to the Philippines, while the latter is also found on Borneo and some of the Indonesian islands.

Philippine Crocodile

Two species of crocodile are found in the Philippines: the Saltwater Crocodile (*Crocodylus porosus*) and Philippine Crocodile (*C. mindorensis*). The former, though now rather rare in the Philippines, is widespread from India to Australia, inhabiting both marine and freshwater habitats. The smaller Philippine Crocodile, by contrast, is endemic to the Philippines, and is now close to extinction, probably with less than 100 adults left in the wild. Not surprisingly, it is classified as Critically Endangered, and is said to be the world's most threatened crocodile, at serious risk of being the world's first crocodilian to suffer global extinction.

Below: The Saltwater Crocodile is far larger than the endemic Philippine Crocodile, and is widespread across Southeast Asia and Australia.

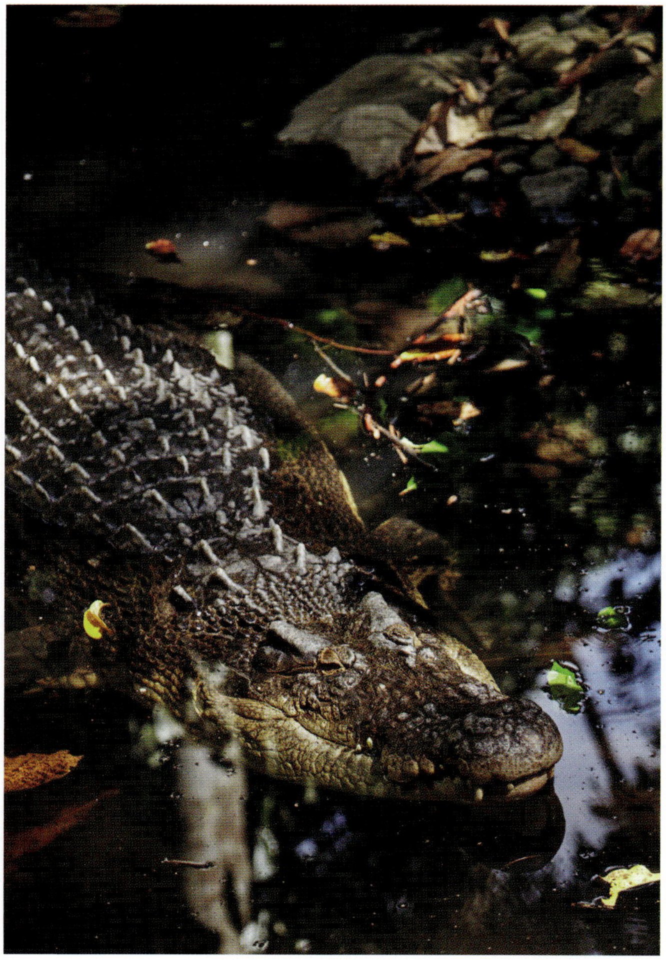

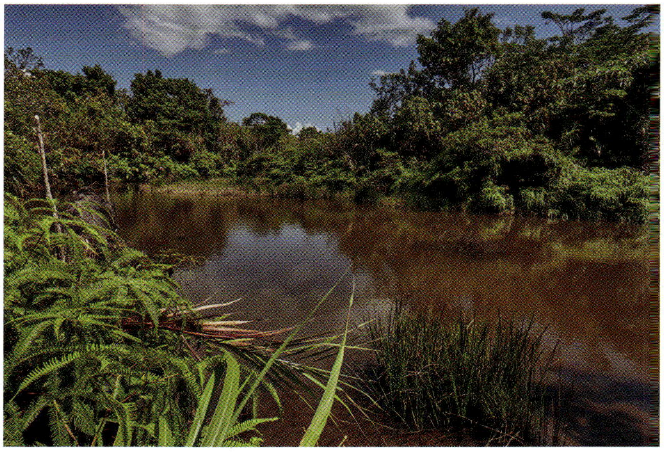

Some biology

Surpisingly little is known about the Philippine Crocodile. It is smaller than the Saltwater Crocodile, reaching a maximum of 3 m (10 ft) in length, compared with the Saltwater Crocodile's 7 m (23 ft). It is also restricted mainly to freshwater habitats, usually consisting of small lakes, ponds, creeks, streams and marshes. Although at one time it was probably spread across the entire country, habitat loss and direct persecution have decimated the population. Sporadic surveys indicate that the last remaining populations are in Mindanao's Liguasan Marsh, on Dalupiri Island off Luzon's northern tip, and in the mountains of Luzon's northern Cordillera and Northern Sierra Madre. The last of these was only discovered in 1999, and seems to be the main hope for the crocodile's survival in the wild.

Crocodile conservation

In the 1980s the DENR launched a captive-breeding programme, the Crocodile Farming Institute, just outside Puerto Princesa, in Palawan. Subsequently renamed the Palawan Wildlife Rescue and Conservation Center (PWRCC), it has been quite successful with its breeding. In 2009, 50 of the captive-bred crocodiles were released into Dicatian Lake near Divilacan, close to the Pacific coast of north-east Luzon. Though a valuable initial experiment, it was not hugely successful: many of the crocodiles did not know how to hunt, leading some to die and others simply to leave the lake.

Opposite, top: Lake Narra, on the western edge of Northern Sierra Madre Natural Park, is one of two lakes where captive-reared Philippine Crocodiles are released back into the wild.

Right: A sub-adult Philippine Crocodile, one of a growing number being released into the wild to try to boost the population of this Critically Endangered endemic species.

Below: A two-year-old Philippine Crocodile at the Mabuwaya Foundation's rearing facility, shortly before its release in the Northern Sierra Madre Natural Park.

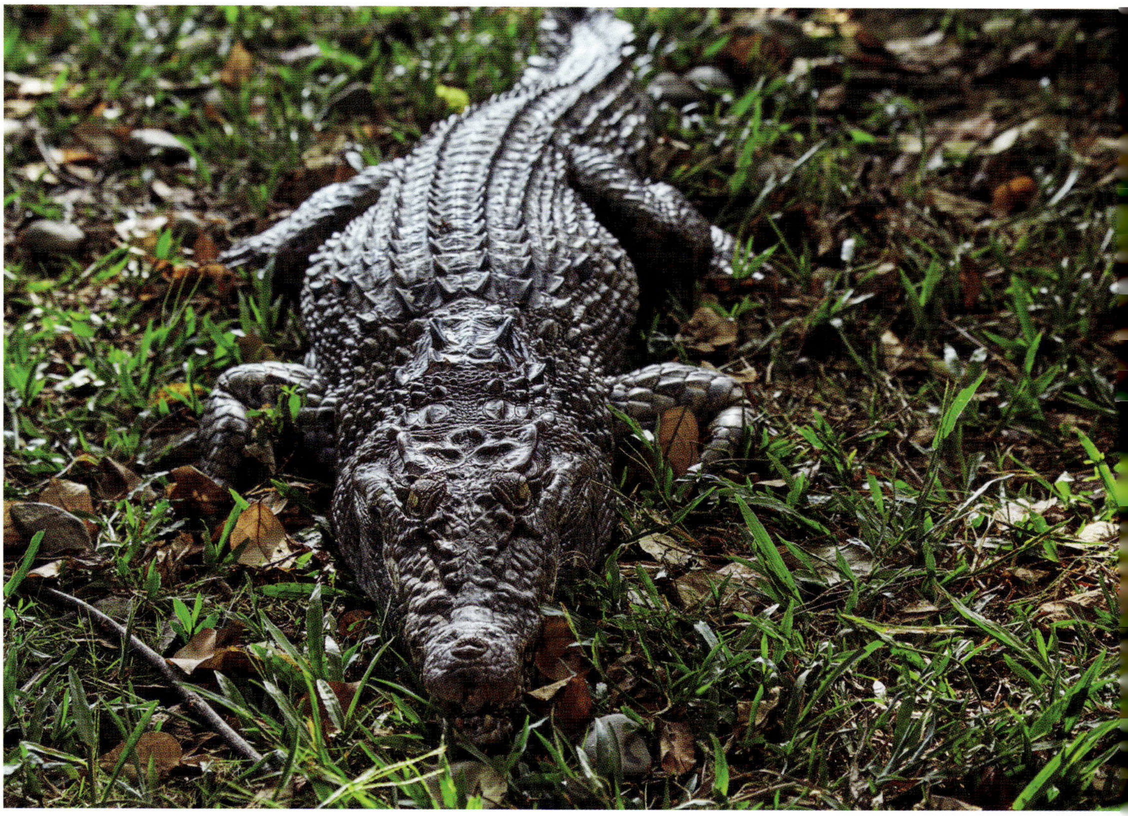

Following on from the discovery of the Northern Sierra Madre wild crocodiles, an NGO called the Mabuwaya Foundation was launched, with the specific intent of saving and expanding this population. This NGO has been very successful with its work, changing attitudes towards crocodiles among rural communities in the Northern Sierra Madre, creating pride in having responsibility for the future of such an iconic Philippine species. Its staff have also been successful with a head-start programme, capturing newly hatched wild crocodiles, then rearing them in safety until they are large enough to fend off predator attacks. At that point, when they are about two years old, the young crocodiles are rereleased into a couple of small lakes well known locally as crocodile reserves.

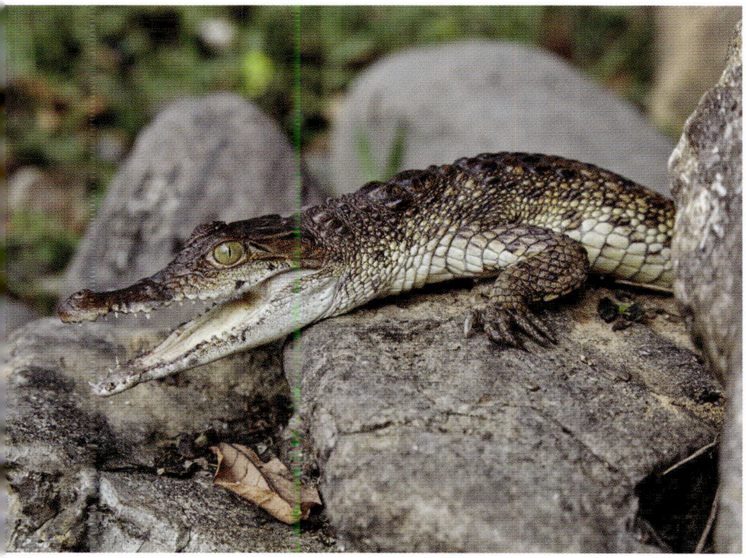

A steady stream of young crocodiles has been successfully released in this way, leading to a slow recovery in the wild Philippine Crocodile population around the western edges of the Northern Sierra Madre Natural Park. The next step for Mabuwaya is to start local captive breeding, moving beyond simply capturing and rearing wild-born hatchlings to having the entire mating, egg-laying and hatching process 'in house', at a newly built breeding facility near San Mariano. Although there is still a very long way to go, with its success to date the work here by the staff of the Mabuwaya Foundation does represent the best chance for the Philippine Crocodile's survival, giving some cause for hope.

Insects: butterflies

The great majority of knowledge about the Philippines' animal wildlife revolves around the vertebrates, or animals with a backbone (including all the terrestrial animals described in this chapter so far). When it comes to the invertebrates, or animals with no backbone, research has barely scratched the surface. As described earlier, a reasonable amount is known about many marine invertebrates, in particular the corals, molluscs and crustaceans, but work on their terrestrial Philippine cousins seems to be, relatively speaking at least, still in its infancy. With approximately 22,000 species, the insects alone outnumber all other terrestrial animals combined, and yet to date far less has been published about them.

*Below: A mating pair of Gray Glassy Tigers (*Ideopsis juventa kinitis*). This nymphalid butterfly is common across Southeast Asia.*

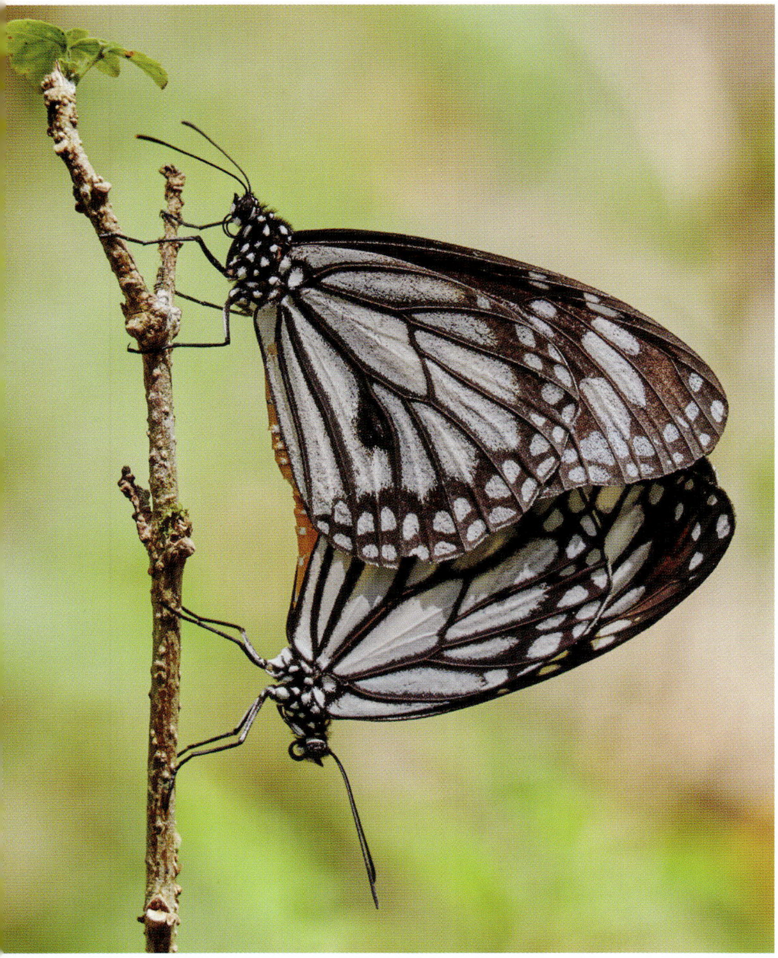

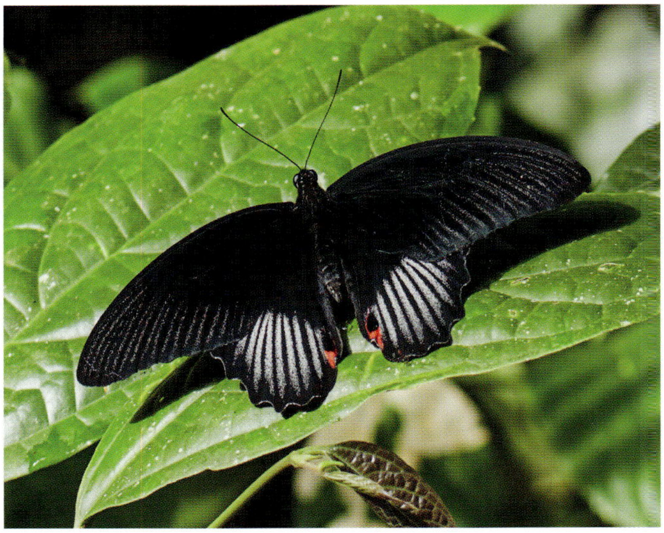

Philippine butterflies

One group of insects with which we are reasonably familiar is the butterflies. More than 900 species are currently known in the Philippines, over four times the number of Philippine mammals, and more are being discovered all the time. Roughly 350 species are endemic to the country, with some – like many other forms of animal wildlife in the Philippines – having quite restricted ranges.

A wide variety of butterflies are common across the Philippines, though mostly in the lowlands, and particularly in forested areas, frequently in forest glades and riversides. Gardens filled with flowering shrubs are, of course, great butterfly locations. Not all butterflies always feed on flower nectar, however: muddy river- or pool-side sites are often frequented by butterflies too, as places where they can pick up valuable salts.

Six of the seven butterfly families are represented in the Philippines, the largest and perhaps most spectacular belonging to the swallowtails (or Papilionidae family). This group includes the enormous birdwing butterflies, some of which can be larger than many small forest birds. One of the most common Philippine swallowtails is the Scarlet Mormon (*Papilio rumanzovia rumanzovia*), a large, mainly black butterfly which in the females also contains several splashes of vibrant red (replaced with white in the male).

All the groups contain a great diversity of colours

*Opposite, top: The female Scarlet Mormon (*Papilio rumanzovia rumanzovia*) is a large butterfly distinguishable from the male by the red wing flashes. Here seen in forest on Mt Makiling, southern Luzon.*

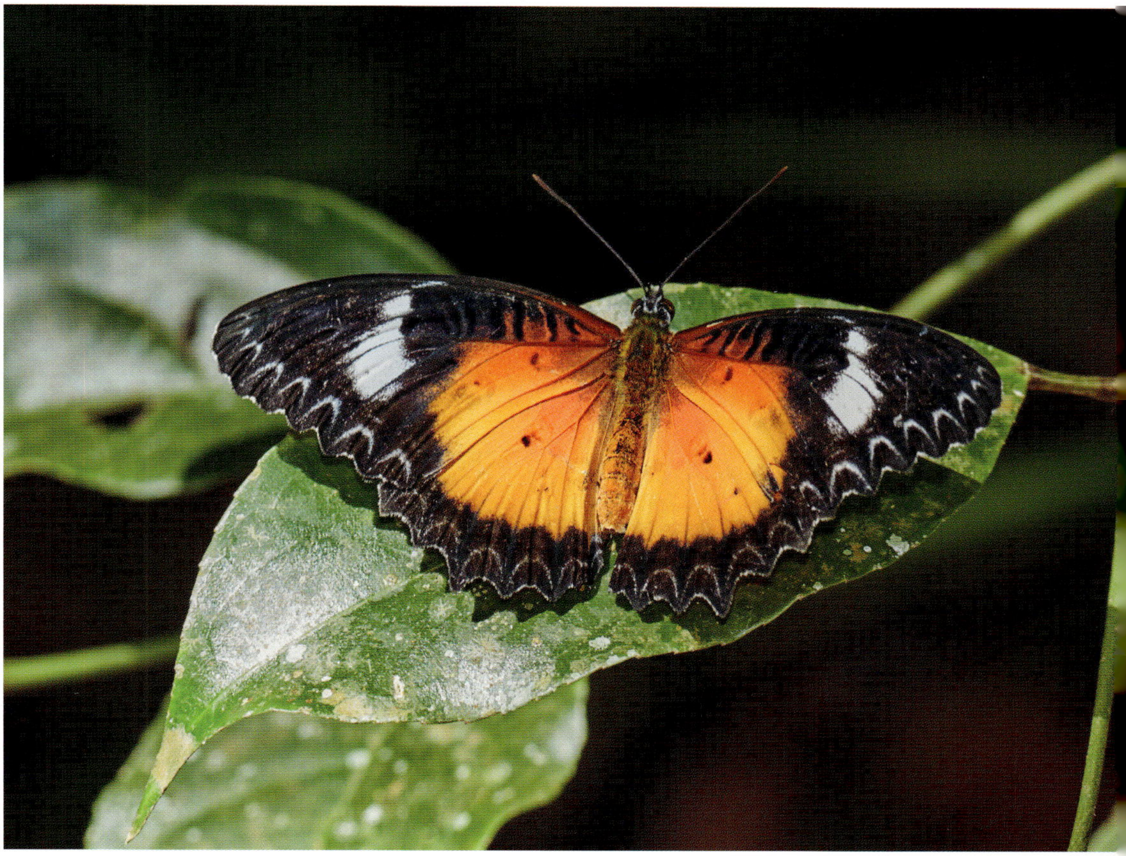

*Right: The Philippine Lacewing (*Cethosia luzonica pariana*) is one of many hundreds of butterfly species endemic to the country. This individual was found in lowland rainforest on the edge of Mt Kanlaon Natural Park, Negros.*

and patterns, ranging from the delicate black-and-white patterns of such common species as the Paper Kite (*Idea leuconoe*), to stunningly vibrant colours, including the beautiful orange of the Malay Cruiser (*Vindula dejone*). Both these species belong to the brush-footed family (or Nymphalidae), so named for the brushes found on their short front legs. Other brilliantly colourful butterflies include the jezebels (*Delias* species), whose wings have a mix of yellow, scarlet, black and white, making for very vibrant butterflies.

Below: The Malay Cruiser is a common forest butterfly across much of Southeast Asia. This individual was found in forest near Davao, Mindanao.

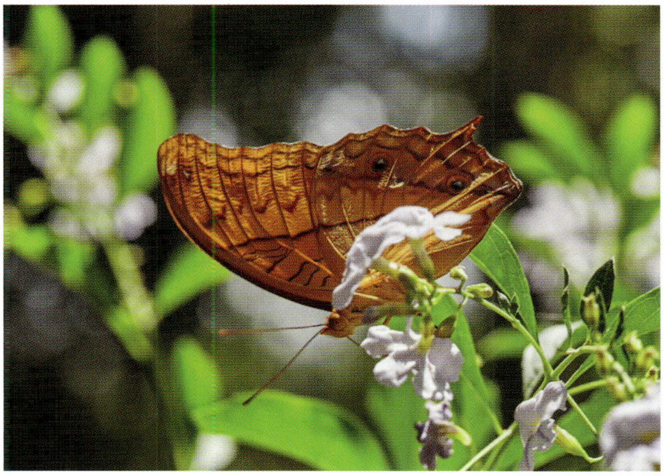

Butterfly conservation

At this stage it seems that little is known about the state of most populations of Philippine butterflies. It is known that the greatest threats for many are loss of habitat (especially lowland forest) and over-harvesting for the butterfly-collector trade. There has been some concern that the great majority of terrestrial protected areas in the Philippines have been established mainly with mammals and birds in mind, and that areas important for their protection are not necessarily the best for butterflies. However, the steady expansion in the number of protected areas, including CHs, and the recognition of a large number of unprotected KBAs, may be able to address this issue.

Orchids

As is the case with insects, not enough is known about Philippine orchids, despite the huge number of species found in the country. Of the roughly 8,500 species of higher plant in the Philippines, over 1,100 are orchids, and more than 900 of these are endemic to the country. This is an incredible record of biodiversity, as well as of species development taking place in isolation from the rest of Southeast Asia.

Not surprisingly, there is a huge range of appearances among the orchids, from tiny plants with equally tiny, rather green and inconspicuous flowers, up to tall plants with large, showy flowers in purple, violet, yellow or white. They can be divided roughly into two groups: epiphytic and ground plants. The latter grow up from the ground, often colonizing damaged areas with compacted and/or very poor soil, such as roadsides and logged sites. The epiphytic orchids grow attached to tree branches, not living there in any parasitic way but using the site as a base from which to grow, off the forest floor and in a wet, yet well-drained environment.

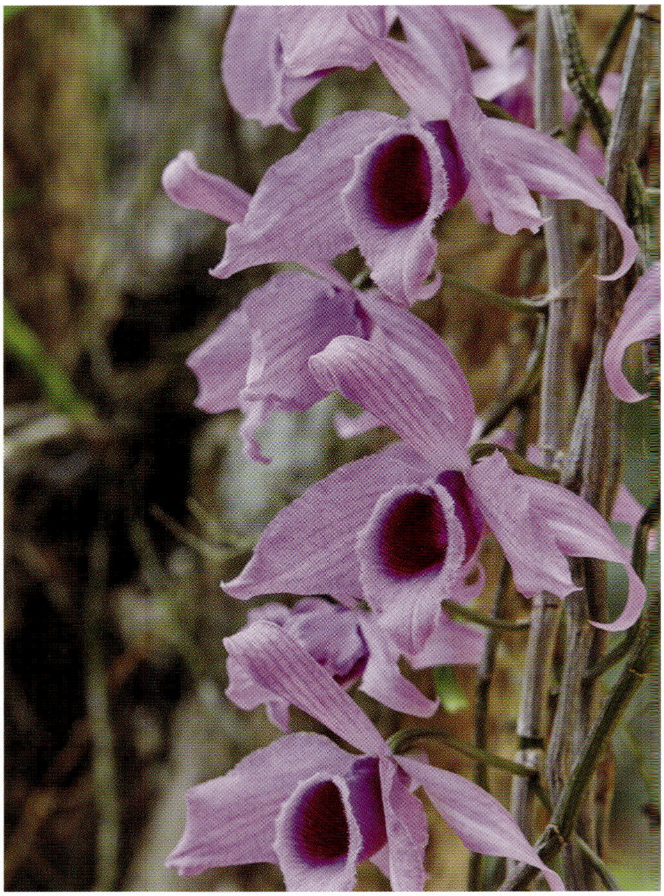

Above: *The Sanggumay Orchid (Dendrobium anosmum), one of 60 Dendrobium species orchids found in the Philippines This one is widespread across Southeast Asia.*

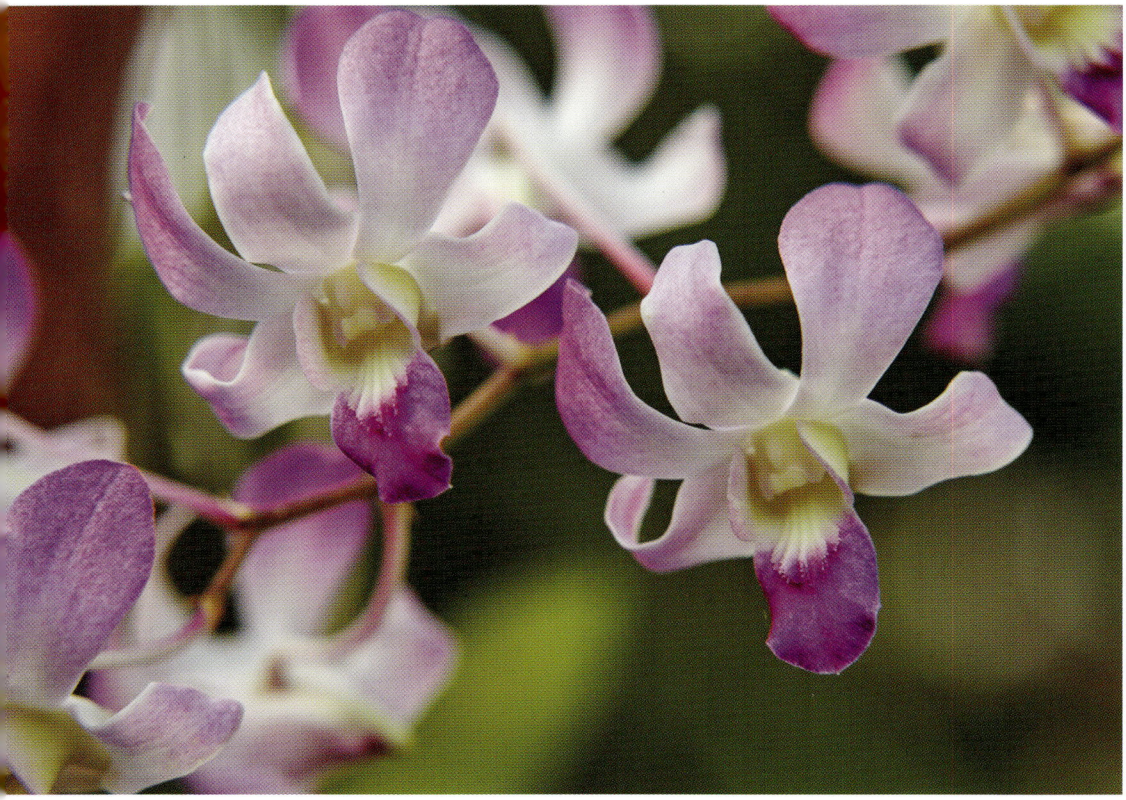

Left: *Orchids are hugely popular as garden plants right across the Philippines, with plants often being taken from the wild, leading an increasing number of species to become endangered.*

Epiphytic orchids

The epiphytic orchids make up by far the largest group of Philippine orchids, their plants festooning trees in most forests, particularly the montane forests, though they are also common in the lowlands. In the montane forests especially, they can be seen crowding tree branches, fighting for space with ferns and mosses. It seems surprising that plants can flourish so well in an environment with so little soil, but the crowded mix of such tree branch-dwelling plants generates just enough soil for their needs from rotting leaf litter and tree bark.

Unfortunately, despite the huge number of orchids in the montane forest habitats, it is unusual to see many in flower. Most flower irregularly or at different times of the year, though with sufficient species diversity in any given location it is usually possible to see at least a few in flower. One of the best-known sites in the Philippines for flowering orchids is Mt Pulag, Luzon's highest peak. The montane forest here harbours an enormous range of orchid species, some of them found nowhere else. It is common to see some orchids in flower from the main Babadak summit trail (see page 147).

Below: A moth orchid (Phalaenopsis *species) growing in its wild home, in forest in Samar Island Natural Park.*

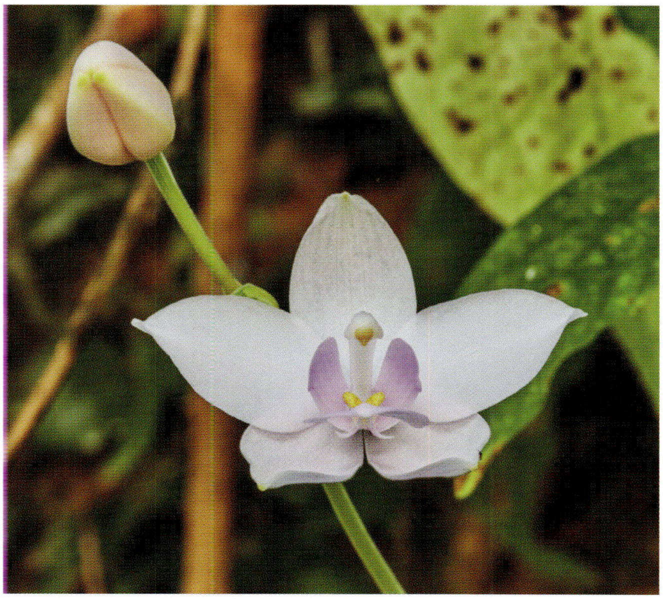

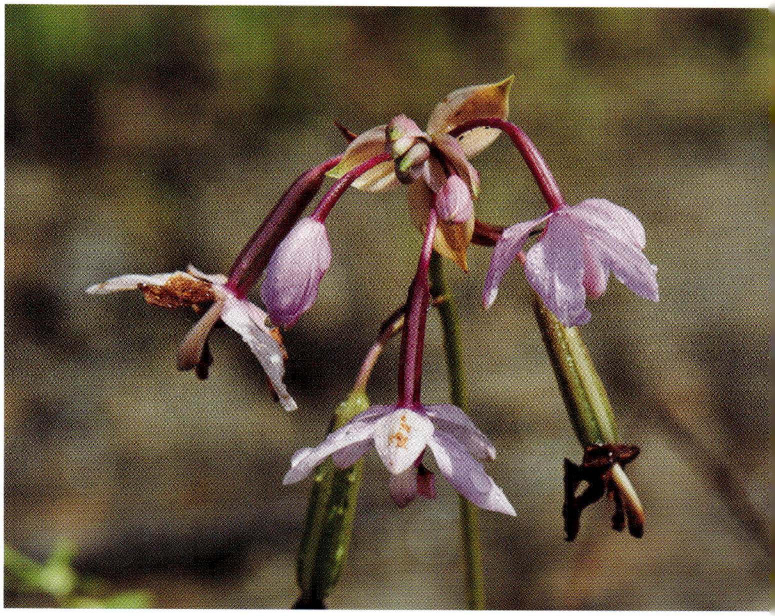

Above: The Philippine Ground Orchid in its wild home, rough ground in Samar Island Natural Park. So widespread has it become due to the garden trade that it is now classed as a pest in some countries.

Orchid conservation

Although there are times in a montane forest when orchids seem to be just everywhere, a number of species are threatened either through loss of forest habitat (particularly for those limited to the lowland forests) or through over-collection for the garden trade. Orchids are hugely popular in the Philippines among gardeners, and many of those that are supplied come directly from the forest, rather than through garden breeding. The latest (November 2017) list of threatened Philippine plants issued by the DENR shows 97 orchid species to be threatened with extinction. The lady's slipper orchids (*Paphiopedilum* species) are particularly badly affected due to massive over-collecting. Also included is the Waling-waling (*Vanda sanderiana*), the Philippines' most famous orchid, with stunning pink or white flowers, which in the wild is limited to lowland forests on Mindanao.

Some Philippine orchids, such as the Philippine Ground Orchid (*Spathoglottis plicata*), have become so popular as garden decoration that they have been disseminated right around the tropical world, and in some countries are now considered invasive pests.

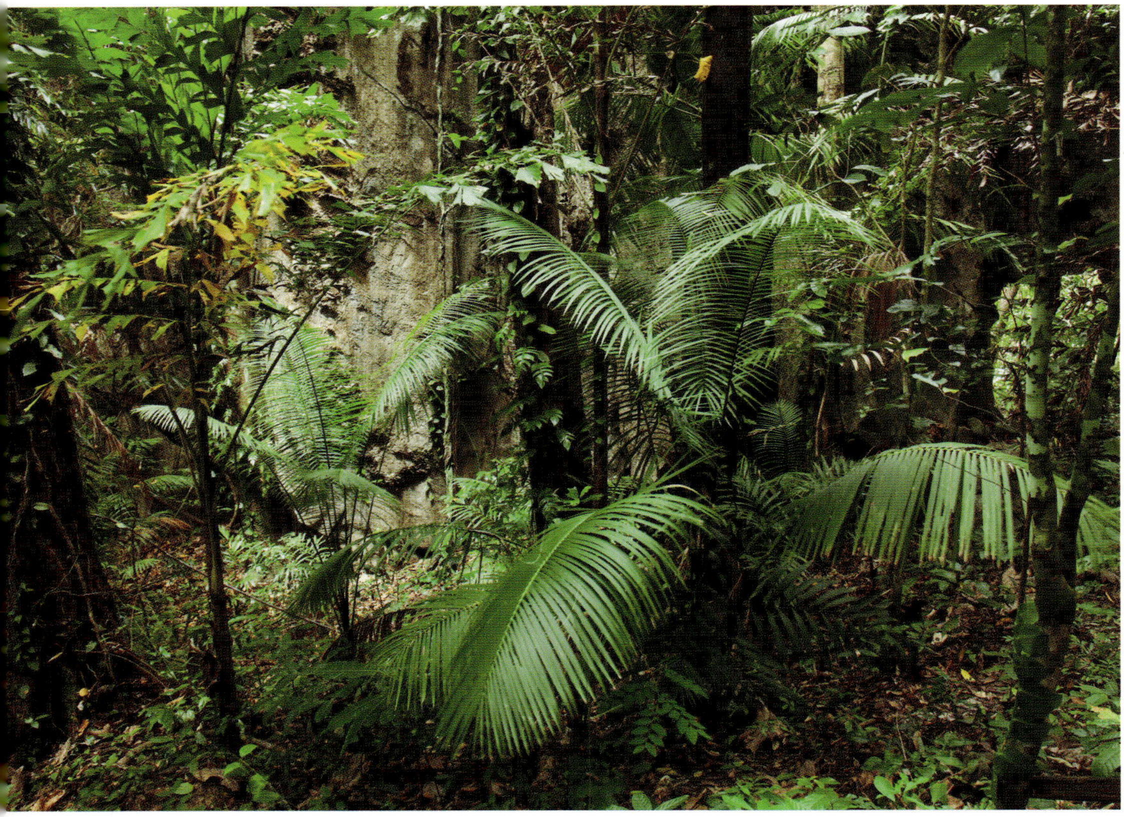

Left: *Rattans (*Calamus *species)* *in limestone forest on Lagen Island, in Bacuit Bay, Palawan.*

Opposite, near right: *Detail of the painful spines on young rattan stems, enabling them to both resist predation and latch on to trees in their climb towards the tree canopy.*

Opposite, far right: *A* Calamus *species rattan climbing upwards towards the light.*

Below: *An* Areca *species palm growing in marshy ground alongside Lake Dunoy, on the western edge of the Northern Sierra Madre Natural Park.*

Palms

This is another hugely important group of plants in the Philippines, amounting to about 150 species across the country, about two-thirds of them found nowhere else. They are widely distributed across a range of terrestrial habitats, from beaches to lowland rainforest and up into lower levels of the montane rainforests. Due to the low light levels beneath the forest canopy, they tend to be less common in closed canopy primary rainforests than in open-canopy secondary forests. In the latter, palms frequently grow profusely, especially the rattans, helping in the forest's natural restoration process.

Along with such trees as the figs (*Ficus* species), palm fruits are a hugely important food source for forest wildlife, especially fruit bats and pigeons and doves, but also wild pigs, deer, civets and some rodents.

Many of the palms are also very useful to the human population, so are widely cultivated. The most famous of these is the Coconut Palm (*Cocos nucifera*), of which there are believed to be more than 200 million in plantations right across the country. Unfortunately,

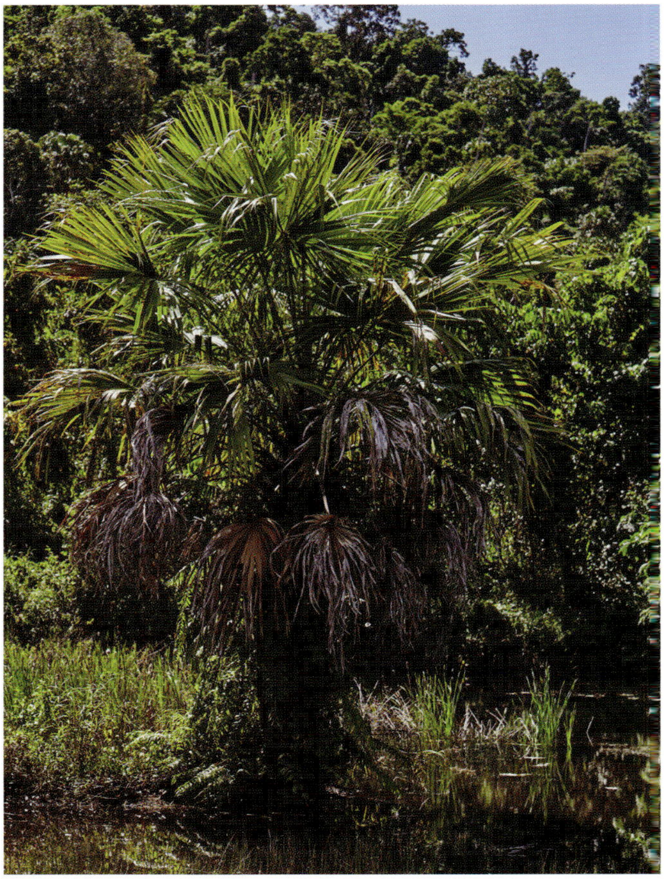

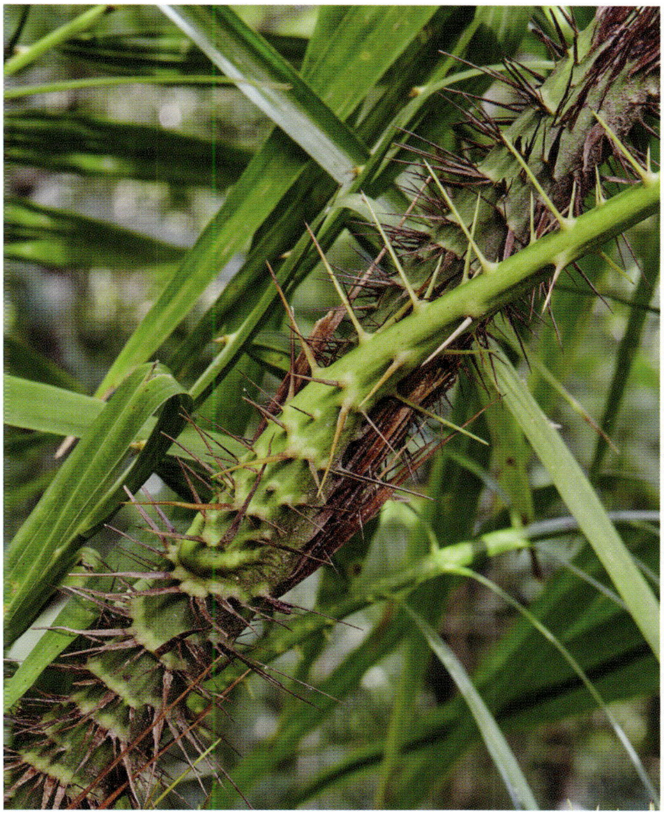

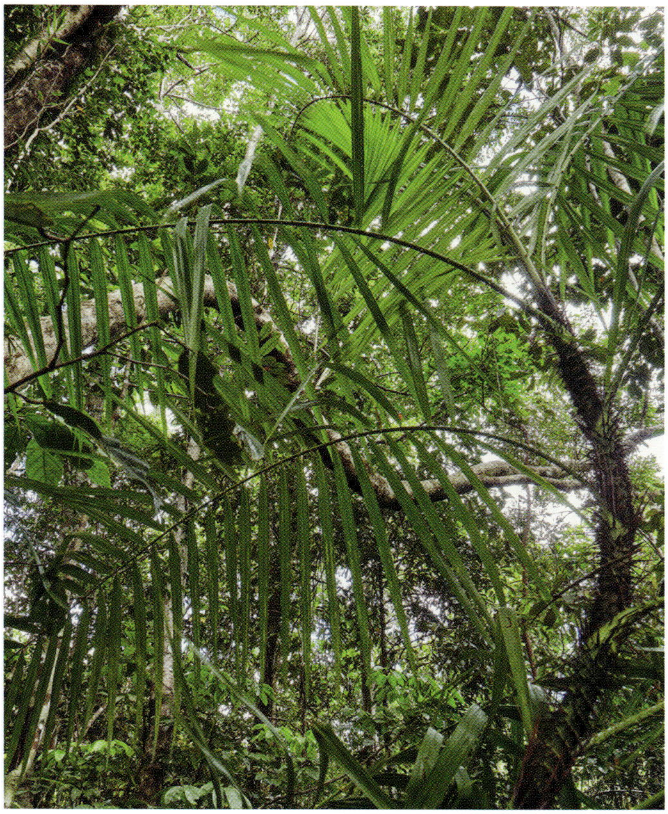

the creation of such plantations in the years following the Second World War played a large part in the destruction of the country's primary lowland rainforests, a tragedy far outweighing the limited economic benefits that came from the plantations.

Widely used palms include the Nipa Palm (*Nypa fruticans*) and Sago Palm (*Metroxylon sagu*). The former is often found in vast swamps consisting of little else at the landward edges of mangrove forests, and is widely used in creating mats and thatching. The Sago Palm forms clusters, the stems of which are an important food source. Other important forest palms include the *Pinanga*, *Areca* and *Heterospathe* species, the fruits of which are often eaten and are an important source of oils, with the leaves being widely used in handicrafts.

Rattans

The single largest group of palms is the rattans, climbing palms that make up roughly half of all Philippine palm species. Most of these are *Calamus* species, and two-thirds of those found in the Philippines are found nowhere else. The rattans are, of course, famous as a major material for furniture, due to the great strength and high flexibility of their stems.

While the non-rattan palms all grow with a single, unbranched erect stem, the rattans produce long, straggling vines that can be as much as 200 m (660 ft) long, which hook themselves on to trees to climb towards the forest canopy. As anyone who has ever hiked in a Philippine rainforest will have discovered, the young stems are covered in very sharp spines, particularly the long, whip-like extensions that are sent out to hook on to tree branches. Becoming caught by these can be a painful experience.

Palm conservation

Although some palms are cultivated, others – especially the rattans – have been traditionally collected from the wild. As a result, a number are threatened, and the DENR's list of threatened Philippine plants includes 57 palm species in total, 22 of them *Calamus* species rattans. Although cultivation of rattans has increased, many remain threatened due to the loss of their lowland rainforest habitat.

Pandans

Often called screw pines due to the spiral arrangement of the leaves as they grow out from the stems, this is another widespread group of plants, with just over 100 species found in the Philippines, most of them endemic to the country. Rather closely resembling palms, they are in fact not related, but instead are a group of tropical or subtropical trees. One of the main distinguishing features is that the pandans have multiple branching stems, while palms have only a single stem, ending in a single cluster of leaves. Like the palms, the pandans can be found in quite a variety of habitats and are also hugely useful, with the fruits of some species being edible, and the leaves frequently being used in weaving and handicrafts, as well as in flavouring foods.

Pandans are enormously versatile plants, able to thrive in a host of conditions. The most familiar

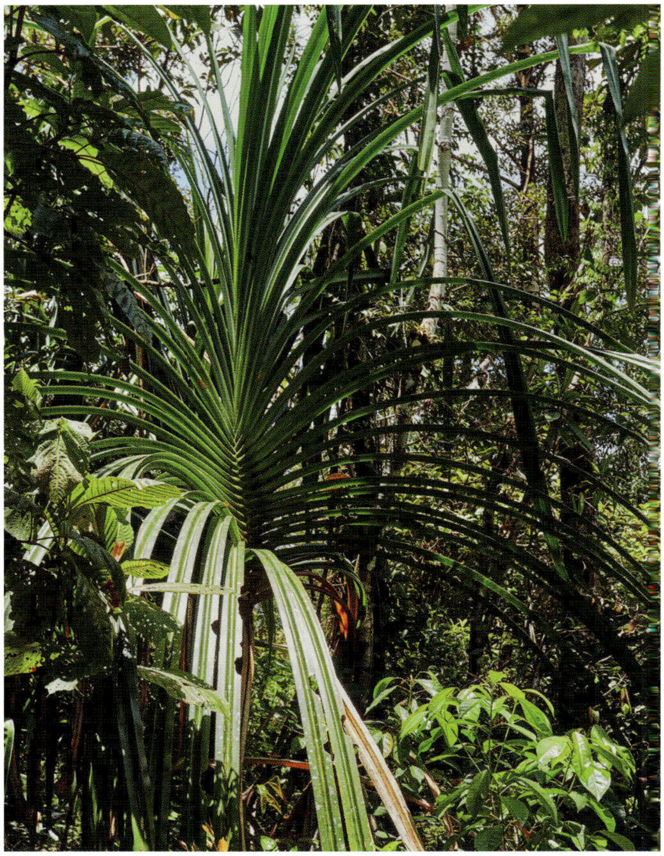

Above: A magnificent Freycinetia *species forest pandan, in lowland forest in Samar Island Natural Park.*

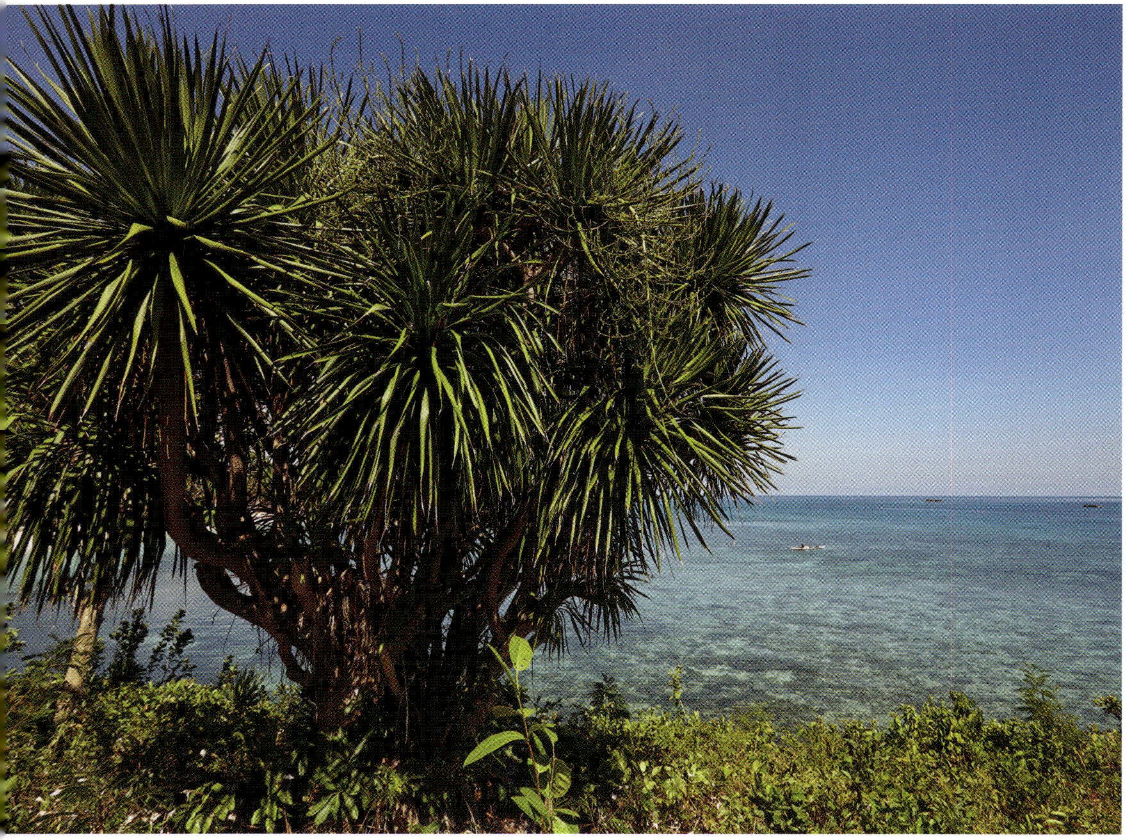

Below: *One of the most common pandan species,* Pandanus tectorius, *a frequent resident of coasts, particularly in beach forest.*

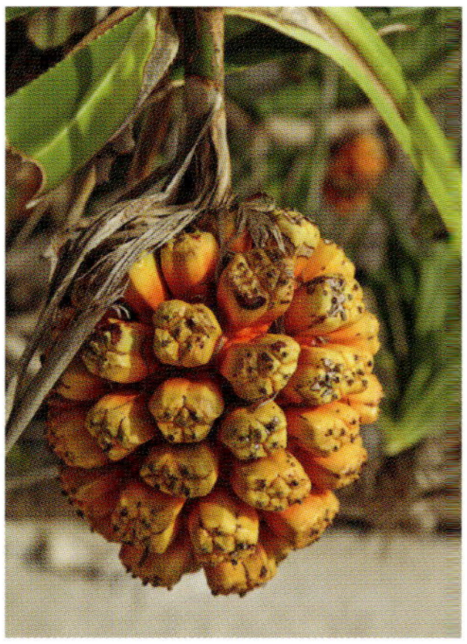

Below: *The easily recognized fruit of* Pandanus tectorius, *on the beach on Balicasag Island, Bohol.*

species are a common component of beach forests and the edges of mangroves, but a large number of virtually unknown species grow in the Philippines' forests, both primary and secondary, up to an altitude of about 1,500 m (4,900 ft). They can often be found along forest streams, but also in swamps and dry areas over limestone.

Species diversity

The Philippines' 100-plus pandan species all belong to a family of plants called the Pandanaceae, which in turn is divided into four genera: *Benstonea*, *Pandanus*, *Freycinetia* and *Sararanga*. The last of these consists of just one species, *Sararanga philippinensis*, an endemic species that grows in forests around the country, and which is the Philippines' only threatened pandan species. The great majority of pandans are either *Pandanus* or *Freycinetia* species, the first of these consisting of the 'classic' pandans with which most people are familiar, while the latter are straggling shrubs or climbers, growing mainly in lowland rainforests.

Pandanus species trees are most familiar from their beachside habitats, where they can provide valuable shade. Looking rather like a strange palm and growing up to several metres tall, they are often held up by prop roots, clusters of branching roots that grow down from the stem a short distance above the ground, anchoring the tree firmly in the sand. They are also noticeable for their bulbous orange- or red-coloured fruits that bear a passing resemblance to pineapples. Growing in such an exposed position, these pandans are able to resist drought, high wind and salt-laden air. Perhaps the most common of these species is *Pandanus tectorius*, whose orange fruits are said to be quite edible. This pandan is widely distributed across Southeast Asia, Australia and the Pacific, spread by its watertight, floating fruits. Forest species, many of which can grow to 15 m (49 ft) or more tall and often have prop roots, include *P. pallidus*, *P. paloerisis*, *P. simplex* and *P. philippinensis*, all of them endemic to the Philippines, and most restricted to specific parts of the country. They are generally found in forests at lower altitudes, but one that grows up to 1,500 m (4,900 ft) is *P. pallidus*.

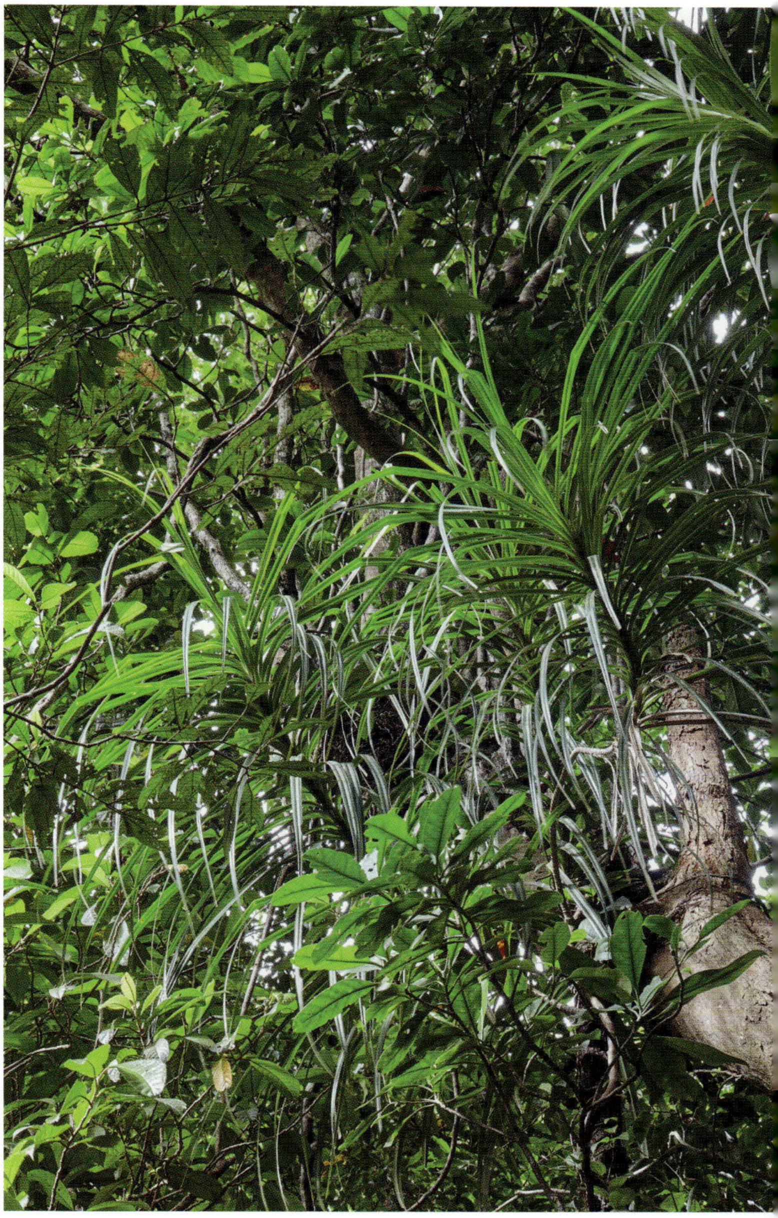

Above: A Freycinetia *species climbing pandan working its way up a tree in forest in Northern Negros Natural Park, Negros.*

The *Freycinetia* species climbing pandans are largely forest dwellers, and consist mainly of shrubs with long, rather spindly branching stems. The great majority are endemic to the Philippines, including such species as *F. apoensis* (endemic to Mt Apo), *F. cumingiana* (endemic to Luzon) and *F. jagori* (endemic to southern Luzon, Samar, Leyte and Mindanao).

Medinillas

The *Medinilla* shrubs are often seen as the quintessential Philippine rainforest plants, if only because of the fact that of the 88 species currently listed, 78 are endemic to the country. They are, in other words, very much a 'Philippine plant', the few that are not endemic being found only in some of the Indonesian islands. Like so many other endemic Philippine plant species, most have quite limited geographical ranges, restricted to just certain parts of the country, and they are nowhere particularly common. They are, however, quite unmistakable when spotted in the forest, and are rather remarkable.

Most *Medinilla* species are epiphytic, even the larger shrubs that reach several metres tall, often seen growing out of branching forks of forest trees. The leaves are ovoid in shape and usually a rich, dark green, with a leathery appearance. It is the flowers, however, that make the *Medinilla* shrubs unmistakable, hanging as large panicles, conically shaped clusters of flowers, each flower within a panicle rarely more than 2 cm (0.8 in) across, and both the flowers and panicle stems coloured red, pink, violet or white.

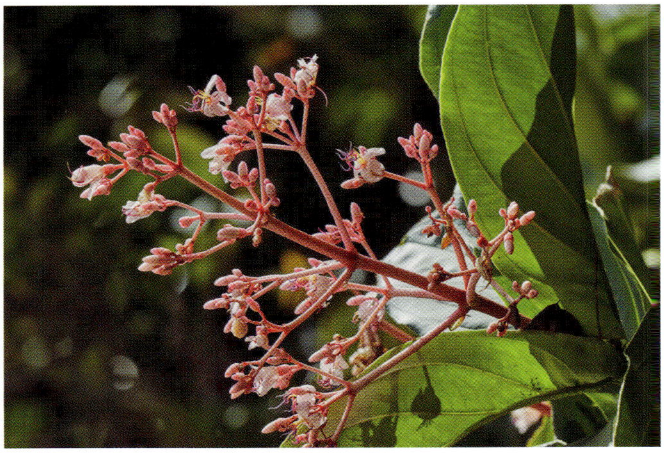

Above: Kapa-kapa, one of the best-known medinillas, in flower in lowland forest in Samar Island Natural Park.

The plants are found primarily in montane rainforests, with some, such as *Medinilla apoensis*, growing at altitudes over 2,000 m (6,600 ft), well into the upper montane/mossy rainforest areas. Most, however, grow in the lower or mid-montane forests, with a few restricted to the upper areas of lowland rainforest. With so much of the Philippines' lowland rainforest having been destroyed, *Medinilla* shrubs that are restricted to this habitat are now very rare.

It is unusual during any forest hike to see more than a handful of *Medinilla* plants, but when spotted it is usually due to their vibrantly coloured flowers, providing a welcome splash of strong colour among the otherwise dark and rather sombre greens.

Threatened *Medinilla* species

At present, 20 *Medinilla* species are listed as threatened according to the DENR's 2017 list of threatened plants, two of them Critically Endangered. These are *M. dalciana* and *M. magnifica*, the former restricted to just one region of southern Luzon and found growing at an

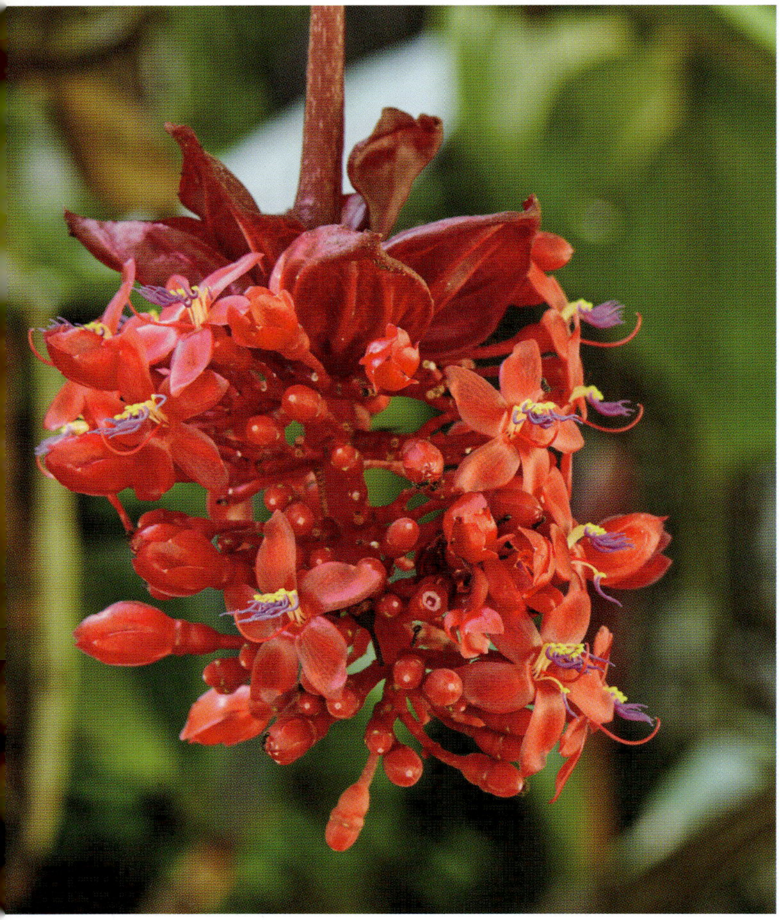

Left: Crimson Medinilla (Medinilla miniata), a Philippine endemic species, here seen in Makiling Botanic Garden.

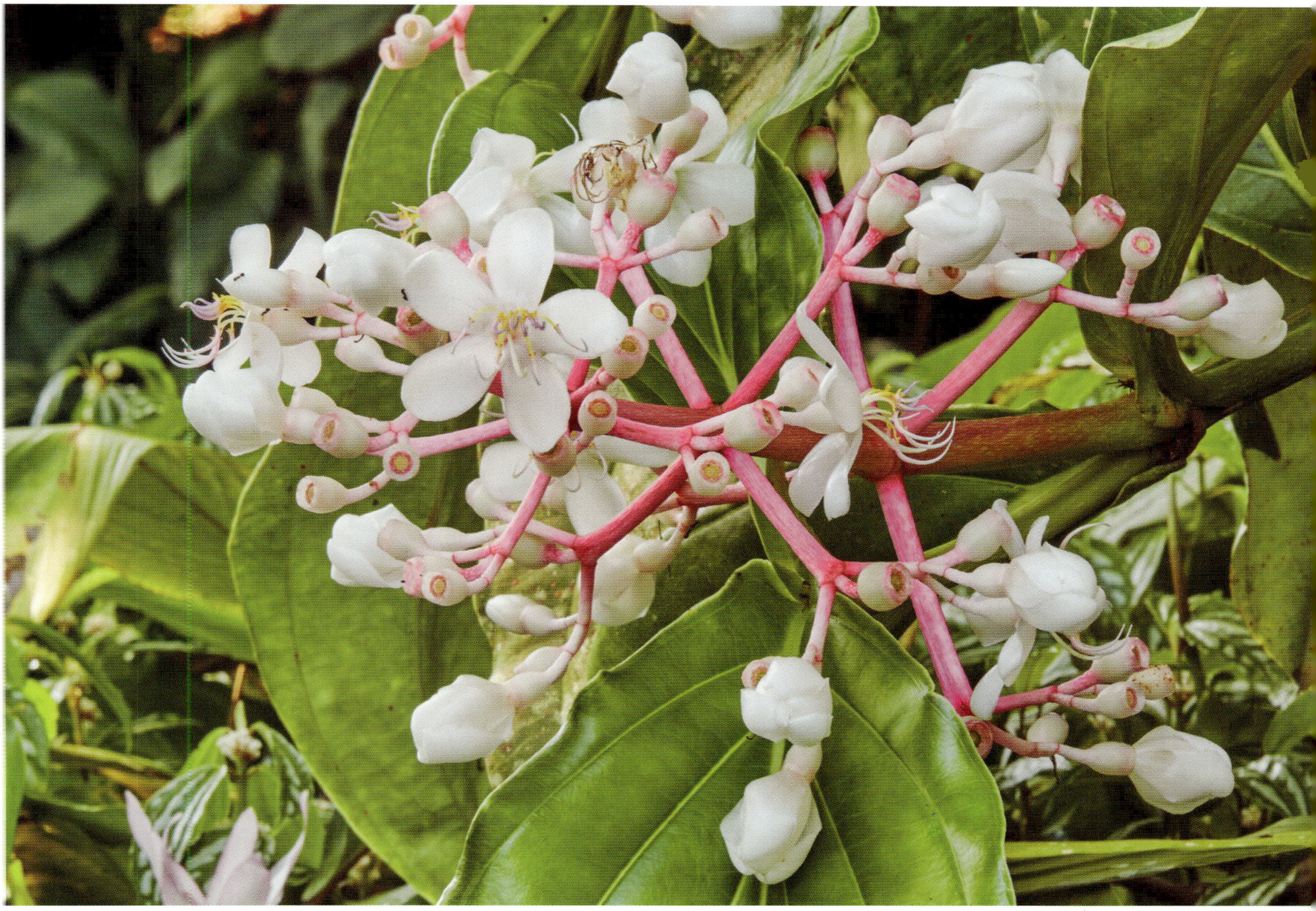

Above and below: *Pear White Medinilla (*Medinilla albiflora*), showing both the flower panicle and a close-up of a single flower. This endemic species is found in southern Luzon, Samar, Leyte, north-east Mindanao and Panay.*

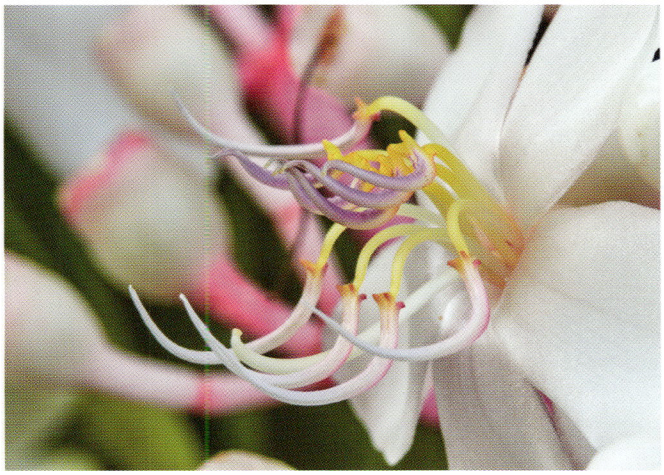

altitude of about 950 m (3,100 ft). *M. magnifica* was previously found in forests around a relatively large part of the Philippines, and across a wide spectrum of altitudes, from lowland rainforest at 300 m (1,000 ft) to montane rainforest at 1,400 m (4,600 ft).

M. magnifica is one of the country's best-known and loved *Medinilla* plants, called locally Kapa-kapa, and in English sometimes Rose Grape. It is one of the few *Medinilla* species to have a Filipino or English name, a mark of its relative familiarity, at least when compared to many of the others. Although now rare in the wild, *M. magnifica* is cultivated extensively, though this may be one of the reasons for its demise in the wild, along with habitat loss. Surprisingly, it is one of the few *Medinilla* species that is not totally endemic to the Philippines, being also found in the rainforests of the Moluccas, in Indonesia.

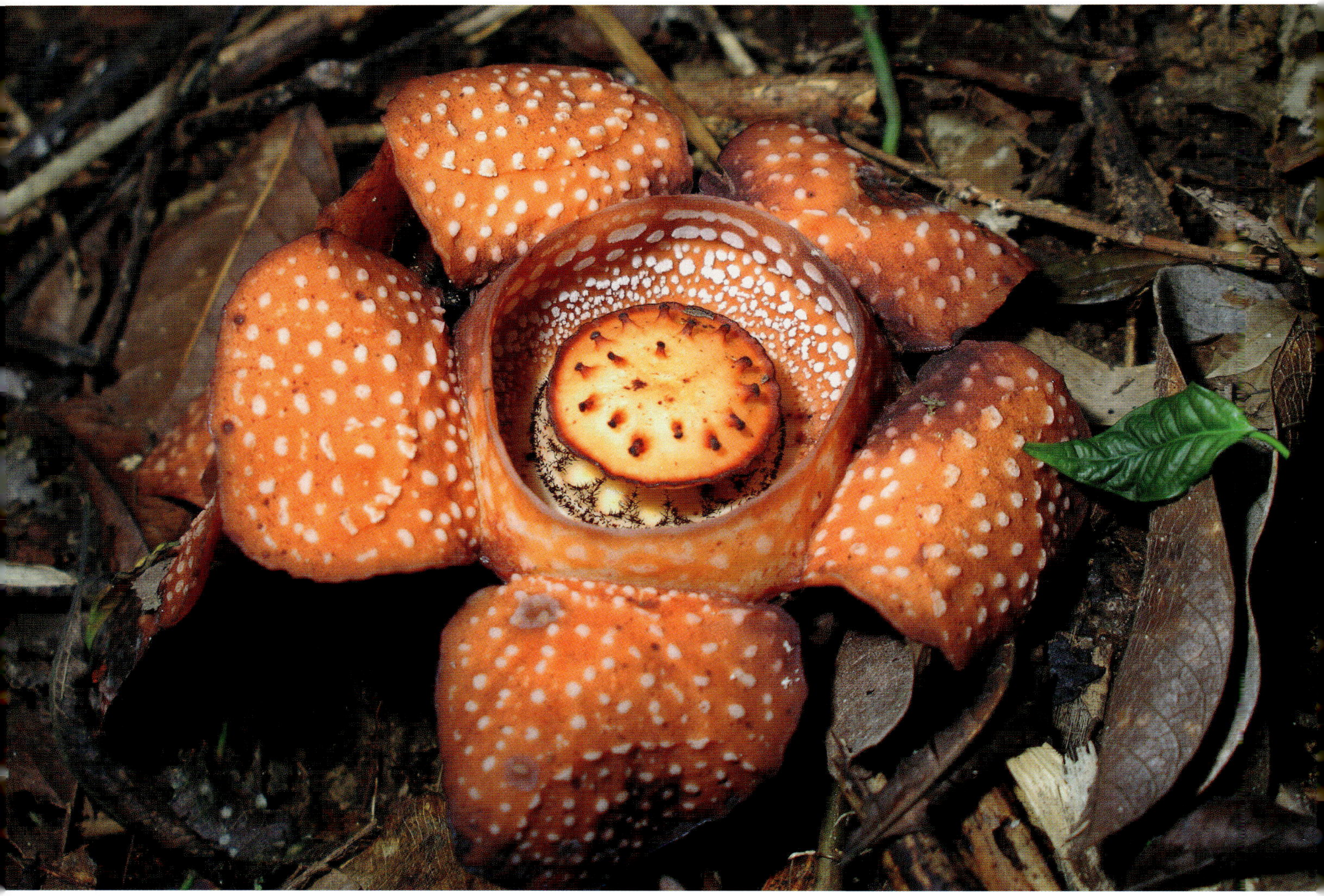

Above: The least rare of the Philippines' rafflesias, Rafflesia lagascae, *whose flowers are 15-20 cm (6-8 in) in diameter, seen in forest on Mt Makiling, Luzon.*

Rafflesias

This chapter is rounded off with an introduction to one of the strangest and most enigmatic of all rainforest plants, the rafflesias. Well known for producing gigantic flowers, the world's largest, up to 1.5 m (4.9 ft) across, rafflesias are truly remarkable plants. Their story in the Philippines is quite instructive, as in many ways it reflects on our lack of knowledge about the country's forest habitats. Until the year 2000 only two species were known in the Philippines, making it something of a sideshow in the rafflesia story internationally, most species being known from Thailand, Malaysia and Indonesia. However, since 2002 a further 11 species have been found in the Philippines, putting the country at the centre of rafflesia biodiversity.

Some rafflesia biology

All rafflesias are wholly parasitic, living inside the roots and stems of several species of a forest vine, *Tetrastigma*. Furthermore, the rafflesias produce no stems, roots or leaves of their own, depending wholly on the vine for nutrition: the only part of the plant ever visible is the flower. So when not in flower, or at least showing flower buds, rafflesias are completely undetectable. Most of the Philippine rafflesias seem to flower through much of the year, though peak time appears to be generally March–May. The record for the world's largest flower is held by *Rafflesia arnoldi*, found in Indonesia, but the second largest is *R. schadenbergiana*, a Philippine rafflesia, whose flower reaches up to 80 cm (31.5 in) across. All other Philippine species have much smaller flowers, the smallest just 9.7 cm (3.8 in) across, belonging to *R. consueloae*.

The growth of Philippine *Rafflesia* species

The Philippines' first two *Rafflesia* species were discovered in the nineteenth century, *R. schadenbergiana* being found in forest near Mt Apo, Mindanao, and *R. manillana* found near the town of Basey on Samar. *R. schadenbergiana* was not found again for more than 100 years and was believed to be extinct, but in 1994 was rediscovered in South Cotabato, then sometime later in forest just outside Mt Kitanglad Range Natural Park, both on Mindanao.

The rush of 'new' rafflesias began in 2002 with the discovery of *R. speciosa* on Panay, then later also on Mt Kanlaon, on Negros. This was followed in subsequent years by *R. baletei* in southern Bicol, *R. leonardi* in Luzon's Northern Sierra Madre, *R. lobata* on Panay, *R. mixta* in north-east Mindanao, *R. verrucosa* in south-east Mindanao, and *R. aurantia* in Luzon's Southern Sierra Madre. *R. lagascae*, initially confused with *R. manillana*, is now considered a separate species, restricted to southern Luzon, but the most widespread of all the Philippine rafflesias. The most recent addition to the list is *R. consueloae*, found in central-north Luzon in 2014.

Below: *The buds of* Rafflesia lagascae *flowers in forest on Mt Makiling. The flowers are the only visible part of a rafflesia, so the buds are an early indicator of the plant's presence.*

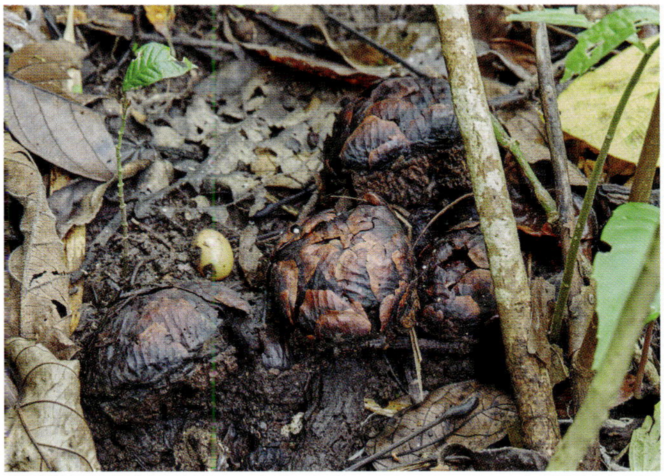

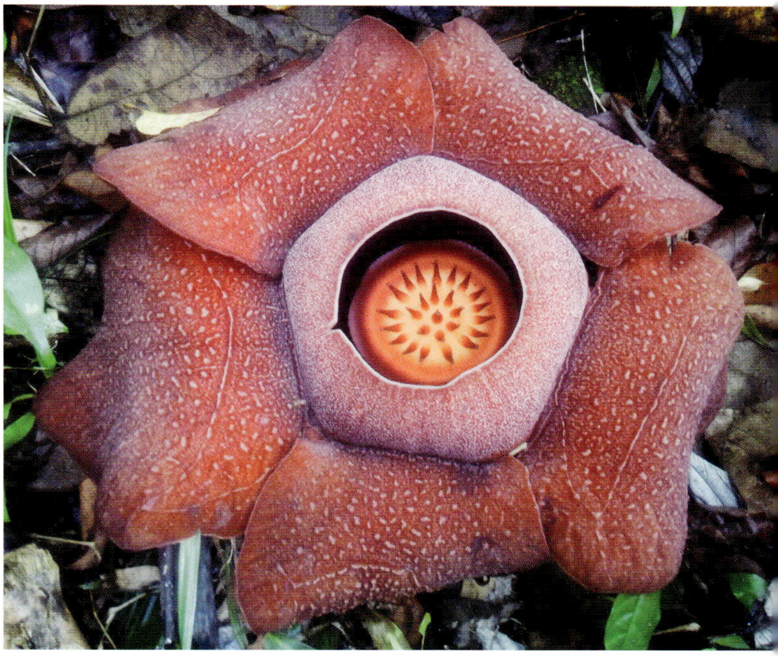

Above: Rafflesia speciosa, *a very rare species found only on Panay and (as pictured) in Mt Kanlaon Natural Park, on Negros.*

Conservation of rafflesias

One of the most distinctive features of Philippine rafflesias is that they are all extremely rare, each one restricted to just a handful of plants at just a couple of locations, or sometimes even only one. Every species is listed as Endangered or Critically Endangered. Worse still, most have been found outside currently protected areas, putting them at high risk of being lost to deforestation or infrastructure development. One of the worst threats seems to be badly managed ecotourism, the great fame of rafflesias in general drawing people to want to see the newly discovered Philippine plants. Unfortunately, trampling of the ground around flowering sites has been found to cause significant damage to the plants, and may be quite a threat to their survival.

There are currently no conservation management plans in place for any of the Philippine rafflesias, though these are urgently needed.

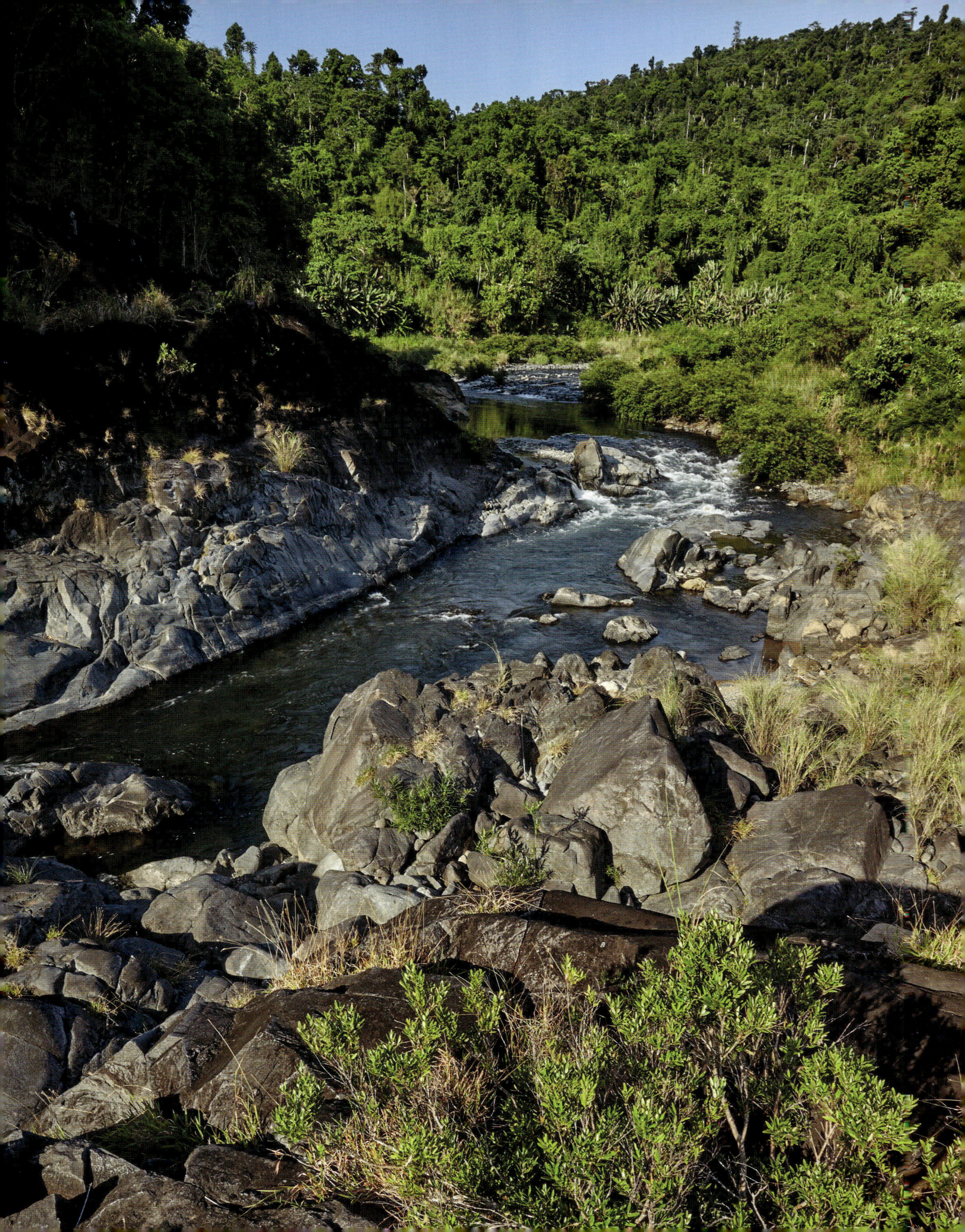

5

Wild Places of the Philippines

The National Parks and other Protected Areas

As described in Chapter 2, there is a growing amount of work going into protecting as much of the Philippines' wild habitats, along with their plant and animal wildlife, as possible, work that is expressed as a whole plethora of projects that are both locally and internationally funded and managed. Underpinning the work is the NIPAS, which currently encompasses some 240 terrestrial and marine protected areas. The aim of the system, of course, is to protect the full range of the Philippines' natural habitats, ranging from coral reefs, seagrass beds and mangrove forests along the coasts, to beach, lowland and montane forests on land, along with the many habitat sub-types, such as swamp and limestone forests, and mountain summit grassland.

While it is well beyond the scope of this book to give a description of every site within the NIPAS, the aim of this chapter is to give a taste of some of the country's most important wild sites, both terrestrial and marine. Although the number of sites described here makes up only a tiny percentage of the Philippines' total number of protected areas, these are among either the most important from the biodiversity conservation perspective, or are part of a growing group of protected areas increasingly popular for ecotourism projects ranging from birdwatching to scuba diving. All the sites encompass KBAs and/or IBAs, areas that have been identified in recent years as critical for the protection of particular plant or animal species.

Most of the sites descibed are specific protected areas. However, in a few instances regions reaching beyond the boundaries of a single protected area are described, as the importance of that region's natural environment reaches well beyond just that one protected area. This is particularly the case for some of the coral reef areas, but also for some terrestrial sites, such as the Sierra Madre mountains of northern Luzon.

Opposite: *The Catallagan River, flowing along the western edge of the Northern Sierra Madre Natural Park in north-east Luzon, is known to be one of the last wild homes of the Philippine Crocodile.*

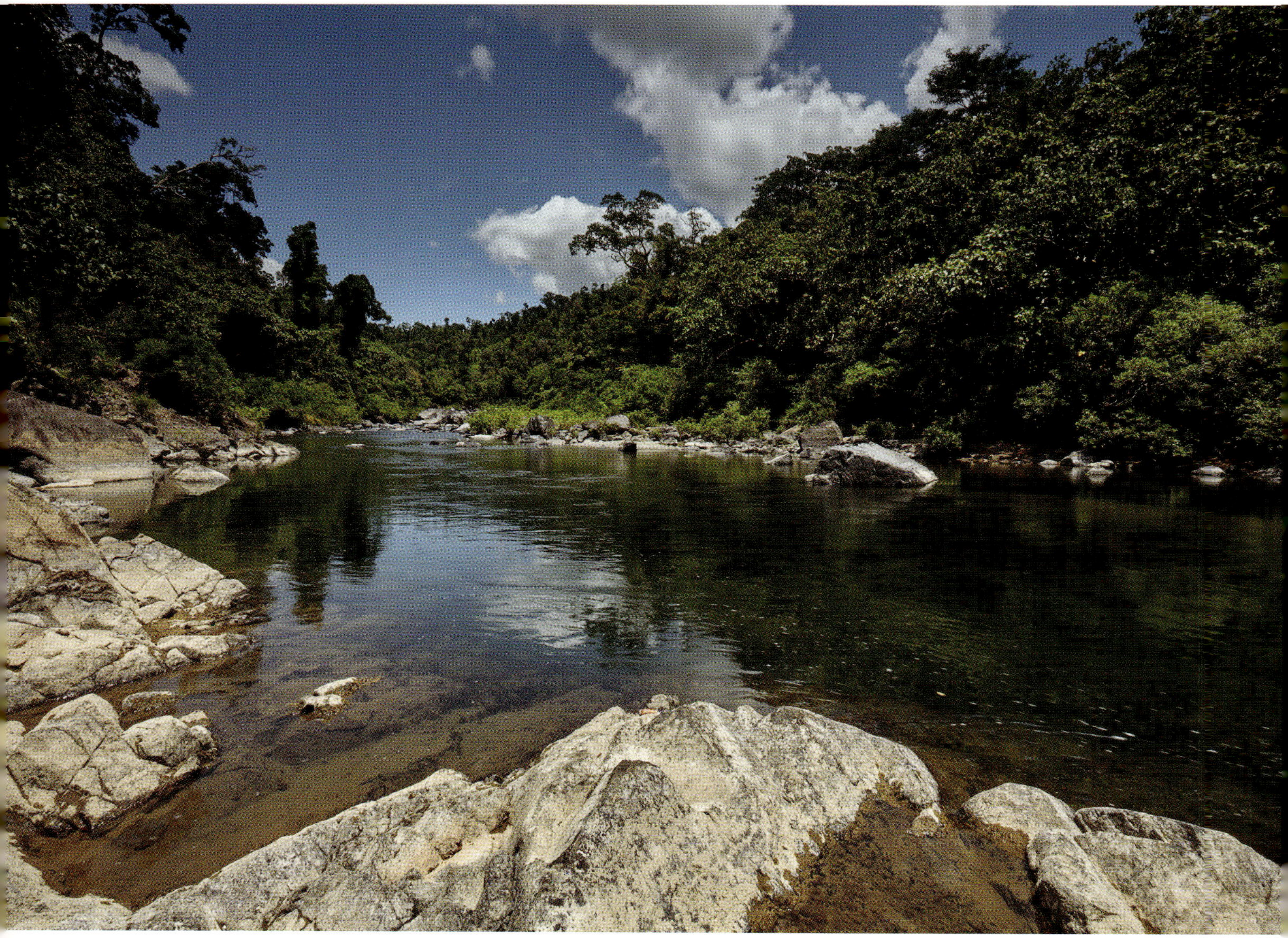

The Sierra Madre Mountains

The Sierra Madre mountains are the Philippines' longest mountain chain, running along much of the eastern flank of northern Luzon for a north–south distance of about 350 km (214 miles). Most of the mountain range is not especially high, though some parts do reach over 1,000 m (3,300 ft), the highest peaks being Mt Cresta (1,672 m/5,486 ft) and Mt Divilacan (1,311 m/4,301 ft). The range is, however, almost universally rugged, making the Sierra Madre a mostly remote and inaccessible place that has enabled much of its forest to survive intact. This mountain range is home to the Philippines' largest continuous area of rainforest, a mixture of old-growth, or closed canopy, primary rainforest in more or less pristine condition, and regenerating secondary forest. Most of the forest is lowland, rich in large dipterocarp trees, though there are some areas of montane forest around the highest peaks.

With such a vast, remote forested area, not surprisingly the region is critically important to the Philippines' biodiversity conservation. These forests provide an increasingly scarce home for endemic species such as the Philippine Eagle (*Pithecophaga jefferyi*), Philippine Crocodile (*Crocodylus mindorensis*) and Golden-crowned Flying Fox (*Acerodon jubatus*), to name just a few.

Furthermore, because few roads cross the mountains, the very thinly populated Pacific coastline also remains largely undamaged. As a result, it is home to a number of important areas of beach and mangrove forest, as well as seagrass beds and coral reefs.

A network of protected areas

More and more of the Sierra Madre is coming under formal protection as a string of protected areas is created along much of the range. The intention, as promoted by the US conservation body Conservation International, is eventually to create a huge biodiversity corridor that will link up all the region's key biodiversity areas, allowing wildlife to move around through a vast protected forest.

The oldest of the protected areas is the Aurora Memorial National Park, lying in Aurora province, in the southern Sierra Madre, just south-west of the Pacific coast town of Baler. Established initially in 1937, it is rather small, covering 5,676 ha (14,133 acres) of rather fine primary lowland rainforest. There are plans to greatly increase the park's size to bring much more of the hitherto somewhat overlooked forests of the southern Sierra Madre into protection.

At the other end of the scale is the Northern Sierra Madre Natural Park (NSMNP), the Philippines' largest terrestrial protected area, covering a vast 356,486 ha (888,310 acres) of forested mountains and coastal habitats. Established in 1997 as one of the country's 10 Priority Protected Areas under the World Bank Global Environment Fund's CPPAP, this park greatly expanded upon the already existing Palanan Wilderness Area. Since then, the NSMNP has been the focus for much of the conservation studies and protection work that have gone into the Sierra Madre region.

More recently created are the Quirino and Casecnan Protected Landscapes, both protecting slices of the southern Sierra Madre, Quirino covering 175,943 ha (434,766 acres), and Casecnan protecting 88,847 ha (219,545 acres). A third site, Peñablanca Protected Landscape and Seascape (covering 118,781 ha/293,595 acres) encompasses more of the northern Sierra Madre and is contiguous with the northern border of the

Opposite: *The Catallagan River surrounded by lowland rainforest, near San Mariano, on the western slopes of the Northern Sierra Madre Natural Park, north-east Luzon.*

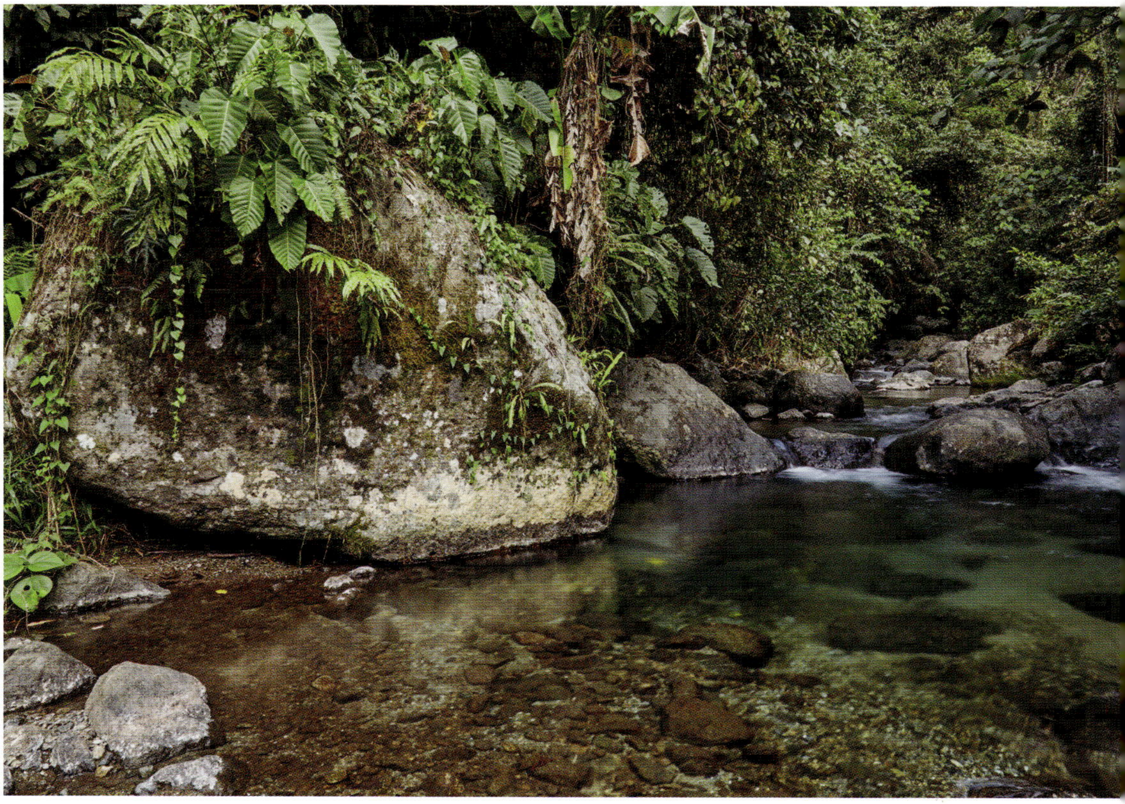

Right: *A stream flowing through lowland rainforest near Baler, close to the boundaries of Aurora Memorial National Park, in the Southern Sierra Madre mountains, north-east Luzon.*

NSMNP. Although the great majority of it covers forested land, a little over 14,000 ha (34,860 acres) is marine, protecting a number of mangroves and coral reefs.

All the protected areas face serious threats. Illegal logging and clearance for farmland are a constant issue, with the NSMNP, for example, losing an estimated 1,400 ha (3,460 acres) of forest annually, mostly along its western fringes.

The NSMNP may also be about to face further environmental threats of this nature. A new road will soon cut right across the mountains, passing through the park, and connecting the currently very isolated Pacific coast communities with Ilagan, the Isabela provincial capital. Inevitably, this will make penetration of the forest by illegal loggers, hunters and would-be settlers much easier unless strong protection measures are put in place.

Other threats include mining, with the Sierra Madre region known to be rich in minerals. Indeed, in the mid-2000s Quirino Protected Landscape had several thousand hectares removed from its protection

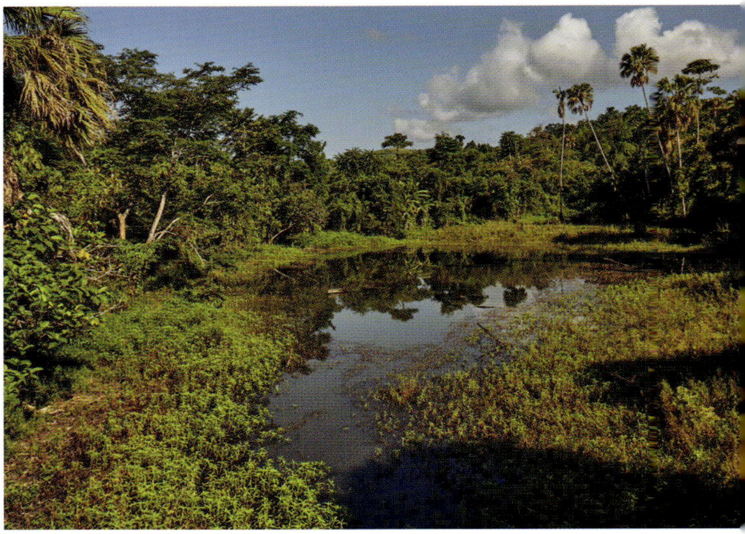

to allow for mining to be established. Furthermore, the environmental value of the Casecnan Protected Landscape has been compromised by construction of a hydroelectric dam within its boundaries.

A biodiversity wealth

Although much remains to be done, studies conducted so far give a vivid picture of just how important the

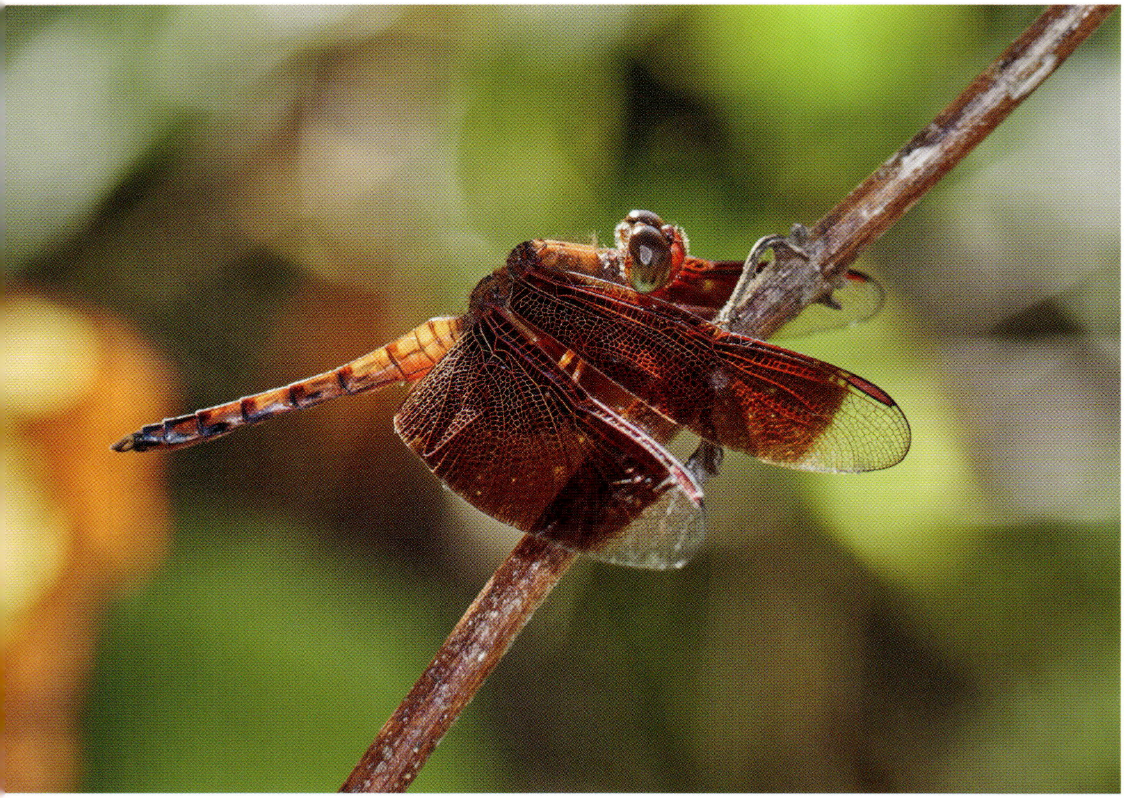

Above: *Dunoy Lake, surrounded by lowland rainforest, is inside the Northern Sierra Madre Natural Park, but is perilously close to the edge of farmland. It is known to be one of the few remaining wild homes of the Philippine Crocodile.*

Left: *The Red Grasshawk dragonfly (*Neurothemis fluctuans*), is widespread across Southeast Asia. Here seen close to the shore of Dunoy Lake, in the Northern Sierra Madre Natural Park.*

Sierra Madre's protected areas are to Philippine biodiversity. About 60 mammal species have been found within the Sierra Madre, roughly 45 per cent of them endemic to the country or even just to Luzon. The results are even more spectacular for birds: of the 600-plus species known throughout the Philippines, almost half can be found within the Sierra Madre, with 30 per cent of these endemic and 19 species classified as threatened.

When it comes to amphibians, reptiles and insects, it is very hard to quote numbers as so many new species are being discovered all the time. However, of the roughly 350 species of amphibian and reptile currently known nationally, 90 have been found in the Sierra Madre, nearly half of them endemic to the country. It is a similar story for butterflies, with almost half of the 128 species found in the region known to be Philippine endemics.

Plants too show a high rate of endemism. Of the nearly 1,100 tree species identified in the NSMNP alone, an estimated 42 per cent are endemic. Among these are some of the largest forest trees ravaged by logging around the country, including the Almaciga (*Agathis philippinensis*), and a number of dipterocarp trees, such as *Shorea* and *Hopea* species.

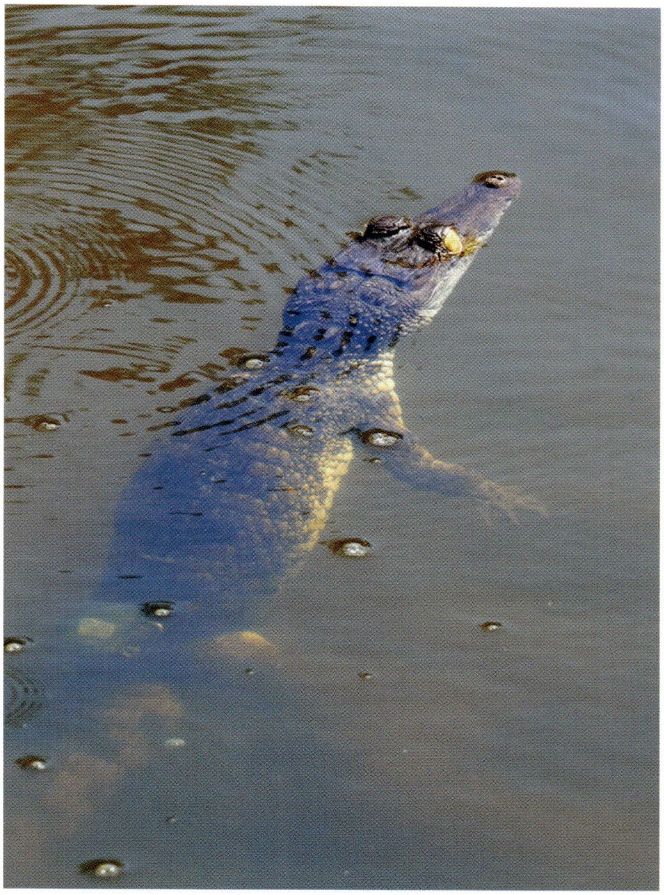

Above: *A two-year-old Philippine Crocodile, shortly after being released into the wild, into a lake in the Northern Sierra Madre Natural Park, following captive rearing by the Mabuwaya Foundation.*

Below: *A Philippine White-throated Kingfisher (*Halcyon gularis*) in rainforest close to the shore of Dunoy Lake. This is an endemic species, only recently split off from the species common across Southeast Asia.*

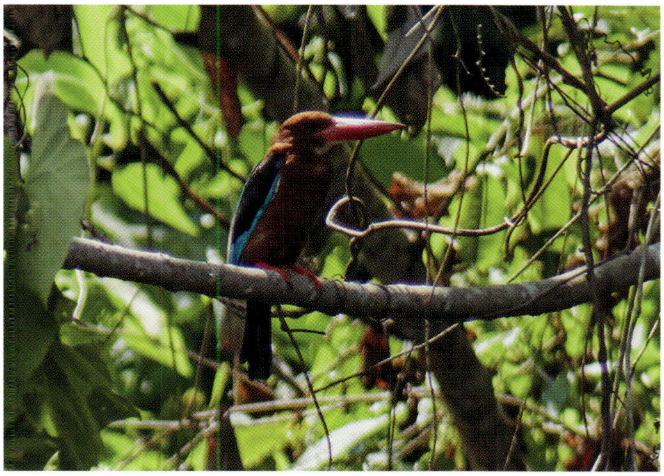

As one of just five remaining known sites for the highly endangered Philippine Crocodile, the NSMNP is a focus for efforts to prevent it going extinct. Central to these efforts is the Mabuwaya Foundation, an NGO headquartered at the Isabela State University, but with a rearing centre in the town of San Mariano, close to the park's western fringe. For the past few years (see page 125), staff have been running a head-start programme, collecting hatchlings found in the wild and rearing them at their San Mariano centre. Once they are about two years old the crocodiles are released back into the wild in two lakes inside the western side of the NSMNP, having reached a size that makes them less vulnerable to predators. As a result, crocodile numbers in the western part of the Park are believed to be slowly increasing.

Mt Pulag National Park

This national park lies across the highest parts of northern Luzon's main mountain spine, the Cordillera Central, protecting forests on Mt Pulag, at 2,922 m (9,584 ft) Luzon's highest mountain. It is a rugged landscape, encompassing not just Mt Pulag itself, but also several other peaks over 2,200 m (7,200 ft), including Mts Tabeyo, Akiki and Panotoan. In total, the park covers 11,550 ha (28,760 acres), its lowest parts reaching down to an altitude of 1,100 m (3,600 ft), all of it steep, mountainous slopes.

Pulag has long been recognized as a very wildlife-rich area, but it was only in 1987 that it fully gained national park protected status. Since then, it has been at the forefront of efforts to protect as much of Luzon's biodiversity as possible. Pulag was one of a tiny handful of protected areas to receive international input in the early 1990s under a WWF-assisted debt-for-nature swap, and from 1995 to 2001 was one of eight sites across the Philippines included in the NIPAP. Since then, the park's conservation status has continued to grow, with its recognition as an IBA and KBA, particularly as a vital site for a wide range of mountain specialists, both plants and animals, that live only in the Cordillera's high altitudes and cool, damp climate.

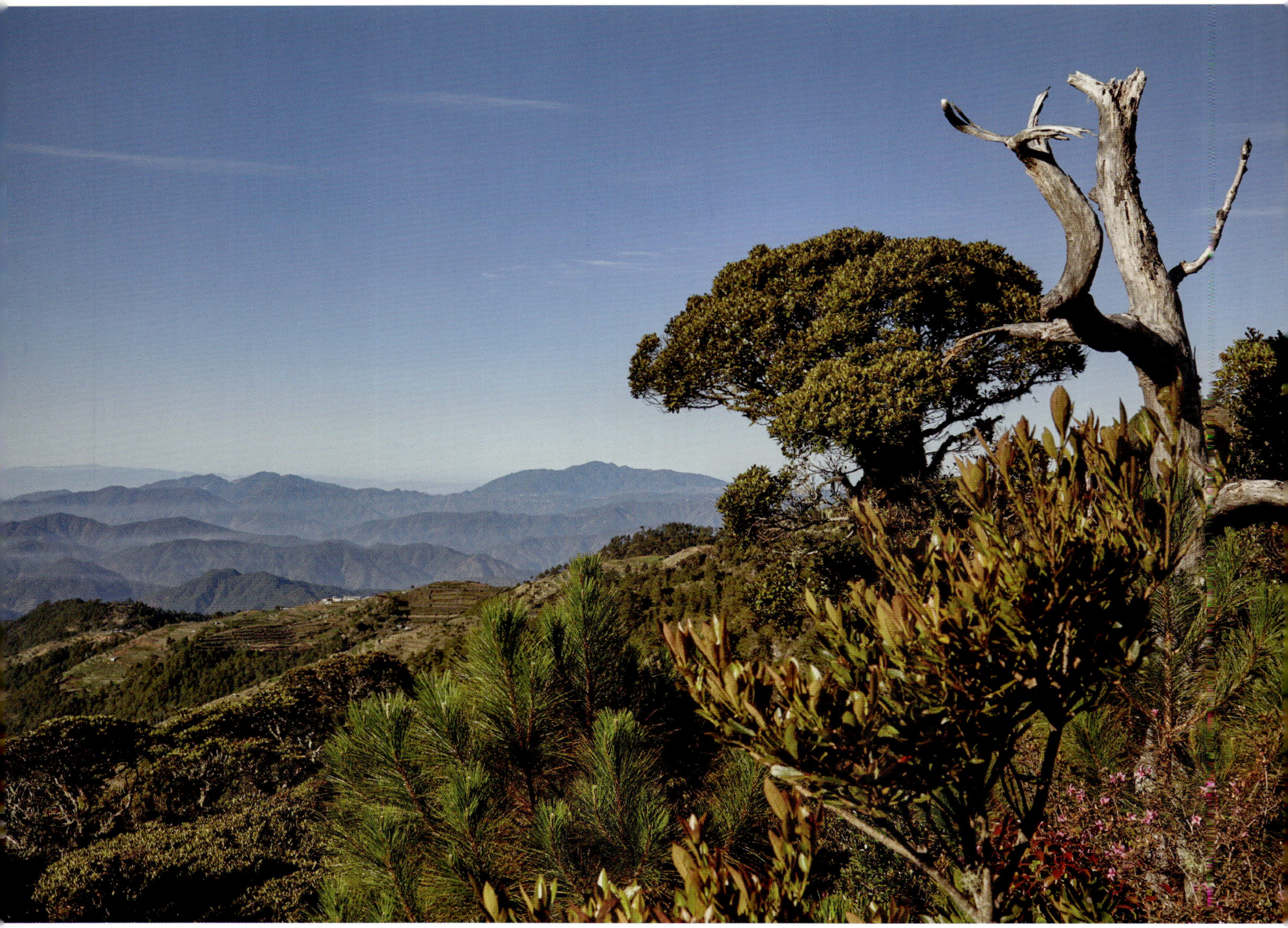

Right: *Latep-Ngapoh Lake surrounded by mist-shrouded mossy rainforest, in the northern part of Mt Pulag National Park, the Cordillera Central Mountains, Luzon.*

Opposite: *A magnificent view across the Cordillera Central Mountains of northern Luzon from a clearing in the montane rainforest of Mt Pulag, at about 2,500 m (8,250 ft) altitude.*

Below: *A typically tangled and chaotic scene in moss-draped upper montane rainforest on the slopes of Mt Pulag at about 2,300 m (7,600 ft) altitude.*

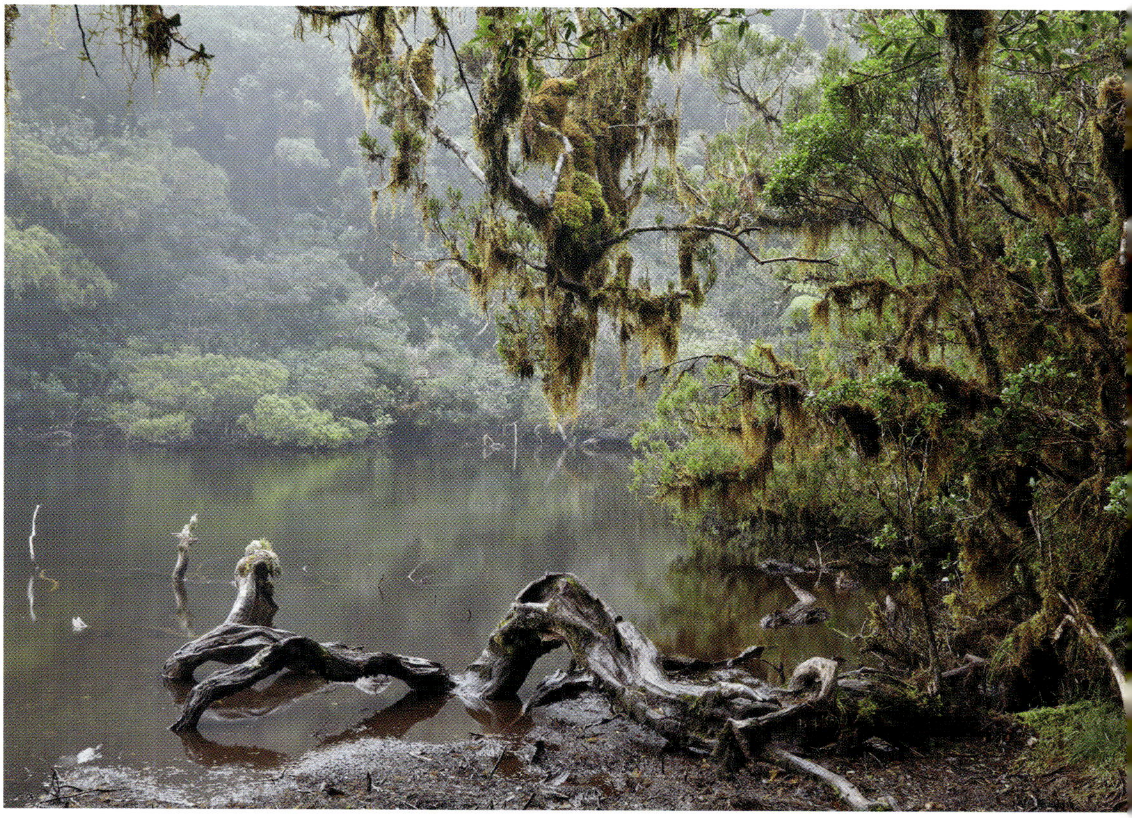

Habitat zones

The park contains essentially three habitat zones, which change with altitude. The lowest, reaching from the park's lowest limits up to an altitude of about 2,200 m (7,200 ft), are pine forests, consisting almost entirely of the Benguet Pine (*Pinus insularis*, also known as *P. kesiya*), along with a number of grasses, shrubs and flowering herbaceous plants scattered among the trees.

Above 2,200 m the forest transitions quite abruptly into upper montane, or mossy, rainforest. Typically for this kind of forest, the trees consist largely of stone oaks and laurels, twisted, gnarled and often very dense, festooned with mosses, orchids and ferns, and constantly dripping wet from the high rainfall and frequent fog. Montane forest takes up about 50 per cent of the park, making it the main habitat.

On Mt Pulag itself, above about 2,600 m (8,528 ft), the forest gives way to open grassland that reaches all the way to the summit, dominated in many areas by the dwarf bamboo *Yushania niitakayamensis*, which grows to a height of about 40 cm (16 in). The other mountains in the park retain forest cover all the way to their summits. There are several, mostly small, lakes, with one on the south side, close to the national park's Babadak ranger station, and four on the north-west side of the park, on the slopes of Mt Tabeyo. Of these four, Lake Tabeyo is now surrounded by farms, even though it is well inside the park's boundaries. The other three, Lakes Ambulalakaw, Latep-Ngapoh and Incoloh, are more remote and are still surrounded by very dense, largely undisturbed montane rainforest.

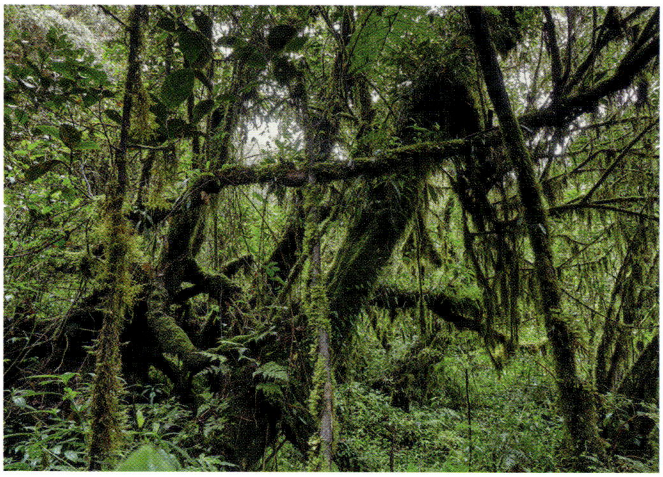

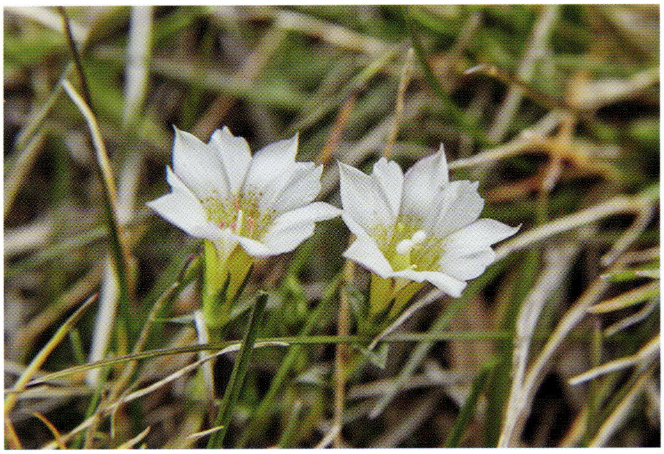

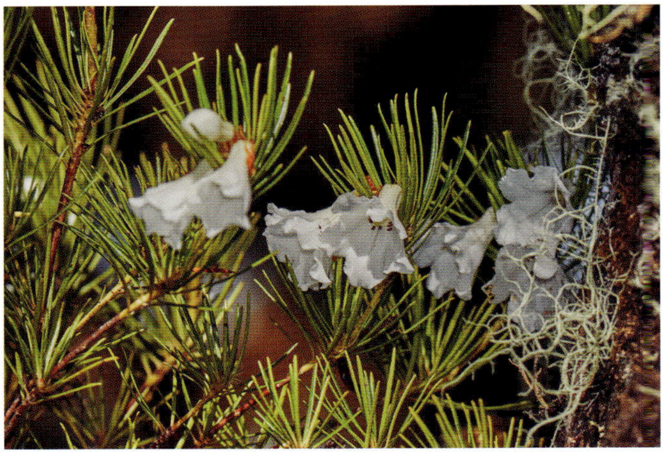

*Above: Tiny gentian flowers (*Gentiana *species) are scattered across Mt Pulag's summit grassland.*

*Above: A very rare plant, the Yew-leaved Rhododendron (*Rhododendron taxifolium*) spotted in Mt Pulag's montane rainforest.*

Below: A rare endemic plant, Loheria bracteata*, actually a type of primula, found in Mt Pulag's mossy rainforest.*

*Below: A wild begonia (*Begonia *species), one of a number of rainforest wild antecedents to the begonia houseplant.*

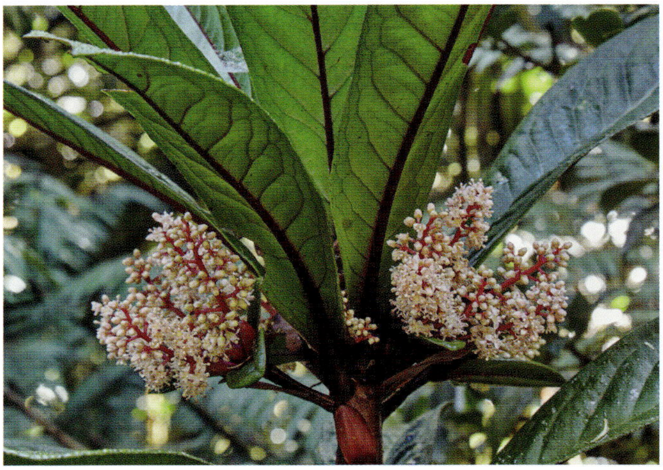

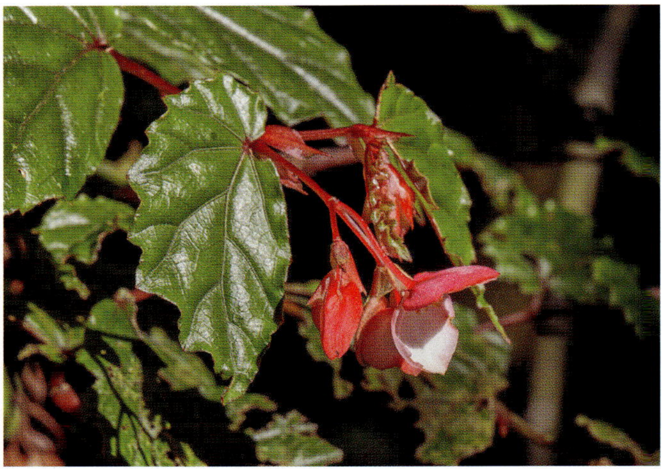

A wealth of wildlife

Studies conducted as long ago as 1910 identified more than 500 plant species on Mt Pulag, but since then the count has grown to a staggering 800-plus. There are over 140 moss species alone, as well as a wide range of orchids and ferns, many of them endemic to Luzon, and a few probably endemic simply to Mt Pulag.

In terms of trees, the pine forests – as already mentioned – consist largely of the Benguet Pine, while the montane forests contain a variety of stone oak species, particularly *Lithocarpus jordanae*, *L. luzoniensis* and *L. woodii*, plus a number of conifers, including a yew, *Taxus sumatrana*. Other trees and shrubs include a *Leptospermum*, a *Viburnum* and two *Medinilla* species.

At the highest altitudes there is a definite link with temperate Asia. This can be seen in the grasslands' dwarf bamboo, *Yushania niitakayamensis,* named after Mt Yushan, Taiwan's highest mountain, where it was first identified. This bamboo is common on the summits of several of Taiwan's mountains, but in the Philippines may be restricted to Mt Pulag. Also on the grassland can be found a tiny gentian, undoubtedly related to Asian temperate gentian species. Most strikingly, several rhododendrons, shrubs more usually associated with the temperate regions of Asia, can be found in Pulag's montane forest.

Of the animals, the mammals and birds have been the most closely studied. At least 22 mammal species have been found in the national park (though more are still being discovered), ranging from the Philippine Brown Deer (*Rusa marianna*) and the Philippine Warty Pig (*Sus philippensis*), to a number of insectivorous and fruit bats, and quite a diversity of rodents. The last of these include several species of earthworm mice, so named because of their diet. There are also a number of cloud rats, including the Luzon Bushy-tailed Cloud Rat (*Crateromys schadenbergi*), Greater Dwarf Cloud Rat (*Carpomys melanurus*) and Northern Luzon Giant Cloud Rat (*Phloeomys pallidus*).

The national park's birds have been extensively studied since the 1990s. To date, 76 species have been found, of which 30 are endemic to the Philippines, making Pulag a home to almost a third of all the country's endemic bird species. Not surprisingly, it is particularly important for high-altitude montane forest specialists, such as the Luzon Water-redstart (*Phoenicurus bicolor*), Flame-breasted Fruit-dove (*Ramphiculus marchei*) and Whiskered Pitta (*Erythropitta kochi*), but the pine forests are also home to a subspecies of the Red Crossbill (*Loxia curvirostra luzonensis*), a Luzon mountain endemic.

Threats to the park

Mt Pulag National Park is under huge pressure from surrounding human communities, especially due to constantly expanding agriculture. Much of the lower pine forest has already been lost, cut down for building timber and to make way for farming. Even in some areas where the pines survive, crops have nevertheless been planted among the trees. The farms continue to creep inexorably up the mountainsides, and in recent years have started to eat into the montane forest, well within the park's boundaries.

A second major pressure is from tourism. There are several good hiking trails leading to Pulag's summit, and in recent years one of these has become hugely popular. This is the Babadak Trail, leading up from the park's ranger station at the village of Babadak, already at an altitude of over 2,400 m (7,900 ft), making it a relatively short hike of about 9 km (5.6 miles). The sheer number of people hiking up this trail is putting a strain on both services and the environment, the latter particularly due to the threat of fire from campers' stoves.

Right: Although Mt Pulag's Benguet Pines grow mostly at lower altitudes, a few lone individuals have managed to colonize the summit grassland at an altitude of more than 2,700 m (8,880 ft).

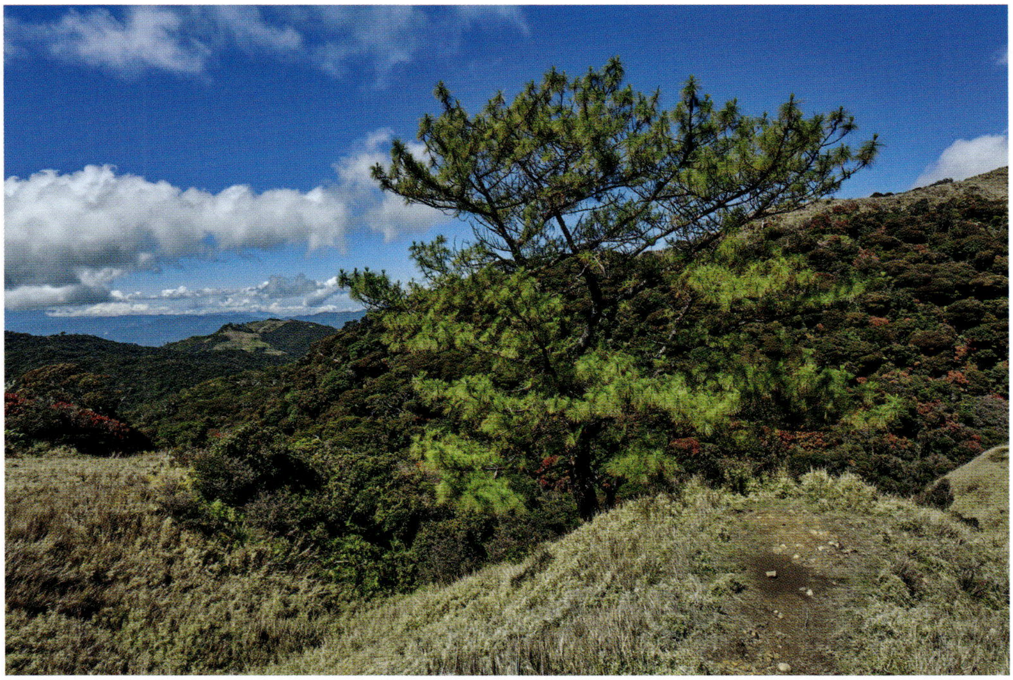

Subic Watershed Forest Reserve

Barely 130 km (80 miles) north-west of Manila sits Subic Bay, until 1992 site of the USA's largest overseas naval base. When the Americans pulled out, the base was converted to an international trade and industrial zone, Subic Bay Freeport Zone. What is of interest to conservation is that the entire place is wrapped up with 10,000 ha (24,700 acres) of lowland dipterocarp rainforest, some of the last forest of its kind on Luzon's west coast.

Throughout the years that the American's held Subic Bay, the forest was protected as a security shield, a vital freshwater source and a training ground for jungle survival. Today, it is a vital piece of lowland rainforest, running contiguously with the forests of Bataan Natural Park beyond Subic's boundaries, a protected area that covers much of the Bataan peninsula. Though the two forests were and still are managed separately, in the 1990s the combined Subic-Bataan Natural Park was one of the 10 priority areas included in the World Bank Global Environment Fund's CPPAP project.

Although the forest is under immense pressure from its industrial and commercial neighbours, much

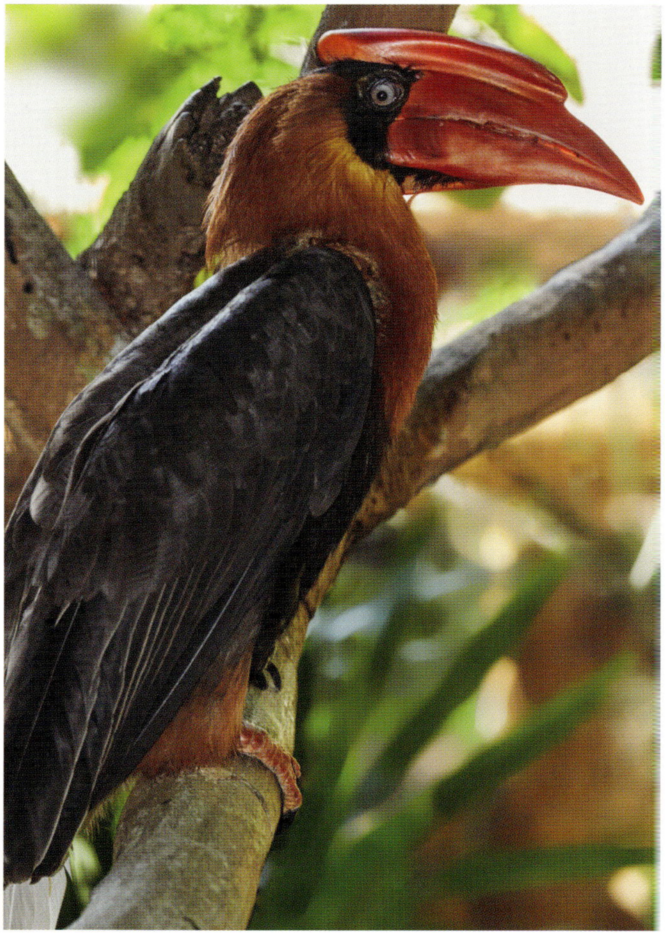

Above: The Northern Rufous Hornbill, though increasingly rare in Luzon, can still be seen in Subic's forests.

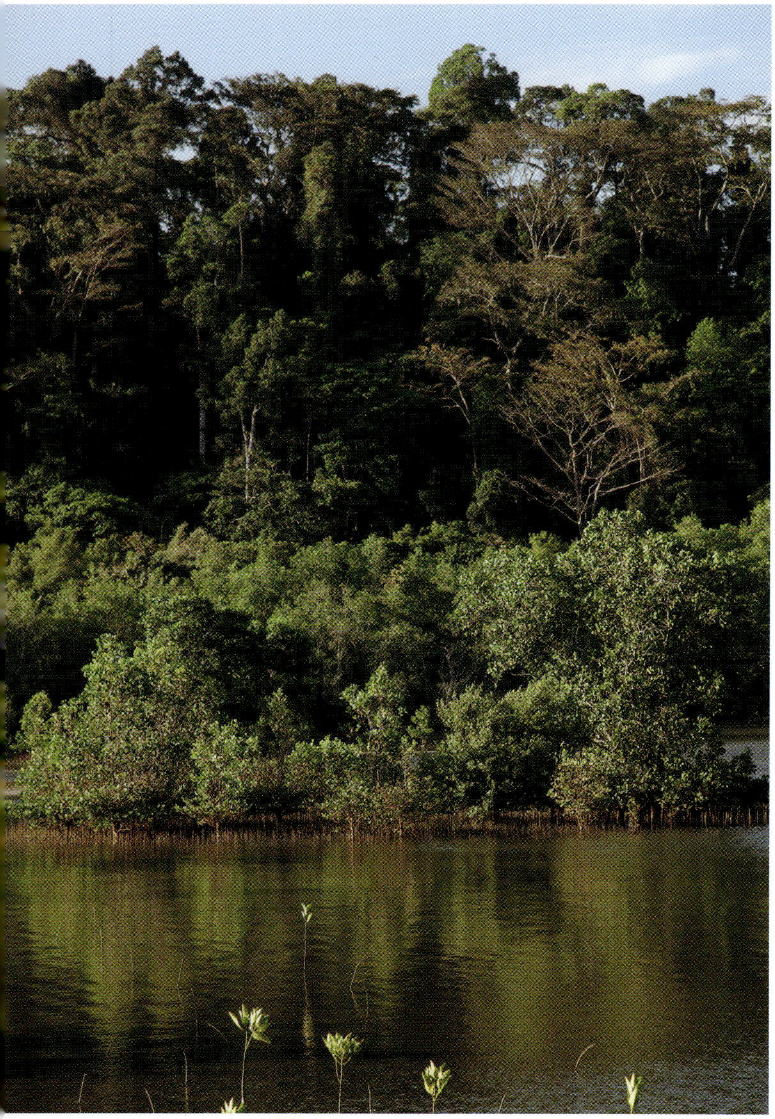

Left: Lowland rainforest along the mangrove-lined shore in Triboa Bay, an increasingly rare example in Luzon of an intact progression from mangrove to terrestrial forest.

of it continues to be in good condition, providing a rare glimpse of what this environment must have been like right across Luzon a century ago. One of the most important features is its habitat continuity, shoreline mangroves passing into beach forest, which then gives way to dense and extensive lowland forest.

Subic's wildlife

Though in good condition, the forest is criss-crossed by a network of roads, created by the American military. Today, those roads provide an easy access to the heart of the forest. As a result, this is perhaps the most popular site for birdwatchers in the Philippines, the roads providing not only access but also clearings that help to make bird life that bit more visible.

Numerous bird species can be found here, ranging from the common Philippine Bulbul (*Hypsipetes philippinus*), Dollarbird (*Eurystomus orientalis*) and White-bellied Sea-eagle (*Haliaeetus leucogaster*), to rarities like the Green Racquet-tail (*Prioniturus luconensis*), Blue-naped Parrot (*Tanygnathus lucionensis*), Luzon Bleeding-heart Pigeon (*Gallicolumba luzonica*), Colasisi (*Loriculus philippensis*) and Northern Rufous Hornbill (*Buceros hydrocorax*). The mangrove areas are a good place to see the endemic Philippine Duck (*Anas luzonica*).

Subic's forests are also home to a number of mammal species, including one of the last really large colonies of fruit bats, which roost during the daytime

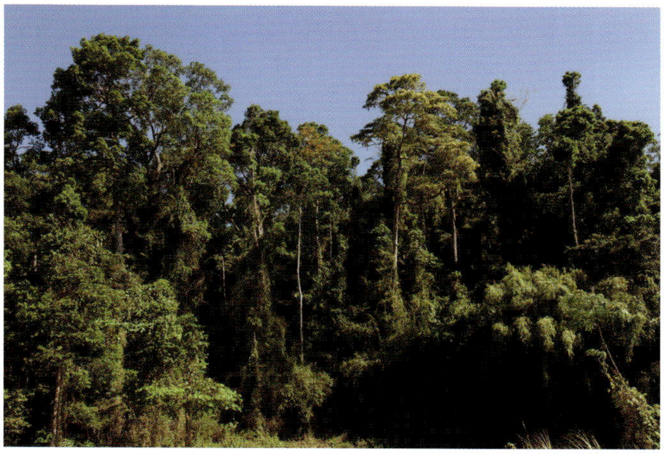

Above: *Lowland rainforest alongside a clearing, crowded with tall dipterocarp trees, a reminder of how much of Luzon once looked.*

in a collection of trees close to the Cubi residential district. Numbering over 20,000 bats, such large colonies were once common across Philippine lowland forests but are now rather rare, making this colony quite special. The colony consists of two species, the Philippine Giant Fruit Bat (*Pteropus vampyrus lanensis*) and the Golden-crowned Fiying Fox. The former is widespread across Southeast Asia, but the latter is endemic to the Philippines and appears to be in serious decline. Fortunately, in Subic these fruit bats have become star attractions, making their protection something of a priority, even though studies into their feeding needs have only recently been undertaken.

Below: *The Philippine Duck can regularly be seen in many of Subic's mangroves and marshes.*

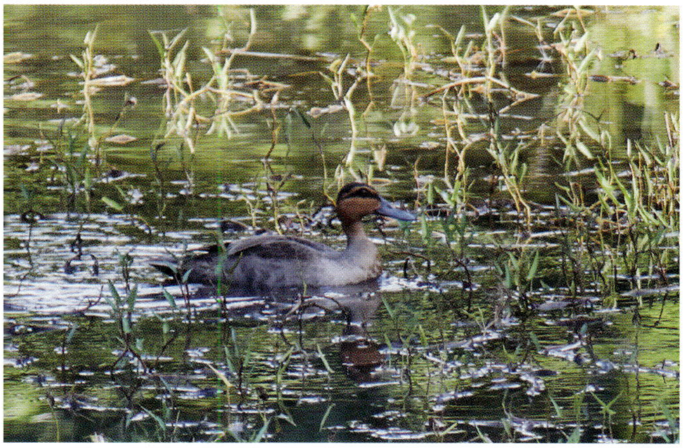

Below: *The Long-tailed Macaque is very common in Subic forests, and is often seen in family groups along the roadsides.*

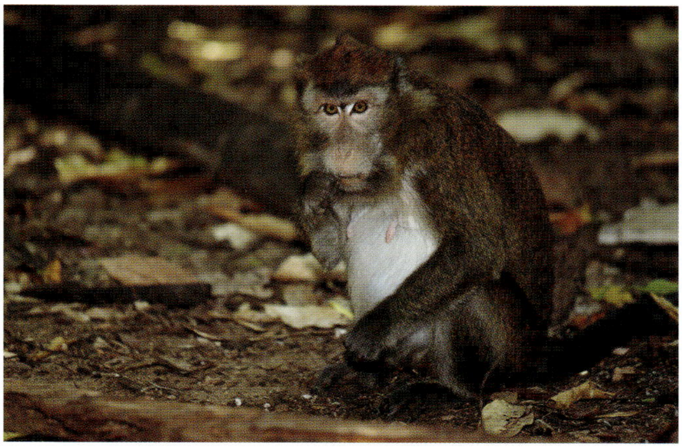

Mt Makiling Forest Reserve

Mt Makiling is a dormant volcano sitting close to the southern shore of Laguna de Bay, the Philippines' largest lake, and barely 60 km (37 miles) south of Manila. Much of the mountain has been protected in one form or another since 1910, largely as a living laboratory and forestry training ground. A campus of the University of the Philippines sits to its north-east, on the edge of the town of Los Baños, which – as its name implies – is well known for its hot springs.

In 1933 Makiling became the Philippines' first national park, though in 1963 it reverted back to being a forest reserve. In 1990 responsibility for management of the entire reserve was handed to the University of the Philippines, and in 2013 it was awarded Association of Southeast Asian Nations (ASEAN) Heritage Park status, confirming Makiling as one of the most important centres for wildlife conservation in Southeast Asia. Today, the Mt Makiling Forest Reserve is officially part of the Los Baños campus, protected very much as a living laboratory for studies into Philippine fauna and flora.

As a result of such prolonged and extensive protection, most of the reserve's 4,244 ha (10,483 acres) is densely forested, with some fine dipterocarp lowland rainforest on its lower slopes, grading into montane forest once above about 600 m (1,970 ft), with some upper montane forest around its three summits. Named rather unimaginatively Peaks I, II and III, all three are a little over 1,000 m (3,300 ft) high, with the highest, Peak III, at 1,109 m (3,637 ft). Together, the three peaks surround what is left of the volcano's crater.

Though well forested, Mt Makiling is not pristine. Like most Philippine protected areas, it suffers from the problem of constant erosion of its edges, both for conversion to farmland and for housing development. These problems are particularly serious around the reserve's western and southern edges, furthest from the university's watchful gaze. Within the reserve, some areas of early forest damage have been restored using foreign species, principally the Large-leaved Mahogany, the result of early experiments in reforestation, but which are not helpful to Philippine native wildlife.

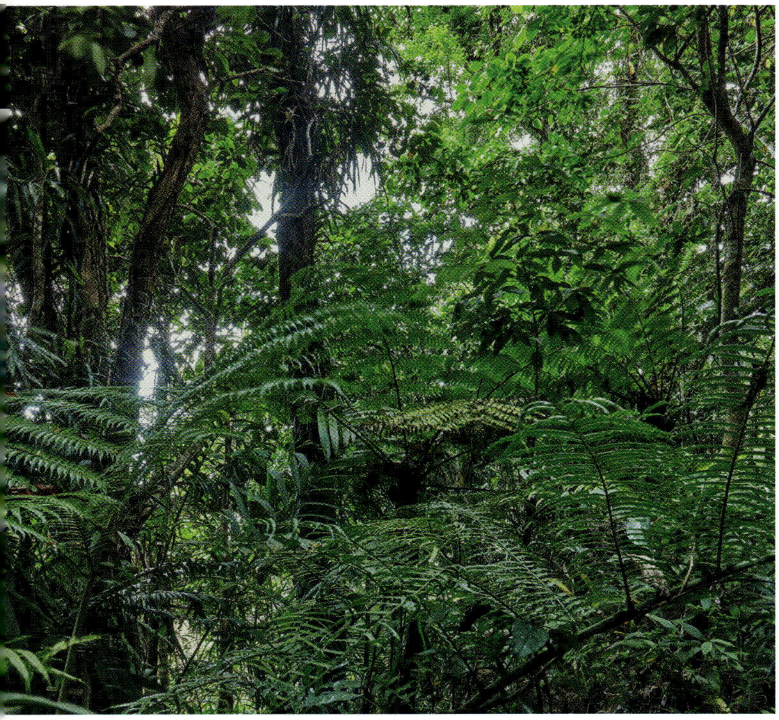

Makiling's flora and fauna

Not surprisingly, Mt Makiling is probably the most closely studied of all the Philippines' protected areas. Much of the early knowledge about the country's incredible biodiversity came from work carried out on this mountain, giving Makiling a very special place in any understanding of the Philippines' biodiversity.

For some time now it has been known that this one reserve contains more than 2,000 species of higher plant, more than the entire flora for many European countries. In addition, nearly 300 fern species and more than 200 moss species have been found here. As well as a number of large dipterocarp tree species, there is also a plethora of small, understorey or ground plants, including several Philippine rarities such as the

Left: *Dense vegetation, with a mix of lianas, tree ferns and vines, typifies Mt Makiling's transition zone from lowland to montane forest.*

Opposite: *Forest crowds in around the Flatrocks area of Mt Makiling.*

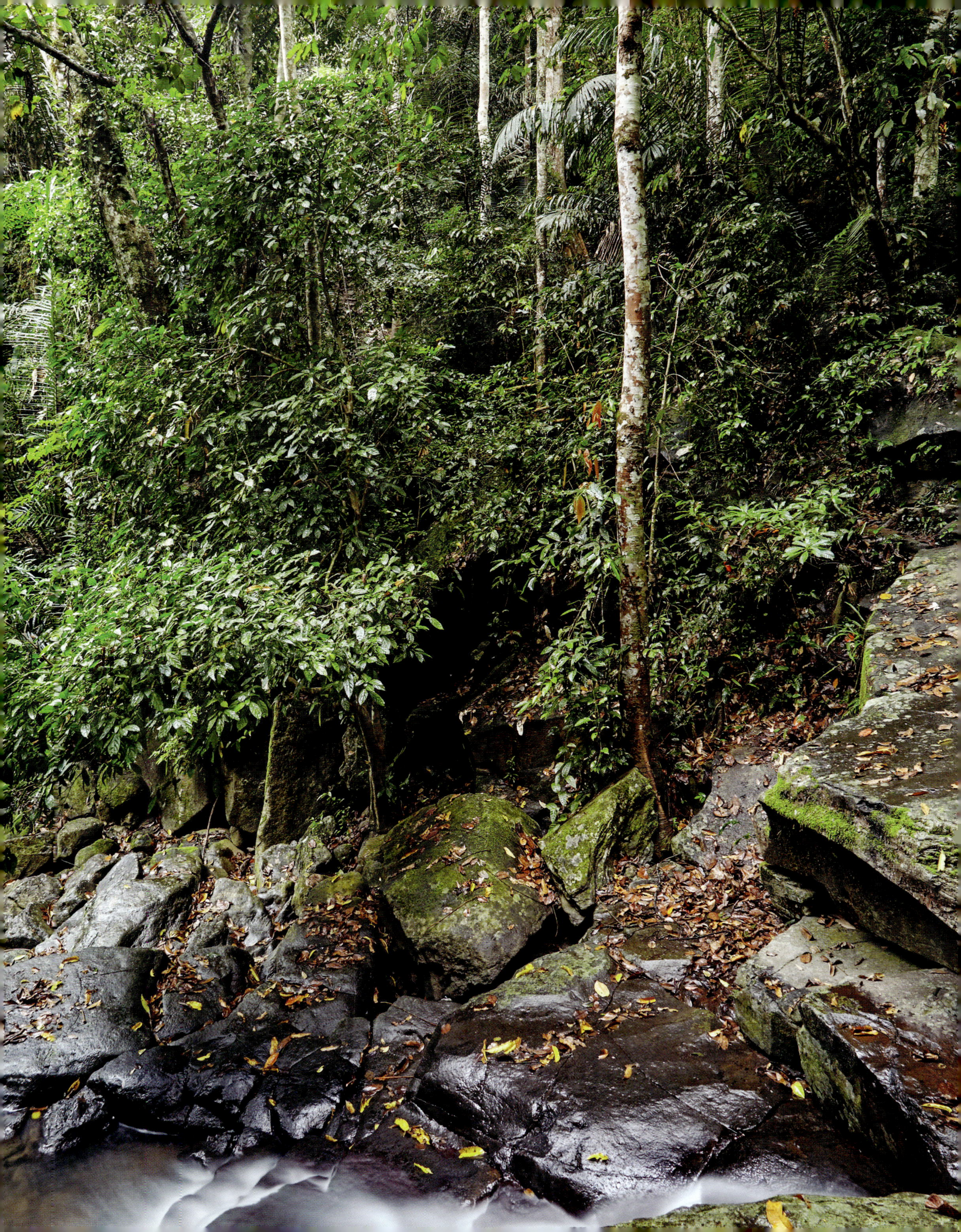

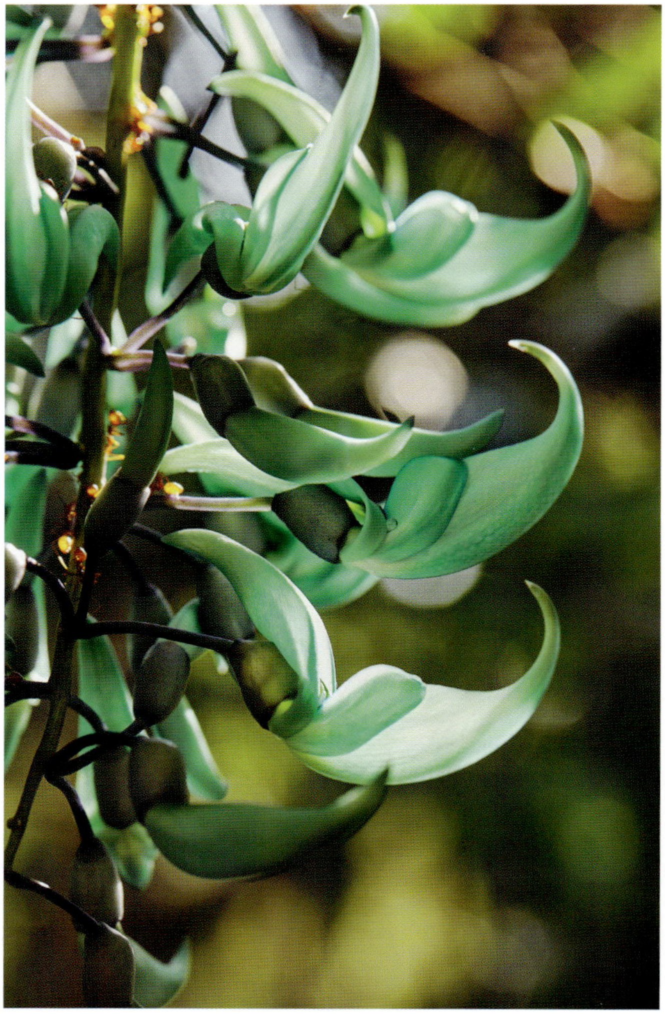

Above: The magnificent Jade Vine is a Philippine endemic that grows on Mt Makiling.

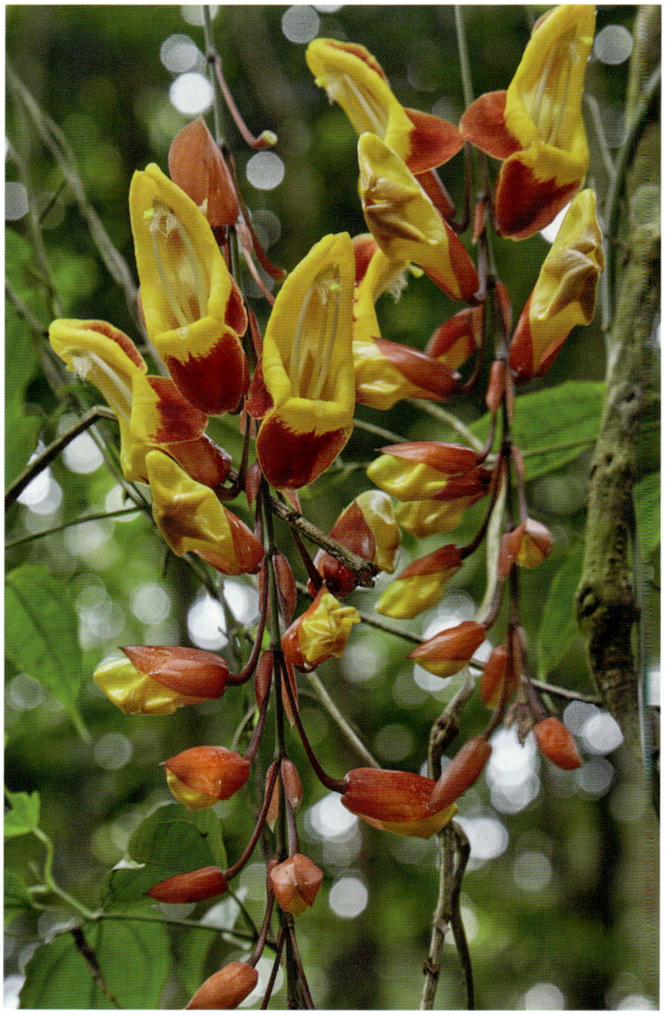

*Above: The Indian Clock Vine (*Thunbergia mysoriensis*), a foreign invasive species taking hold in Mt Makiling's lowland forest.*

Kapa-kapa (*Medinilla magnifica*), an epiphytic shrub that produces spectacular red and pink flower clusters (see page 135). There is also the stunningly beautiful Jade Vine (*Strongylodon macrobotrys*), a climber that is normally rather inconspicuous until it bursts into flower, hanging streams of bright jade-green flowers. Then there is at least one *Rafflesia* species, *R. lagascae*, the least rare of the Philippine rafflesias (see page 136).

Ficus trees are quite common, their fruits vital to fruit bats, including the Philippine Giant Fruit Bat (*Pteropus vampyrus*), Golden-crowned Flying Fox and Pygmy Fruit Bat (*Otopteropus cartilagonodus*). Other mammals include a number of endemic rodents, such as the Northern Luzon Giant Cloud Rat, as well as the more widely known Long-tailed macaque (*Macaca fascicularis*), Philippine Brown Deer and Philippine Warty Pig.

Birds are probably, from the public's perspective at least, the star attraction on Mt Makiling, with nearly 200 species identifed here, most of them forest specialists and many endemic to the Philippines or even just Luzon. An example of the latter is the Luzon Bleeding-heart Pigeon, while rather rare Philippine endemic species that can sometimes be seen include the Philippine Falconet (*Microhierax erythrogenys*) and Philippine Eagle-owl (*Bubo philippensis*). The endemic Luzon Hornbill (*Penelopides manillae*) is surprisingly common in the lowland forests, its mocking call regularly teasing birdwatchers while it remains remarkably well hidden.

Ecotourism on Mt Makiling

Being so close to Manila and quite accessible, Makiling sees a lot of visitors, ranging from local school parties to international birdwatchers. Enjoyment of the forest via the university campus has been greatly facilitated by the Mt Makiling Botanic Gardens, created at the point where campus and forest meet. Though containing some formal arrangements, including an arboretum of dipterocarp trees and a display of native Philippine plants, much of the garden consists largely of natural lowland dipterocarp forest through which good-quality paths have been laid to ease the exploration.

For anyone wanting to penetrate further into the forest reserve from the campus, there is a trail all the way to the summit, called the Mariang Makiling Trail. For the first few kilometres this is in fact a road, cut in the 1970s as part of early plans to build a geothermal energy plant, though this eventually disintegrates into a path that continues to the summit. A kilometre or two into the forest a side path leads off the road to an area called Flatrocks, a set of boulders in the midst of a very steep section of river, creating a natural clearing in the forest. Higher up the road another side path leads to Mudspring, an area of bubbling, boiling mud in the midst of the forest, a reminder that though dormant, this volcano is certainly not dead. In recent years the spring has been expanding, leading to concerns that Makiling's volcanic activity might be on the increase.

Finally, along with Subic Watershed Forest, Mt Makiling is considered to be one of the finest places in the Philippines for birdwatching. Not only has the habitat and its bird population been well protected for many years, but the clearings created by the road and the streams provide areas in which birds are more visible than they would otherwise be. Moreover, the campus itself is known to be a good place for birdwatching due to its many, though relatively dispersed trees.

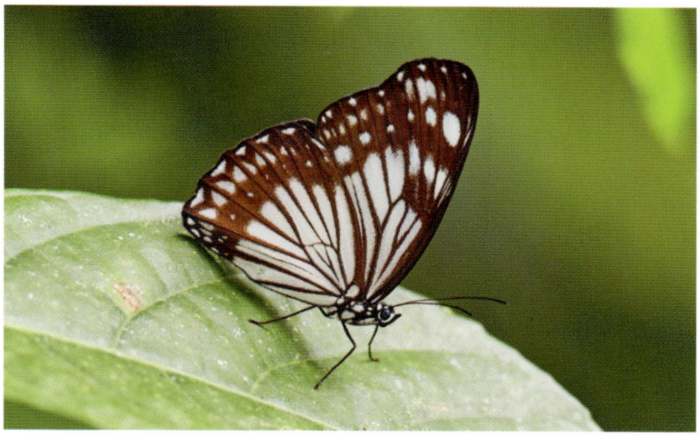

Above: *A Philippine Brown Butterfly (*Zethera pimplea pimplea*) in Mt Makiling's lowland rainforest.*

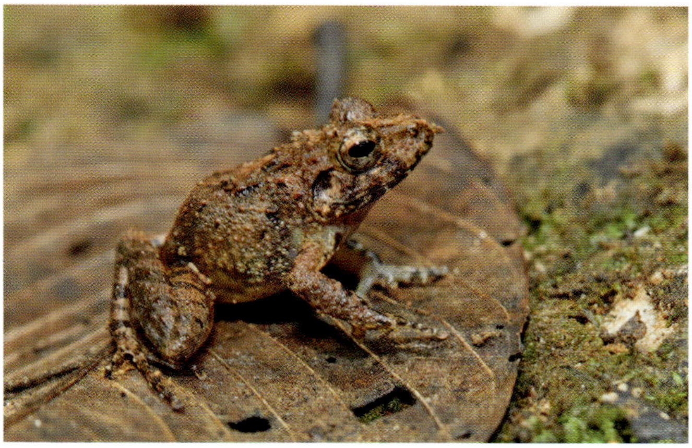

Above: *The Diminutive Forest Frog (*Platymantis mimulus*), a Philippine endemic species, at Mudspring, on Mt Makiling.*

Below: *The steaming, boiling pool that is the volcanic Mudspring, in rainforest on Mt Makiling.*

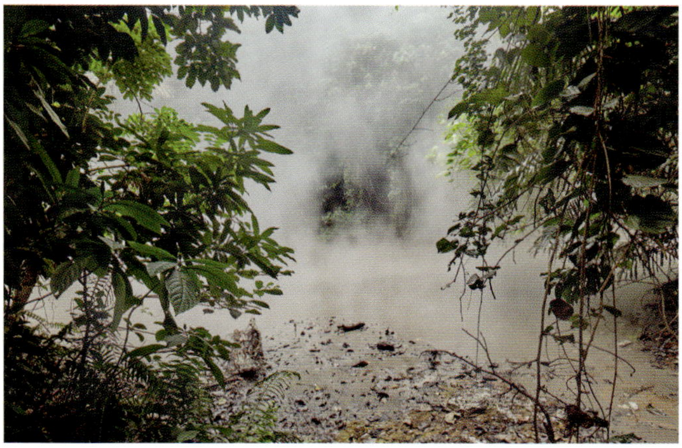

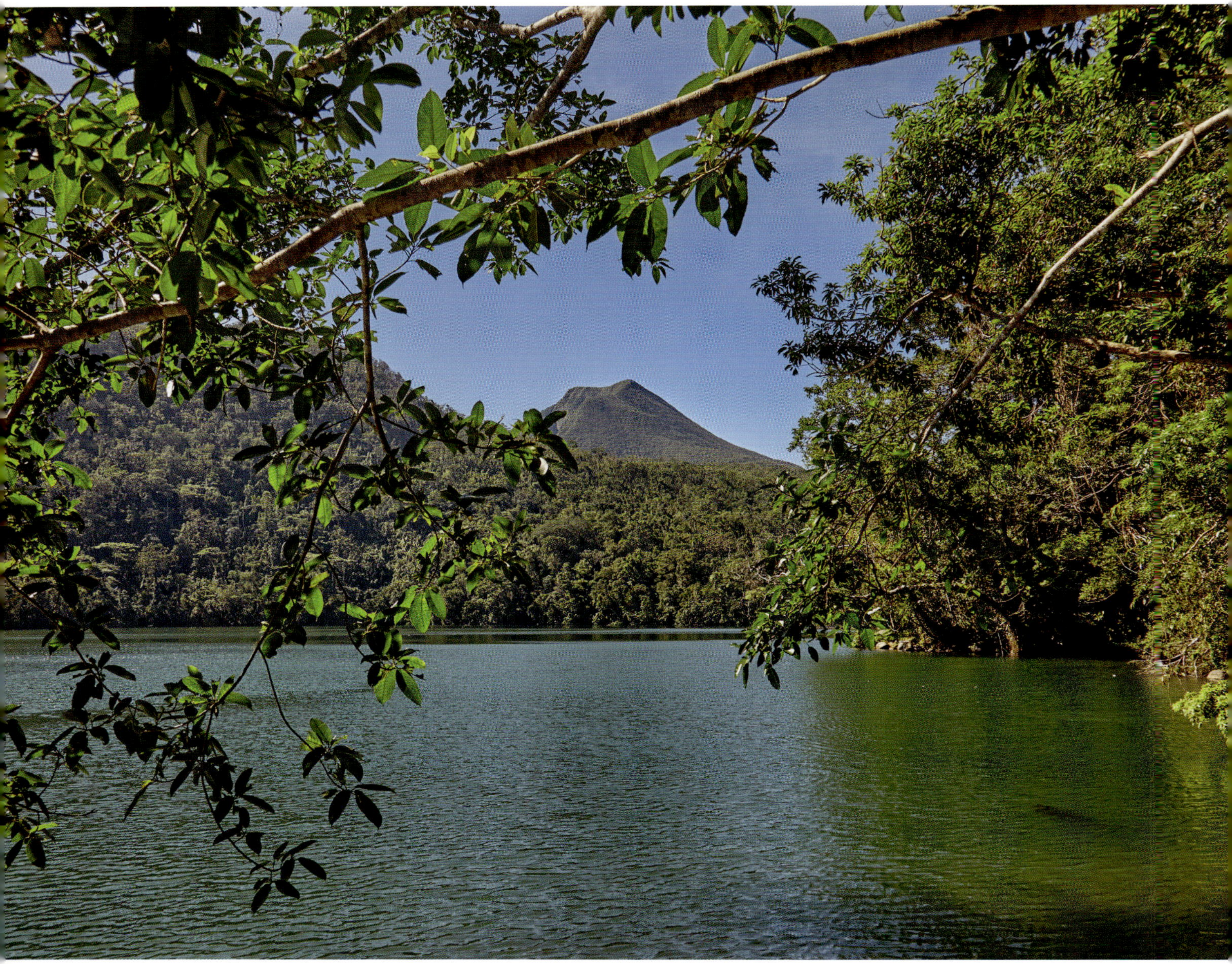

Bulusan Volcano Natural Park

This natural park protects an area of rainforest in southern Bicol, close to the southernmost tip of Luzon. As the name suggests, the park encompasses a volcano called Mt Bulusan, one of the Philippines' most active volcanoes, closely monitored by the PHIVOLCS. It is the southernmost volcano in the Bicol Volcanic Chain, which consists of a number of inactive and three active volcanoes, including Bulusan. The other two active volcanoes are Mts Iraga and Mayon.

Covering an area of 3,673 ha (9,080 acres), the park is relatively small, but much of the forest within it is still in quite good condition, consisting mainly of lowland rainforest, with some montane forests on the upper slopes. The highest slopes are covered mainly with grassland and bare rock, areas where poor soil and frequent volcanic activity prevent tree growth. The mountain consists of three peaks: Bulusan itself, which holds the active crater, plus Sharp Peak and Mt Hormahan. Between the peaks sits a poorly drained area, flooded for much of the year and called Lake Aguingay.

Due to its ongoing volcanic activity, much of

Opposite: *The lovely Bulusan Lake, surrounded by healthy lowland rainforest, with Sharp Peak, one of Mt Bulusan's three peaks, in the background.*

Right: *The Philippine Sailfin Water Lizard is one of the many Philippine endemic reptiles found in the forest around Mt Bulusan.*

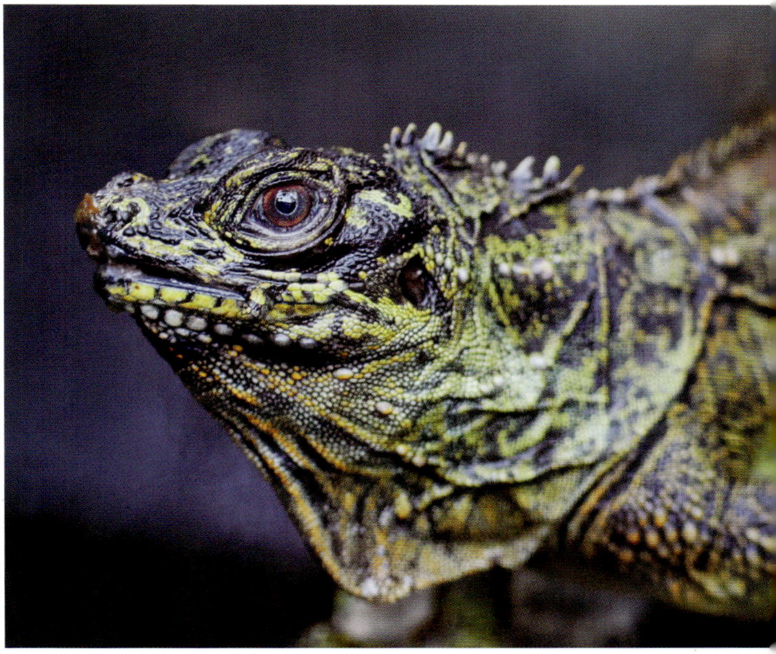

Bulusan is closed to both hikers and researchers most of the time. However, its lower slopes, particularly around the lovely Bulusan Lake, are usually open, and it is in this area that a few studies on the park's wildlife have been carried out. The lake sits to the south of the mountain's summits, at an altitude of about 600 m (1,970 ft). It is easily reached by a motorable road from the small town of Irosin to the south, and there is an excellent trail right around the lake, giving not only fine views of Sharp Peak, but also a chance to experience some lowland rainforest. Unfortunately, the smouldering Bulusan summit is not visible from here, but it can be seen from the main road, which runs north–south just west of the mountain.

The forest and its wildlife

Although the forest around Bulusan Lake is disturbed, it is nevertheless very attractive and in good condition. However, being at a rather elevated altitude the trees here do not reach the huge size commonly seen in the forest at Subic Bay, but the forest is rich in a variety of tree ferns.

It is probable that most of the larger mammals, such as the Philippine Brown Deer and Philippine Warty Pig have been hunted to extinction here, but it is believed that the endangered Southern Luzon Giant Cloud Rat (*Phloeomys cumingi*) still survives. Early morning and late afternoon are good times for birdwatching around the lake. Although the forest is quite dense in places a good range of birds can be seen, including some endemics.

Even a casual observer will quickly realize that the forest around Bulusan Lake is very rich in reptiles and amphibians, with skinks and other lizards regularly seen basking on logs or scurrying away into the undergrowth. This has been confirmed by scientists: one study conducted in 2015 and 2016 found eight species of frog, 10 lizards and eight snakes, including the Spiny Waterside Skink (*Tropidophorus grayi*), Philippine Pit Viper (*Trimeresurus flavomaculatus*) and Rough-backed Forest Frog (*Platymantis corrugatus*). The endangered Philippine Sailfin Water Lizard (*Hydrosaurus pustulatus*) is also known to live here.

Below: *The Philippine Pit Viper, an endemic snake commonly found in Bulusan's rainforest.*

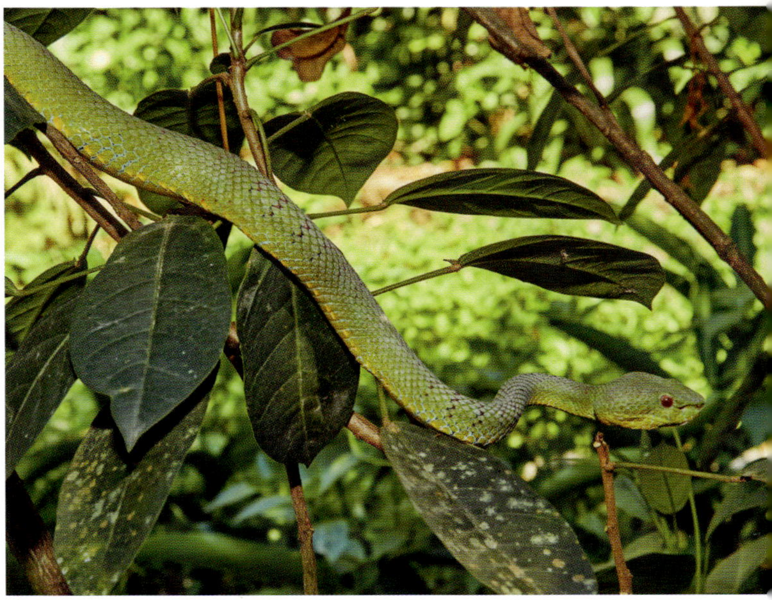

Puerto Galera

Sitting close to the northernmost tip of Mindoro, just 130 km (80 miles) south of Manila, and reached by ferry from the Luzon port of Batangas to the north-east, Puerto Galera is one of the Philippines' best-known and most popular tourism destinations. Visitors pour in for its sandy beaches and superb scuba diving, not really for its wildlife value, though in fact the area is hugely important to the conservation of Mindoro's natural environment. Not only is Puerto Galera blessed with some of the Philippines' most beautiful, most varied and best-preserved coral reefs – to which the town owes most of its fame – but also inland it is backed by mountains that still host some reasonably good rainforest. Mts Malasimbo and Talipanan tower over the town and its western beaches, their upper slopes still green with trees.

It is the coral reefs that get all the attention, however. Puerto Galera town sits at the innermost end of a beautiful natural harbour, a very enclosed bay created by a peninsula to the east and a cluster of islands to the west. The coral reefs are strung out along the outer seaward end of the peninsula, as well as along the seaward shores of the islands. This string of reefs contains a variety of habitats, ranging from shallow, sheltered coral gardens mainly in the inter-island channels, to steep walls and drop-offs exposed to the sea and the strong currents of nearby Verde Island Passage (which separates Mindoro from Luzon). With such a variety of habitats, these reefs are home to

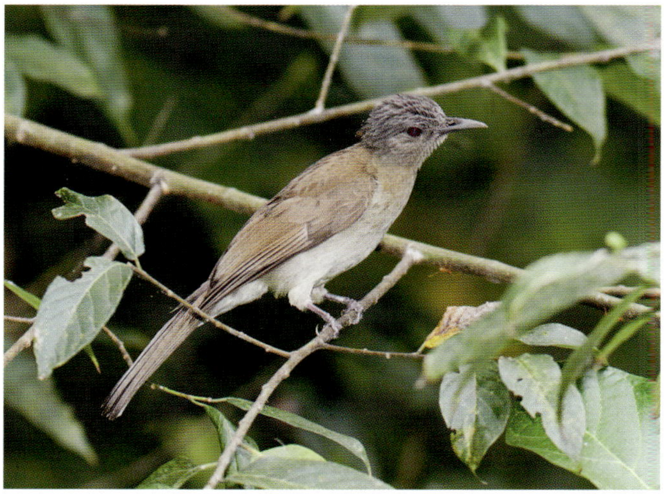

Above: *The Mindoro form of the Philippine Bulbul in a patch of remnant forest near Sabang. This is a widespread endemic bird, though with several forms.*

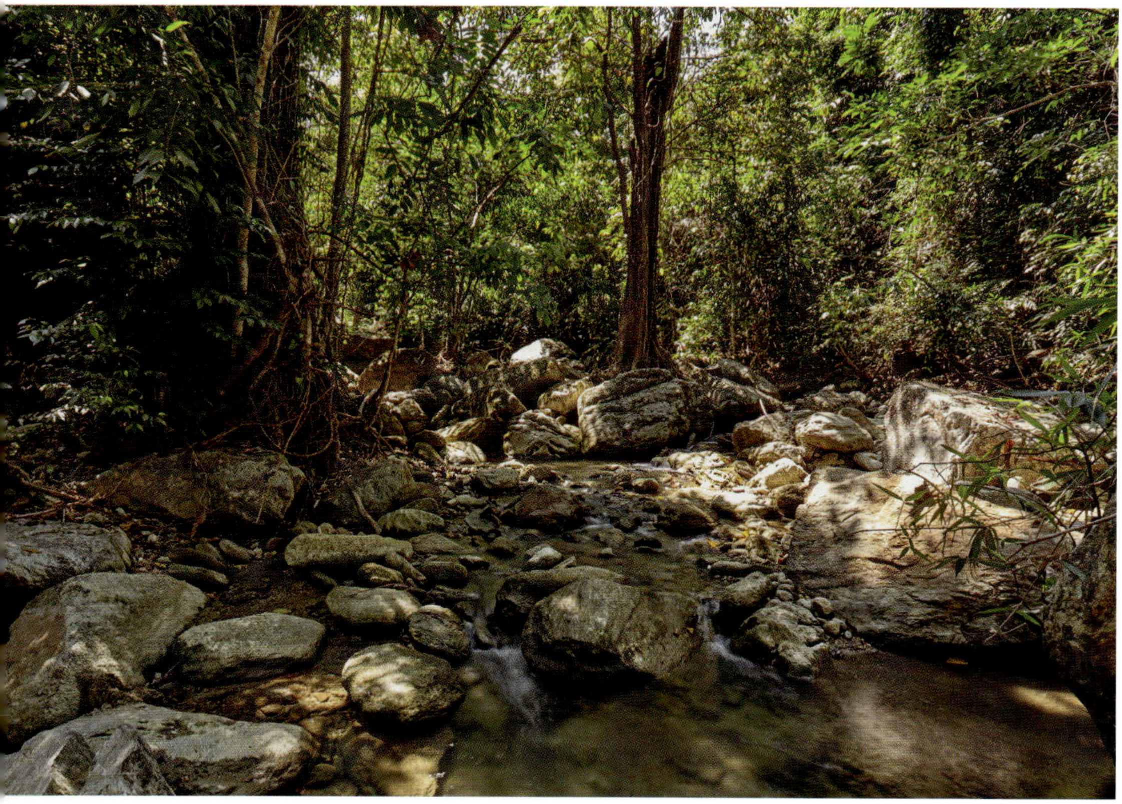

Left: *Tangled vegetation in dense secondary lowland rainforest, on the lower slopes of Mt Talipanan, near Talipanan village.*

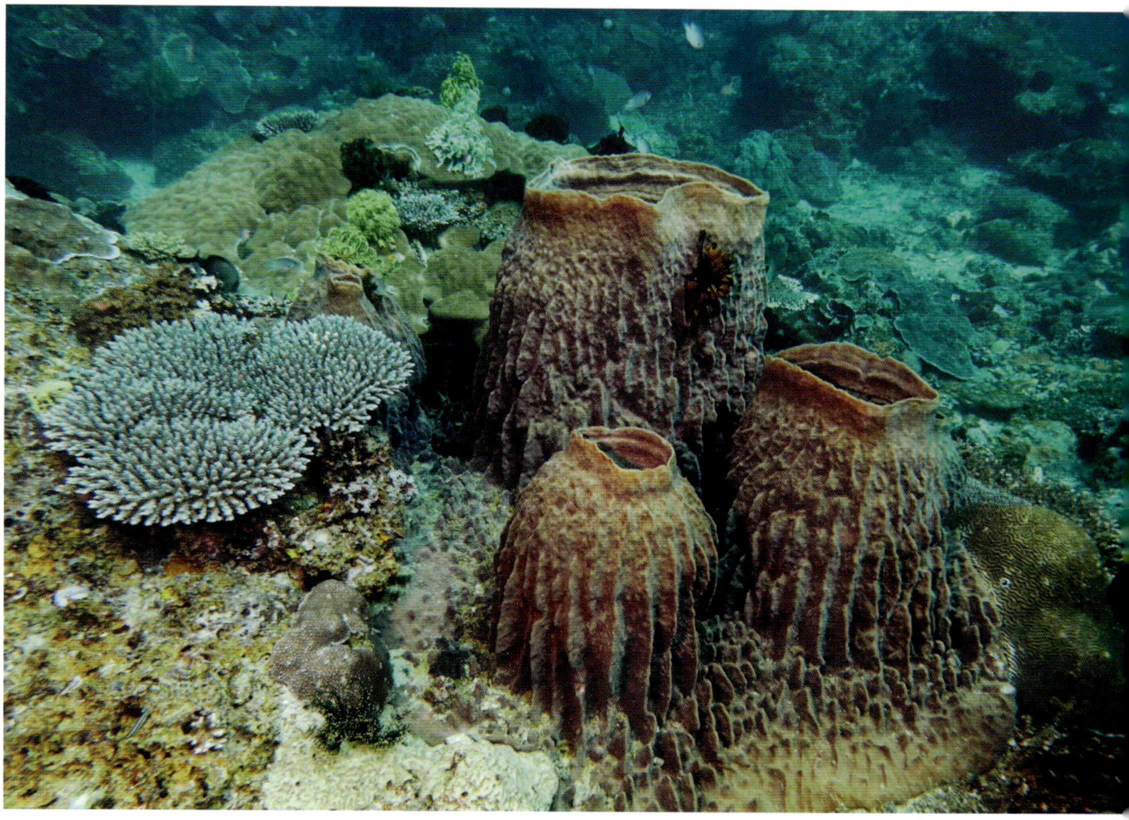

Right: *A group of barrel sponges (*Xestospongia* species), the largest more than 1 m (3.3 ft) tall, with an* Acropora *species hard coral to the left, a classic Puerto Galera coral reef scene.*

Below: *A mushroom leather coral (*Sarcophyton *species), a very common soft coral across open areas of Puerto Galera's reefs.*

hundreds of coral types, ranging from *Acropora* species table and staghorn corals, especially in the shallow coral gardens, to the delicate filaments of *Dendronephthya* species soft tree corals and gorgonian sea fans, which make their homes on the exposed steep slopes and walls.

Sponges (including huge barrel sponges; *Xestospongia* species), sea anemones (anthozoans), feather stars (crinoids) and reef fish abound, the last of these including such stunningly colourful delights as Moorish Idols (*Zanclus cornutus*), angelfish, damselfish and butterflyfish. The deeper and more exposed areas are also likely to include some of the larger fish, including barracuda (*Sphyraena* species), Rainbow Runners (*Elagatis bipinnulata*) and the occasional shark, mainly Whitetip Reef Sharks (*Triaenodon obesus*).

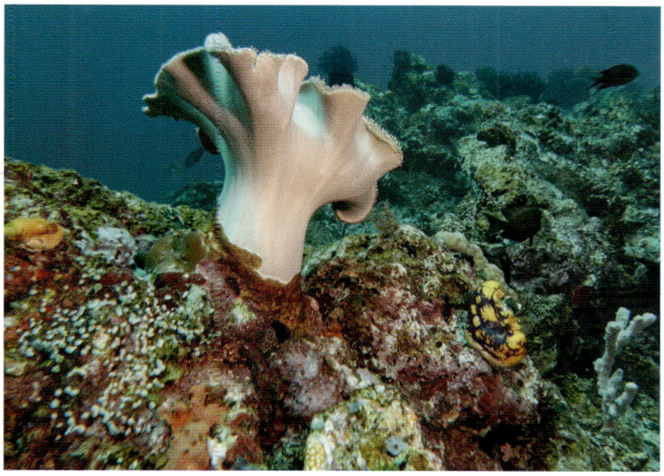

So important is Puerto Galera's marine and terrestrial environment that in the 1970s UNESCO awarded the area Man and the Biosphere (MAB) reserve status. Covering about 23,200 ha (57,304 acres), the reserve takes in not just the coral reefs but also some parts of Mt Malasimbo. Unfortunately, booming tourism, coupled with poor planning and control, has caused rampant resort and housing development that has damaged the shoreline, mangroves, seagrass beds and reefs, through both direct impact and untreated sewage pollution.

In 2014 UNESCO threatened to take away the MAB status, but it was reconfirmed in 2017 following approval of a government action plan. As of 2018, it looks as though implementation of that plan is starting.

El Nido–Taytay Managed Resource Protected Area

Taking in the island-studded Bacuit Bay and the town of El Nido, this is one of the Philippines' most beautiful locations. Sitting on the north-west coast of Palawan, facing the South China Sea, it is protected for its hugely varied and rich natural environment, both marine and terrestrial, and for the conservation of commercially important fish stocks. Its history as a protected area began in the early 1980s as a marine reserve aimed at protecting the bay's fish and marine turtles. In the early 1990s, El Nido's importance was increasingly recognized when it was included in a WWF-organized, Haribon Foundation-managed debt-for-nature swap. Between 1997 and 2001 it was one of eight protected areas included in the NIPAP. Since 1998, the site has been called the El Nido-Taytay Managed Resource Protected Area (ENTMRPA).

Landscape, seascape and habitats

The protected area covers 90,391 ha (223,266 acres), of which 36,018 ha (88,964 acres) are on land and 54,303 ha (134,128 acres) are marine. It encompasses the whole of Bacuit Bay, and all the surrounding land, taking in the entire El Nido region and part of neighbouring Taytay. One of the critical features of the ENTMRPA is that in many areas its habitats run the full range from open water, to coral reef, to seagrass and seaweed beds, to mangrove and beach forests and

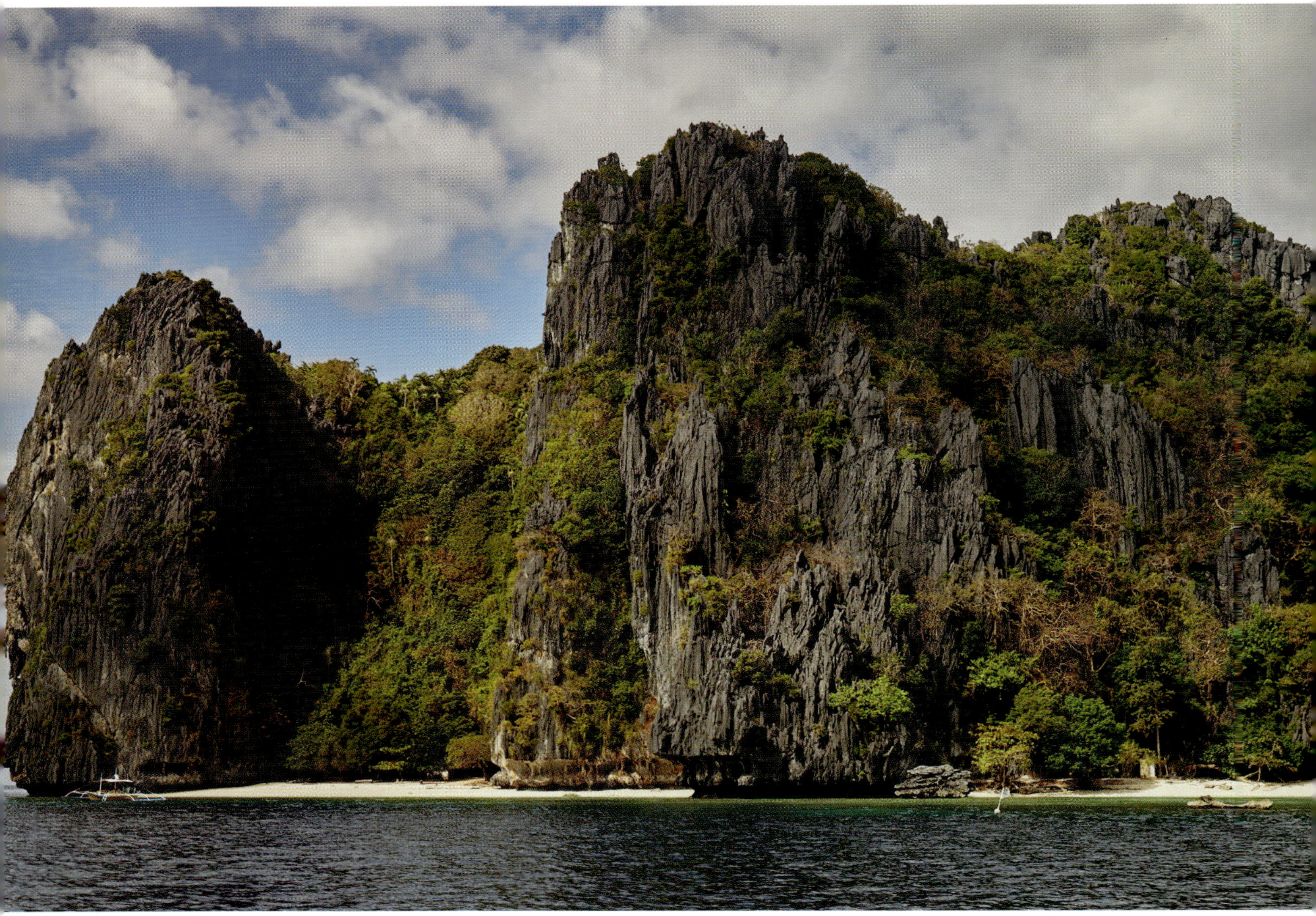

Right: *A lowland rainforest detail on Lagen Island, showing buttresses at the base of a large dipterocarp tree, surrounded by a tangle of verdant rattan palms (*Calamus *species).*

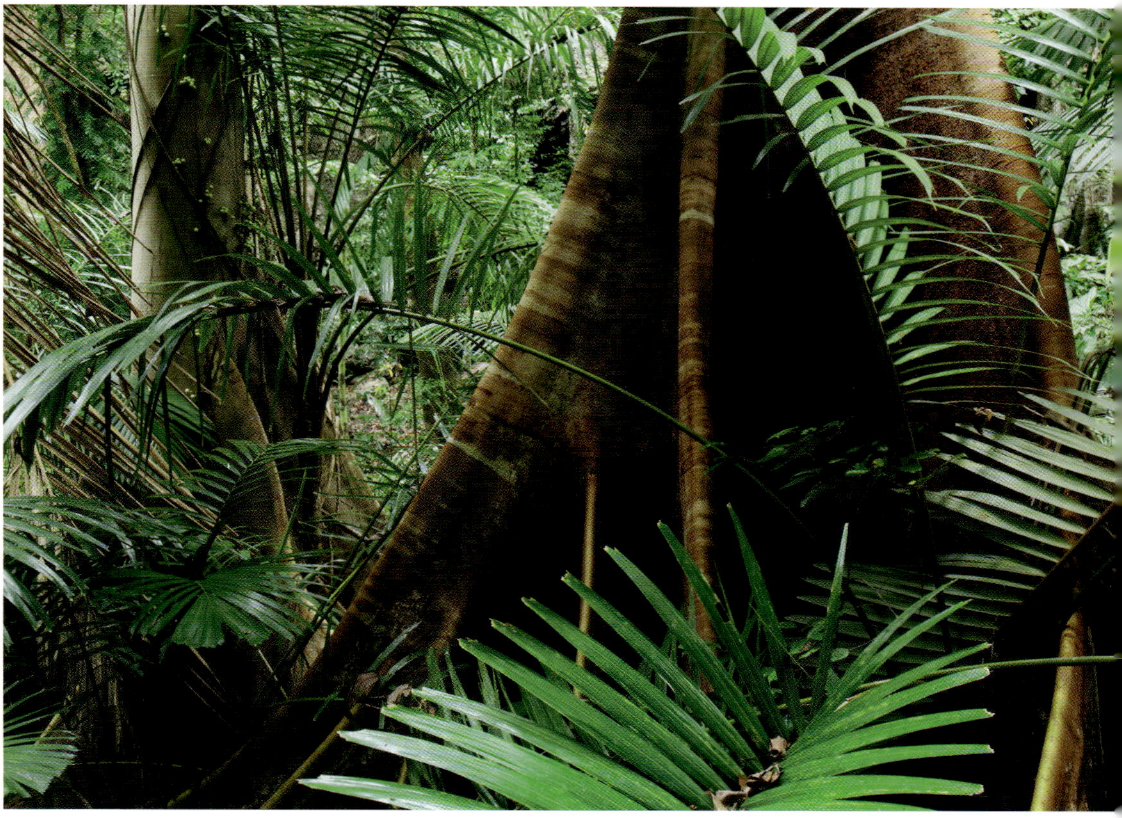

Opposite: *Lagen Island, one of Bacuit Bay's many islands, has a classic karst limestone landscape: jagged sheer cliffs, interspersed with pockets of limestone rainforest, mainly on the gentler, more fertile slopes.*

finally to lowland rainforest. With El Nido's highest point at 640 m (2,100 ft), on Cadlao Island, none of the land is high enough for montane rainforest.

Bacuit Bay contains some 45 islands, consisting mainly of jagged karst limestone cliffs interspersed with small white-sand beaches and mangroves. Much of the mainland is similar, though also with a number of coastal plains, river estuaries and extensive mangrove forests included. Taken together, it is a very beautiful scene that has been attracting rapidly increasing numbers of visitors since the 1980s, making El Nido now one of the country's main must-see places.

Many of the islands and parts of the mainland have associated coral reefs that include both shallow, sandy coral gardens, and steep drop-offs into relatively deep water. The limestone cliffs have also been eroded to create a number of highly sheltered lagoons that include many corals, and are places where turtles are commonly encountered.

The shallowest sandy or muddy areas are sites of extensive seagrass and seaweed beds, particularly around Cadlao Island, and along the mainland coast north of El Nido town, areas that are important to the young of many fish and shellfish, as well as to the local population of the Dugong (*Dugong dugon*).

Immediately along the shores mangroves are quite common. On the islands these are usually very small patches and narrow bands of trees, but along the mainland shore there are several rather large mangrove areas, particularly in the estuary of the Aberawan River, which sits in Bacuit Bay's south-east corner. This is quite a wild, primeval place, very good for wildlife watching.

On land there are four types of forest: lowland evergreen, lowland semi-deciduous, limestone and beach forests. The last of these is frequently found as a narrow band of highly specialized trees and shrubs growing along the landward edge of the beaches and some of the mangroves.

The lowland evergreen forest grows in areas where there are few or only gentle slopes, and where soils are relatively deep and water supply is consistently available, even during rainless periods. It consists of the larger tropical rainforest trees, including a range of

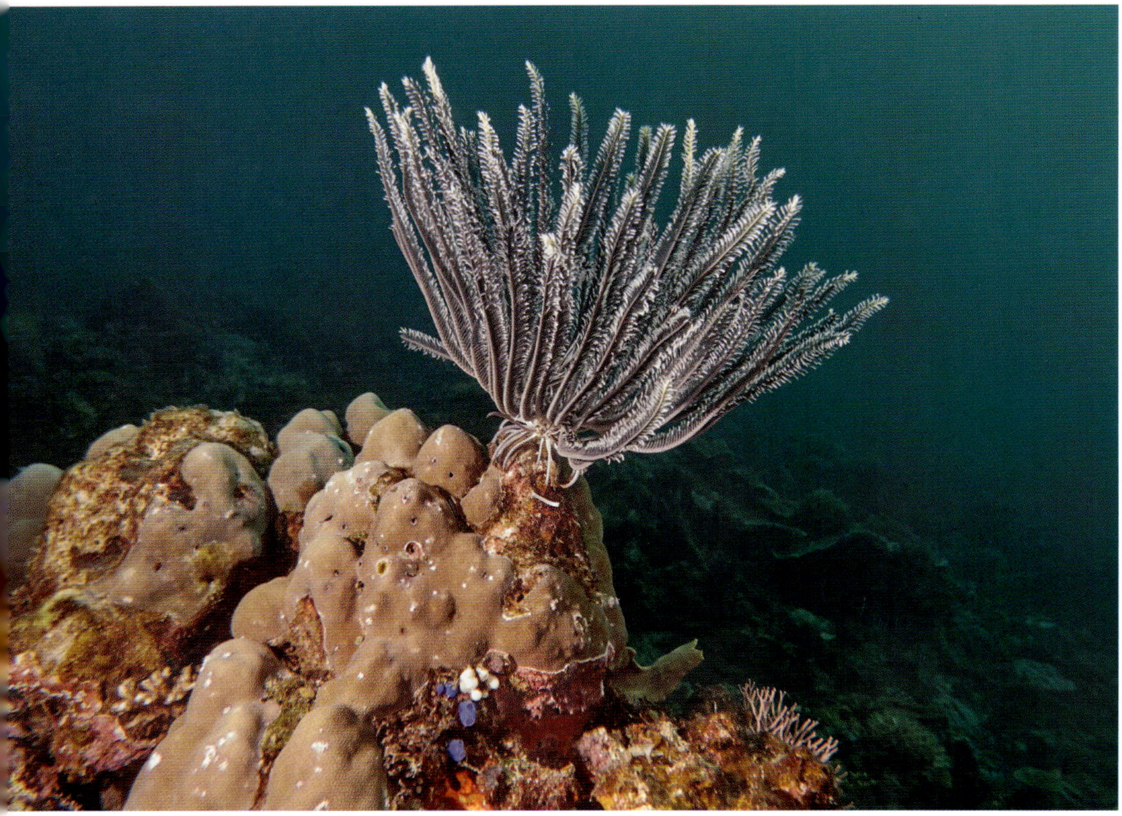

Left: *A magnificent feather star, or crinoid, species unknown, off Pangalusian Island, in Bacuit Bay.*

Opposite: *The Green Imperial Pigeon, a large pigeon that is commonly seen across El Nido and Bacuit Bay, mostly in forest.*

Below: *A Blue-spotted Stingray (*Neotrygon kuhlii*), an occasiona sight on Philippine reefs, here seen swimming across a sea bed littered with broken corals, off Entalula Island, in Bacuit Bay.*

dipterocarp trees, as well as a host of palms (including rattans), gingers and orchids. In El Nido, this evergreen forest includes, along one coastal strip, a few areas of floodplain wetland, in which the forest is very boggy and contains some bog specialists.

The semi-deciduous forest consists of plants that are able to adapt to the thinner soils and eratic water supplies, particularly during the dry season, which are typical of the gentler slopes on the limestone islands. The trees here cope with water shortage by being at least partially deciduous, reducing their water needs. The limestone forest consists of plants that are adapted to the dry conditions found on many of the karst limestone islands, where the soil is thin or almost non-existent, and slopes can be very steep. El Nido's forests are not pristine, those on the mainland having been logged over some time before protection arrived. Much of the forest has recovered, however, and is now good-quality secondary forest. On the islands, there was less logging, though they were heavily harvested for such plants as rattans. In recent years these have recovered and are once again healthy.

Biodiversity in El Nido

The wealth of plant and animal species is quite staggering. The coral reefs, for example, are believed to be home to an estimated 855 species of coral (both hard and soft) and reef fish. Four species of marine turtle – Green (*Chelonia mydas*), Hawksbill (*Eretmochelys imbricata*), Olive Ridley (*Lepidochelys olivacea*) and Leatherback (*Dermochelys coriacea*) – are

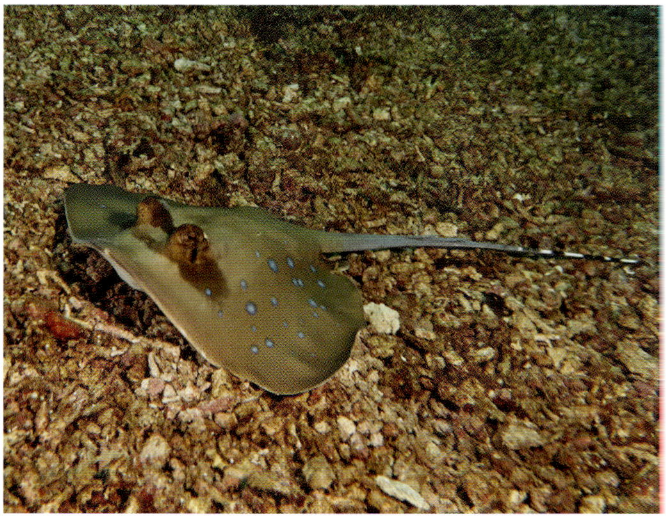

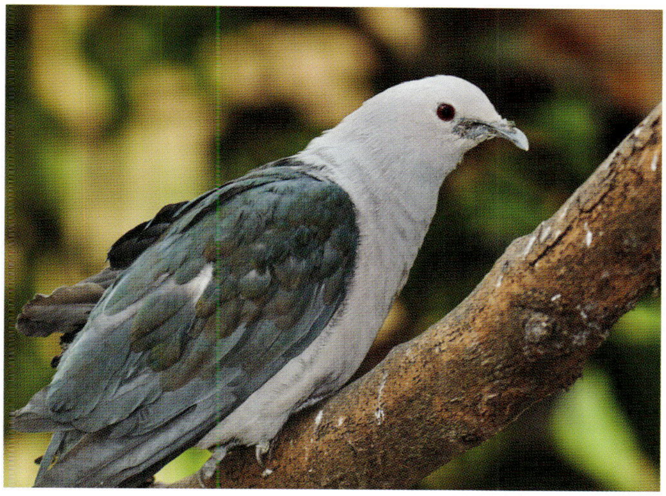

number of mammals that are found in Palawan and other Southeast Asian countries, but not other parts of the Philippines, for example the Asian Short-clawed Otter (*Amblonyx cinerea*).

For birds, all 18 species currently known to be endemic to Palawan are found in El Nido. These include such species as the Palawan Peacock Pheasant (*Polyplectron emphanum*), Tabon Scrubfowl (*Megapodius cumingii*) and Palawan Hornbill (*Anthracoceros marchei*). The latter two are commonly seen on a few of the islands. The Critically Endangered Philippine Cockatoo (also known as the Red-vented Cockatoo; *Cacatua haematuropygia*), a species endemic to the entire country but which today can be found mainly in Palawan, is occasionally spotted here.

regularly seen in Bacuit Bay, though only the first two are believed to nest here. Marine mammals are known to include two dolphin species, two whales and the highly endangered Dugong, which feeds in El Nido's seagrass beds.

On land, the forests – particularly the evergreen lowland forests – are home to an array of mammals, including six of the large species endemic to Palawan, such as the Palawan Pangolin (*Manis culionensis*), Palawan Binturong (*Arctictis binturong whitei*) and Palawan Porcupine (*Hystrix pumila*). There are also a

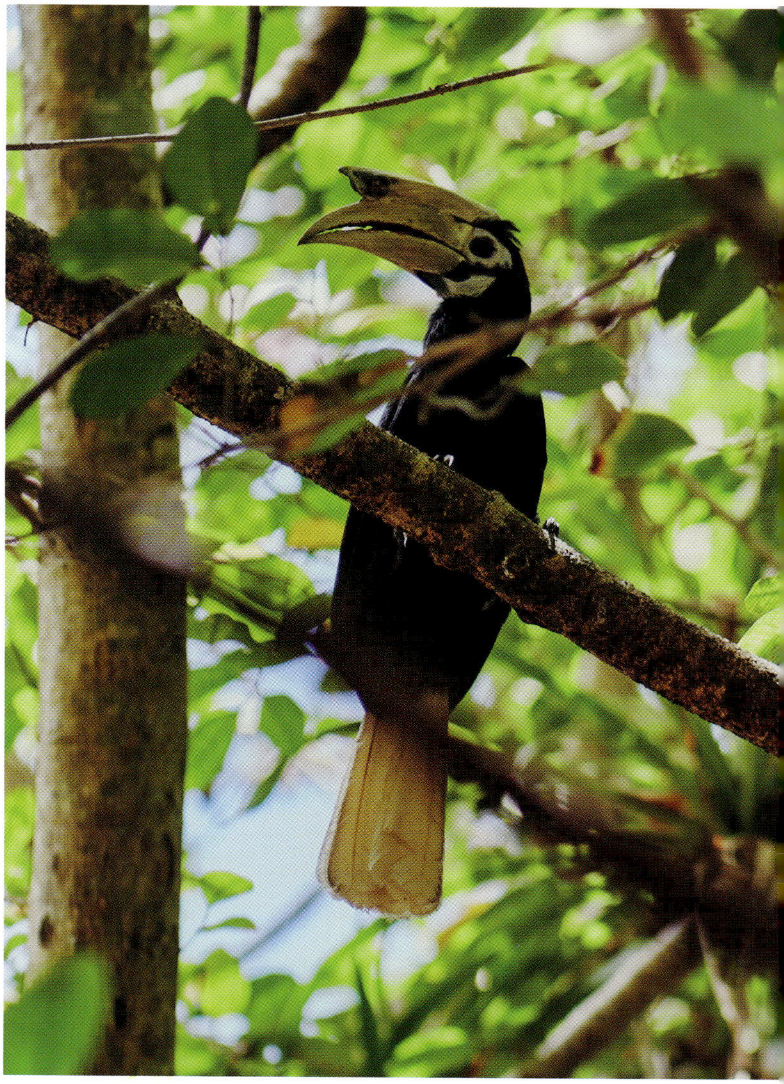

Right: *A Palawan Hornbill in forest on Pangalusian Island.*

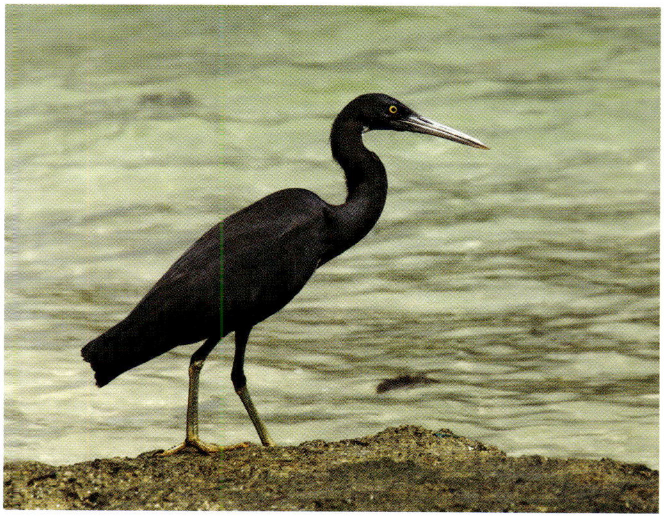

Below: *The Eastern Reef Egret (*Egretta sacra*), is a common coastal bird, on this occasion on the shore near El Nido town.*

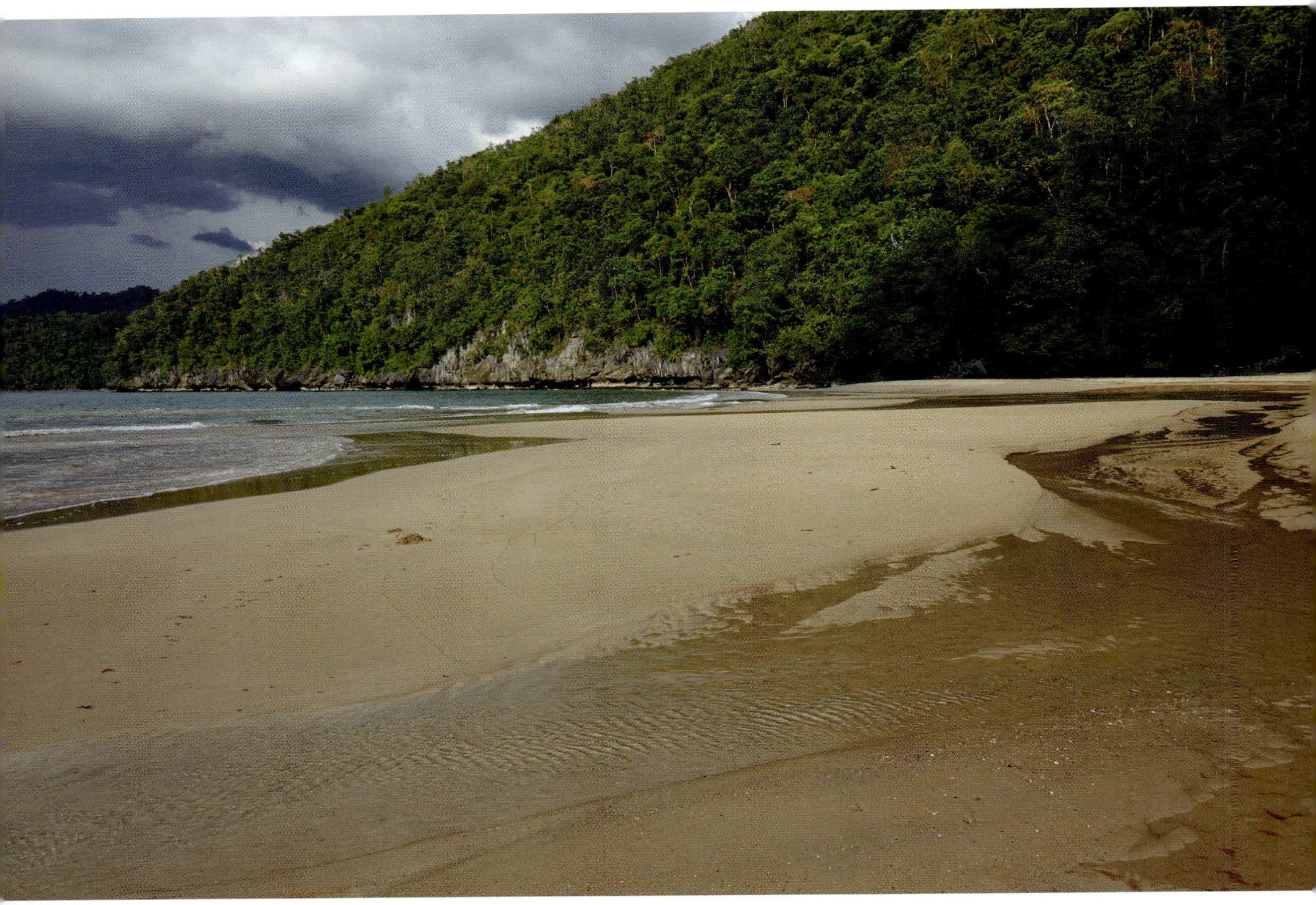

Puerto Princesa Subterranean River National Park

In a province filled with as much natural beauty as Palawan, Puerto Princesa Subterranean River National Park manages to stand head and shoulders above everything else. Not only is the park critical to the conservation of Palawan's rich biodiversity, but also the landscape is quite stunning, and has rightly been rewarded (since 1999) with UNESCO World Heritage Site status.

The park sits on the west coast of Palawan, facing the South China Sea, about 80 km (50 miles) north of the provincial capital, Puerto Princesa. These days it is easily accessible, leading it to have become Palawan's biggest tourist attraction, if not one of the biggest in the whole of the Philippines. Visitors are not drawn by the park's dense forest, or by its rich wildlife, however but by the underground river that flows through its karst limestone mountains before seeing the light of day in a lagoon that, though surrounded by forest, is just barely 200 m (722 ft) from the sea. From the lagoon the river drains to the sea through the sand.

The landscape

The park covers 22,202 ha (54,839 acres) of steep, rolling karst limestone hills, most of it densely forested, topped by dome-shaped Mt St Paul's at 1,028 m (3,382 ft) high. Being limestone, any rainfall percolates down through the rocks, forming water sources deep underground. As a result, the Cabayagan River flows for at least 8 km (5 miles) through

Opposite: The forested mountains reach down to the beach and sea, creating an increasingly rare complete environment from coral reef to lowland forest.

Right: An outrigger canoe carrying visitors through the cave on the underground river.

Below: The underground Cabayagan River sees the light of day as it flows out from its low cave entrance and into a lagoon a short distance behind the beach.

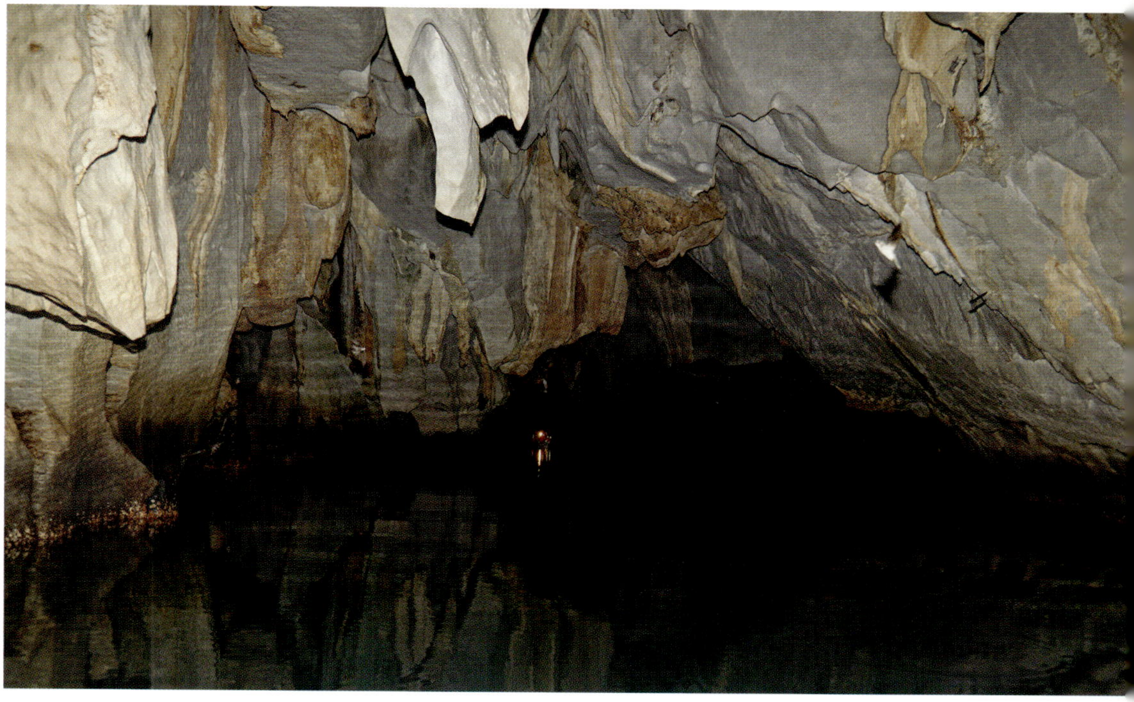

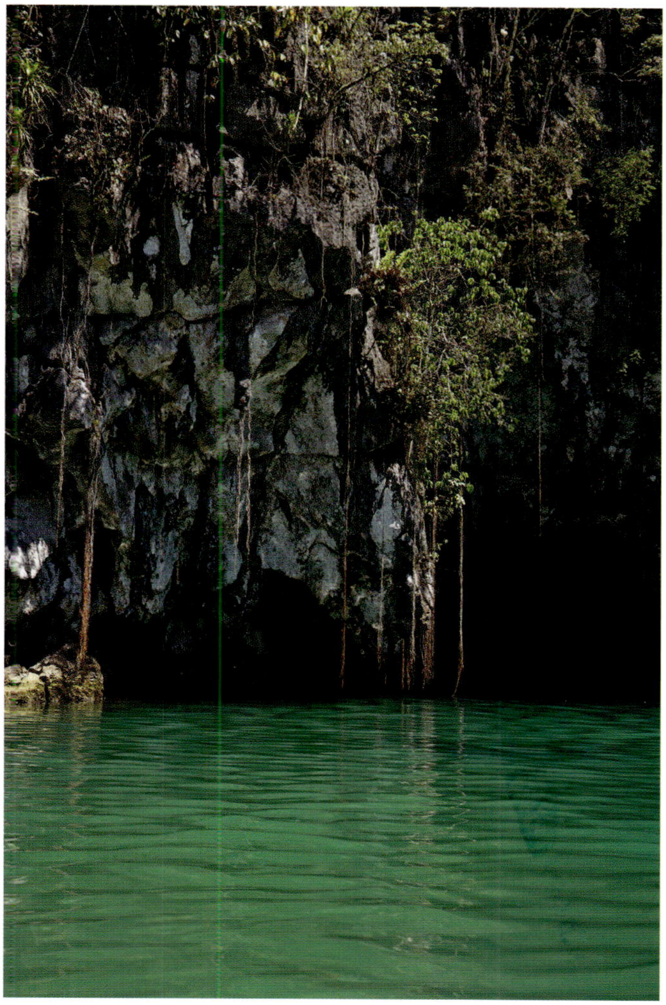

underground caverns, making it one of the world's longest subterranean rivers, before reaching that coastal lagoon. Another river, the Poyuy-Poyuy, flows along the park's southern boundary, reaching the sea just east of the beautiful beach resort of Sabang. Along the coast, the park consists of a small marine element, protecting some important seagrass and seaweed beds, as well as a healthy coral reef.

Habitats, fauna and flora

About two-thirds of the park is forested, with eight types of forest recognized. Probably the most extensive is lowland limestone forest, consisting of trees that are adapted to life in the thin, poor soils that overlie karst limestone. There are also some areas of evergreen lowland rainforest, consisting of a number of dipterocarp species as well as other typically large lowland rainforest trees (such as Ipil, *Intsia bijuga*), though here naturally smaller, due to the limestone base, than is common in most lowland rainforests. Other forest types include some freshwater swamp forest in the lowland forest areas, montane forest, mainly on the upper slopes of Mt St Paul's, and a narrow strip of beach forest along the coast, particularly noticeable along the shore in front of the

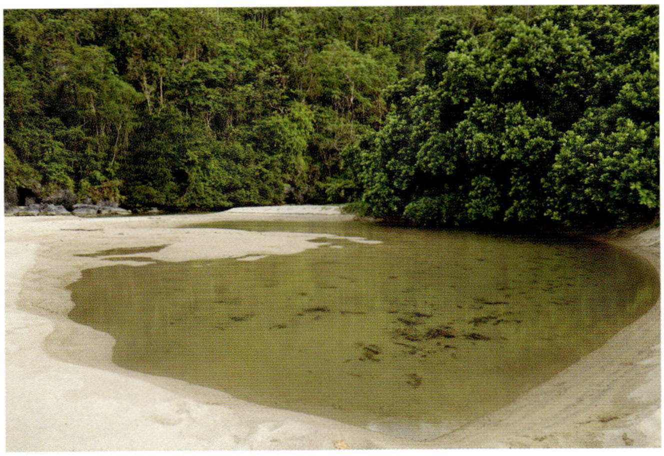

lagoon. The Poyuy-Poyuy River is lined with an extensive mangrove forest that feels quite wild and primeval.

The park's plant life consists of at least 800 species, including about 295 trees, many of which are dipterocarp species. Other ecomonically important wild plants here include a large mix of rattans (*Calamus* species), the coniferous Almaciga tree (*Agathis philippinensis*) and Rambutan (*Nephelium ramboutan-ake*).

For animals, there are more than 250 species of vertebrate, at least two-thirds of them birds. The park's bird count varies depending on the information source, but it comprises roughly 170–200 species, constituting about 67 per cent of all bird species found in Palawan. That includes all 18 species currently recognized as endemic to Palawan, such as the Palawan Scops Owl (*Otus fuliginosus*), Tabon Scrubfowl (*Megapodius cumingii*; a rather strange, flightless, chicken-like bird that creates huge mounds of earth and litter to pile over its eggs), Palawan Flowerpecker (*Prionochilus plateni*) and Palawan Tit (*Pardaliparus amabilis*).

Most of the birds are usually quite difficult for the casual visitor to see. Visitors are much more likely to encounter the numerous Long-tailed Macaques and Water Monitor Lizards (*Varanus palawanensis*) that frequently hang around the park ranger station close to the entrance to the underground river. Though the monitor lizards look ferocious, they are actually quite docile, and are generally safe to approach. Mammals apart from the macaques include eight bat species that live in the cave of the underground river, as well as the Asian Short-clawed Otter and Palawan Stink Badger (*Mydaus marchei*).

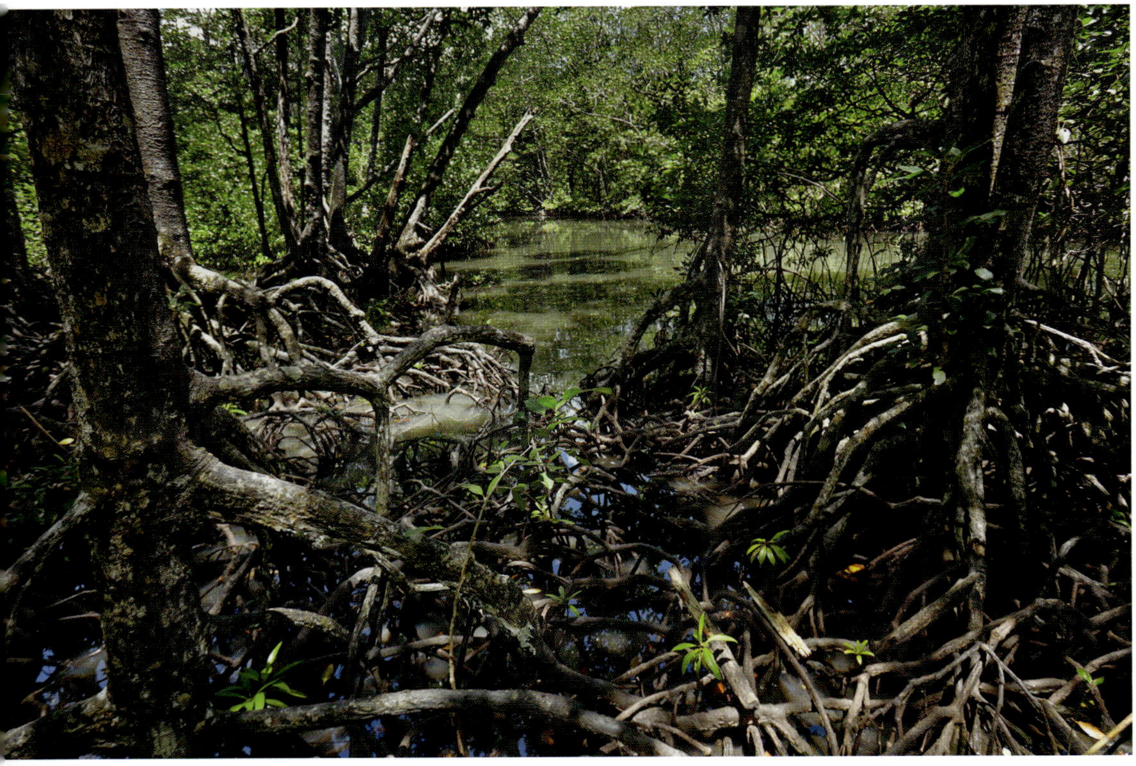

Above left: *The Cabayagan River's lagoon, surrounded by lowland rainforest, just behind the beach. At this point the river drains through the sand and into the sea.*

Left: *A tangle of aerial roots in dense mangrove forest along the Poyuy-Poyuy River, on the southern edge of the national park.*

Opposite below: *The shy Palawan Peacock Pheasant is rarely seen, but is known to live within this national park.*

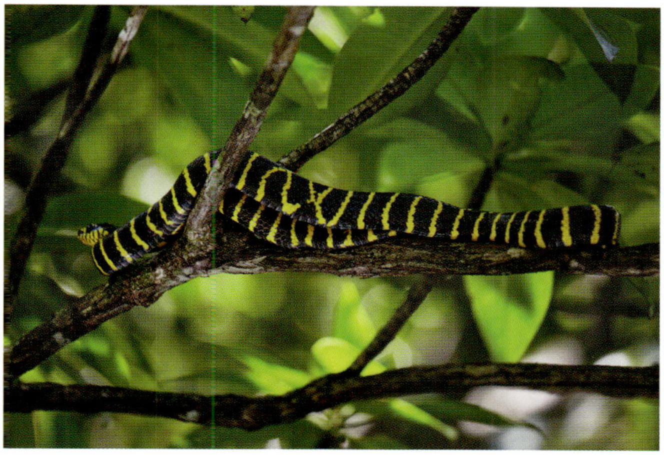

Above: *A Mangrove Snake spotted resting in the branches of a mangrove tree along the Poyuy-Poyuy River.*

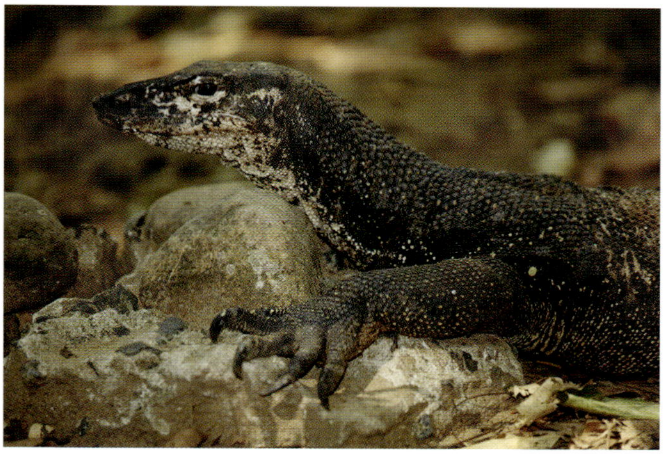

Above: *A Palawan Water Monitor Lizard in forest close to the national park ranger station.*

Apart from the Water Monitor Lizard, the park's 19 reptiles include the Reticulated Python (*Python reticulates*), a freshwater turtle (*Caura ambionensis*), and Green and Hawksbill Turtles. In the mangroves of the Poyuy-Poyuy River it is not uncommon to see the black-and-yellow banded Mangrove Snake (*Boiga dendrophila*) coiled up in trees beside the river.

Visiting the park

Although there are a couple of good hiking trails within the park, including one that follows the coast, linking the park with the Sabang beach resort to the west, very few people explore the park beyond the underground river. The latter has become hugely popular in recent years, leading the area around the departure and arrival point for tours to become very crowded and noisy. Access to the park from Puerto Princesa is initially via road to Sabang, followed by a 15-minute *banca* boat ride into the park, finishing at a beach landing point close to the underground river's entrance.

Trips into the underground river, covering just the first kilometre underground, start and finish from the lagoon shore, and use small outrigger canoes that are paddled by a single oarsman. No engines are allowed in the cave due to the noise and pollution. With the help of a powerful lamp, it is possible to see an assortment of stalactites and stalagmites, as well as a regular stream of bats and swiftlets coming and going.

Trips into the Poyuy-Poyuy mangroves also make use of small outrigger canoes, starting and finishing from a landing run by a local community business on the park's western border close to Sabang. This really feels like a wilderness experience, with wildlife regularly seen, ranging from Long-tailed Macaques, to various kingfishers and/or egrets, to Mangrove Snakes.

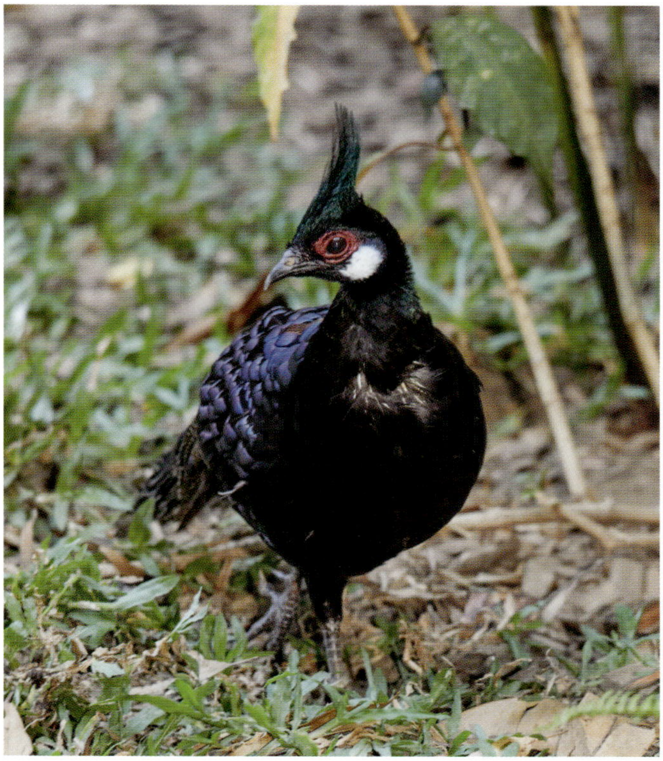

Tubbataha Reefs Natural Park

The jewel in the crown of the Philippines' marine protected areas system is undoubtedly Tubbataha Reefs Natural Park. Located in the midst of the Sulu Sea, 150 km (93 miles) south-east of Puerto Princesa, this is one of the country's largest marine protected areas, and one of the most important in the entire Indo-Pacific region. The park sits literally at the heart of the Coral Triangle, its huge coral reef system serving as a distribution centre for coral and fish larvae for the entire Sulu Sea and probably well beyond. It is a very remote location, surrounded by seas that are frequently rough, and exposed to the south-west monsoon in June–October, and to the north-east monsoon in November–February. Only during the relatively calm March–May months is the reef accessible. Covering an area of 97,030 ha (239,664 acres), the park protects three reef systems. Two of them – Tubbataha Reef itself – consist of atolls, while the third is a submerged reef 20 km (12 miles) to the north, known as Jessie Beazley Reef. All three rise up from the submarine Cagayan Ridge, which runs north–south through the Sulu Sea. The two atolls, called North and South Atolls, are about 16 km (9.9 miles) and 4.5 km (2.8 miles) long respectively, and 4 km (2.5 miles) wide.

A few small, very low-lying islets lie scattered along the reefs, but the entire reef system has no human settlement, apart from a tiny base on North Atoll, home to the handful of rangers who guard the park throughout the year. The atolls are surrounded by deep water, so although there are expanses of shallow, gently sloping coral gardens, the atolls are quite famous for their huge and almost sheer submarine walls dropping 100 m (329ft) or more.

Below: *Gorgonian sea fans such as this one (possibly an* Echinogorgia *species) are quite numerous and grow to great sizes on the sheer submarine walls typical of the Tubbataha reefs.*

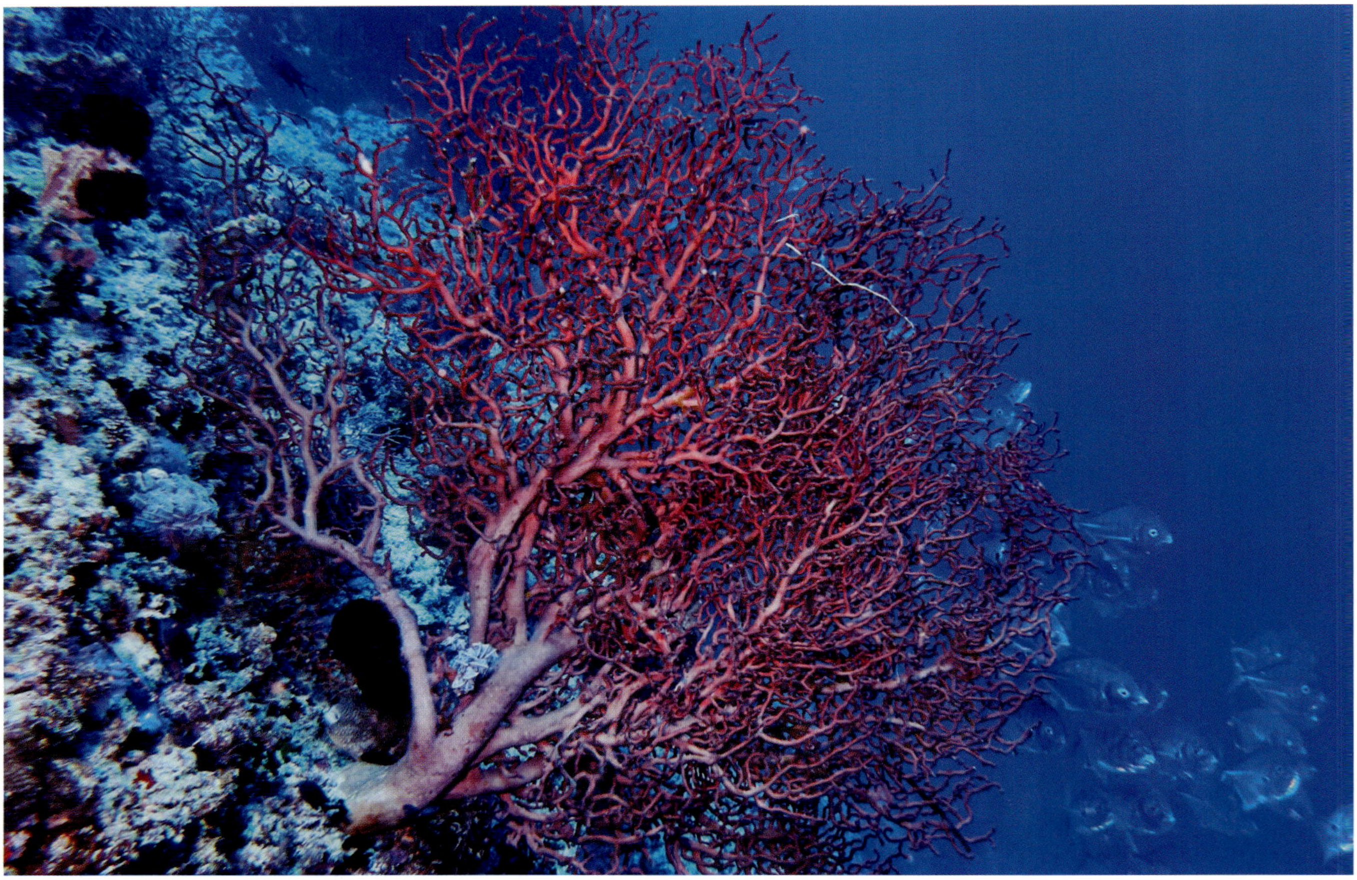

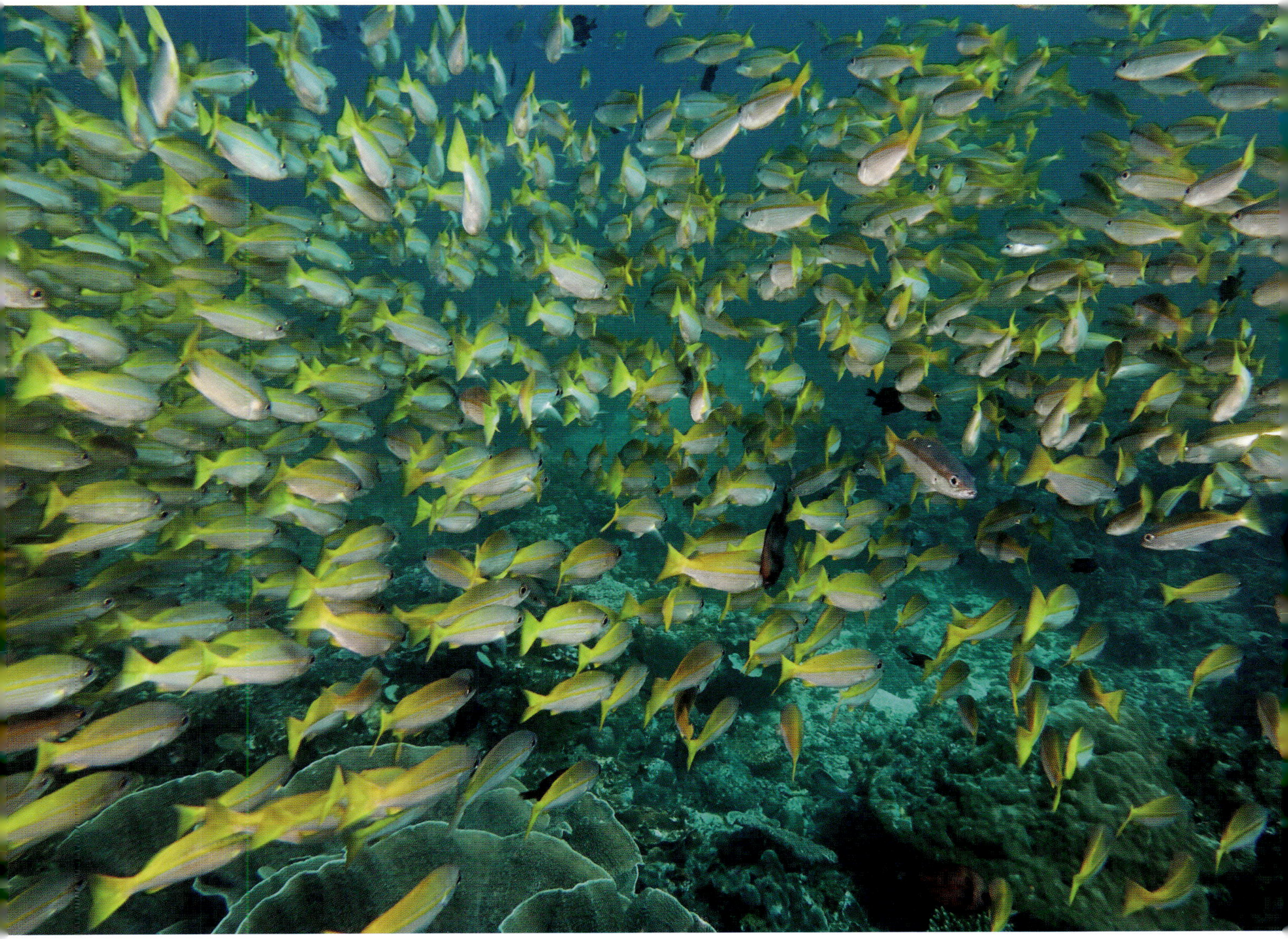

Above: A shoal of Bigeye Snappers (Lutjanus lutjanus), *one of the many fish species commonly seen across Tubbataha's coral reefs.*

Below: Green Turtles are a common sight across Tubbataha Reefs, which they use as an important feeding site.

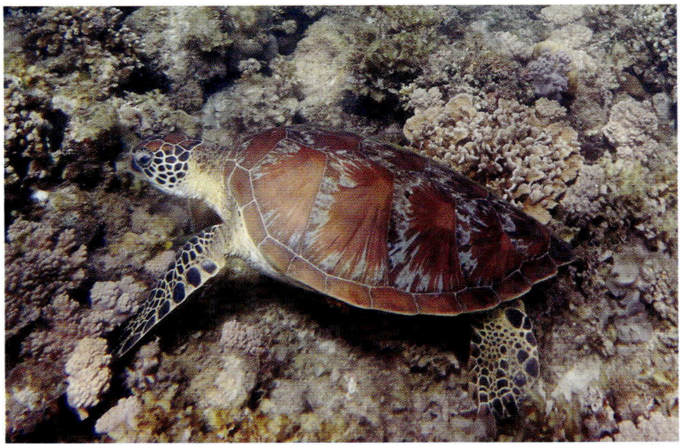

A hotbed of biodiversity

Just a few numbers are all that is needed to drive home the importance of this park. The reefs contain at least 360 species of coral, a large proportion of all corals found in the Philippines, and half of the entire global coral species count. The reefs they have built support a staggering 600 fish species, ranging from such reef types as a host of clownfish, triggerfish, Moorish Idols, damselfish, angelfish, and so on, right up to shoals of the larger deep-sea fish, including tuna (*Thunnus* species), barracuda, trevally (*Caranx* species)

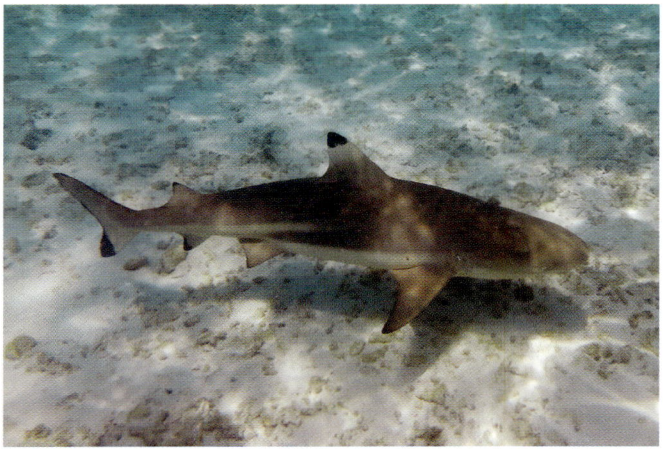

Above: *A Blacktip Reef Shark swimming across sandy shallows. Sharks are very common across Tubbataha.*

and Reef Manta Rays (*Manta alfredi*), to name just a few. Tubbataha is one of the few remaining sites where it is possible to see the Humphead Wrasse (or Napoleon Wrasse; *Cheilinus undulatus*), a large reef fish that has been fished close to extinction in other parts of the Philippines due to its high demand at restaurants around Southeast Asia.

Thirteen species of dolphin and whale have been reported around Tubbataha, along with 11 shark species, most commonly thresher (*Alopias* species), and Grey Reef Sharks (*Carcharhinus amblyrhynchos*) and Whitetip Reef Sharks. The few beaches are important breeding grounds for Green Turtles and Hawksbill Turtles. Both are regularly spotted swimming around the reefs.

The tiny specks of land are vital sites for an estimated 38,000 marine birds, consisting of about 100 species. The great majority use the islets simply as staging grounds during migrations or as resting and roosting sites during feeding forays around the reefs. Only six species breed here, mostly on Bird Islet, one of the park's largest islands (though still only a little more than 1 ha/2.47 acres), at the northern end of North Atoll. These six species are the Red-footed Booby (*Sula sula*), Brown Booby (*S. leucogaster*), Great Crested Tern (*Thalasseus bergii*), Sooty Tern (*Onychoprion fuscatus*), Black Noddy (*Anous minutus*) and Brown Noddy (*A. stolidus*).

North Islet used to be covered with scrubby vegetation, but in recent years this has mostly died away, poisoned by the huge amounts of bird guano, an indication that more and more birds are crowding on to the islet. This could be taken as an indication of the success of protection here, or as a worrying sign pointing to the loss of suitable habitats elsewhere in and around the Sulu Sea.

Protecting the reefs

Until the 1980s Tubbataha's own isolation was its greatest protection. Up until then fishing boats were largely dependent on sail, and the reefs were visited only occasionally during the calm months by sea nomads and fishermen from the Cagayancillo Islands, 130 km (80 miles) north-east. But then sail gave way to engines and fishermen from far and wide were able to reach Tubbataha, bringing with them destructive cyanide and dynamite fishing methods. Alarm bells were rung, and in 1988 the government declared the two Tubbataha atolls a national marine park, the country's first. The reefs' isolation initially made protection difficult, but in the 1990s a permanent ranger station was set up on North Atoll, today manned throughout the year by men from the Philippine Navy, the Coastguard, the park's management office and the Cagayancillo municipal office.

Below: *Two different gorgonian sea fans on a sheer wall,* Acabaria *species on the left and* Subergorgia *on the right.*

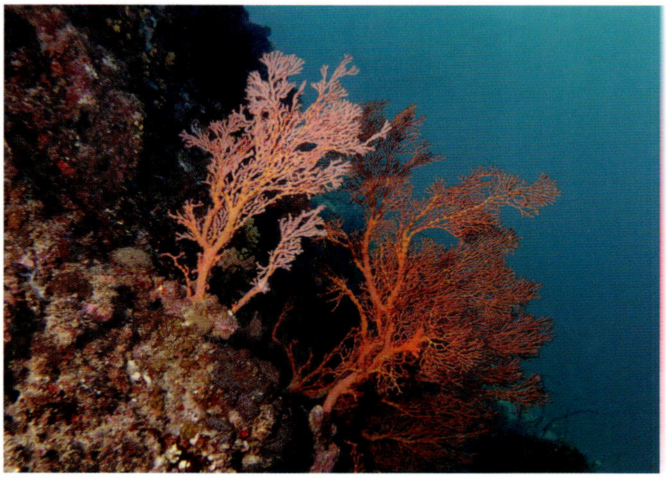

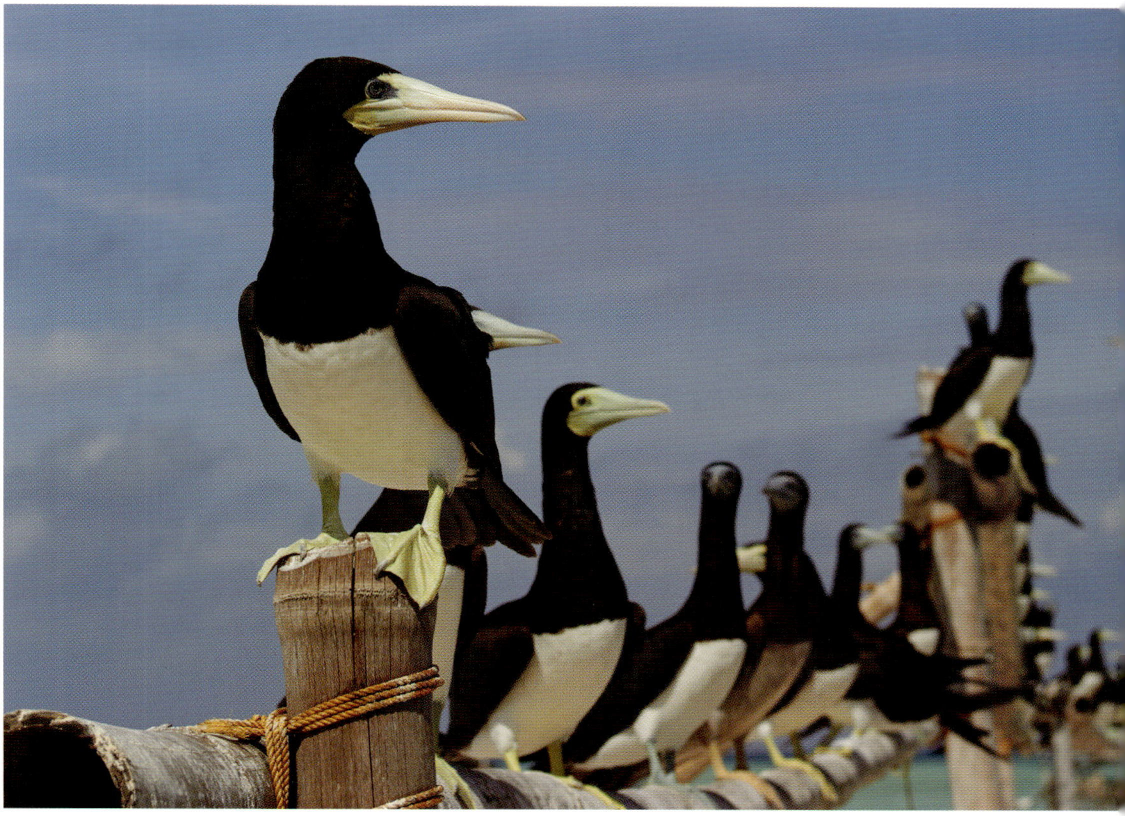

Right: *Brown Boobies on Bird Islet in Tubbataha's North Atoll. They are one of the natural park's most common birds, one of six species that breed here.*

International accolades started to arrive. In 1993 UNESCO declared Tubbataha a World Heritage Site, in 1999 Ramsar declared it a vital wetland area, and in 2015 the park became the 35th ASEAN Heritage Park, giving it recognition as one of Southeast Asia's most important protected areas.

In 2006, the park was greatly expanded from its then 33,200 ha (82,000 acres) to its current 97,030 ha (239,664 acres), with the inclusion of Jessie Beazley Reef. At the same time the park's name was changed from Tubbataha National Marine Park to Tubbataha Reefs Natural Park.

Today, the park is strictly protected, with all forms of fishing and collection prohibited, and entry fees charged to visiting divers. Boats are not allowed to anchor, but must use mooring buoys, and visitors are not allowed to land. When the fishing ban was initially introduced there were regular prosecutions of fishermen, which did eventually result in a steep drop in illegal fishing. As a result, since the creation of the park damage inflicted in the 1980s and '90s has steadily healed.

The park is still surrounded by threats, however. Foreign fishing boats, mainly from Vietnam and China, still try to fish illegally around Tubbataha, while oil and gas exploration in the Sulu Sea increasingly threatens the marine environment generally. At present, despite these threats, Tubbataha Reefs Natural Park represents a beacon of success for Philippine marine protection, hopefully pointing a way forwards for protection of more reefs in the future.

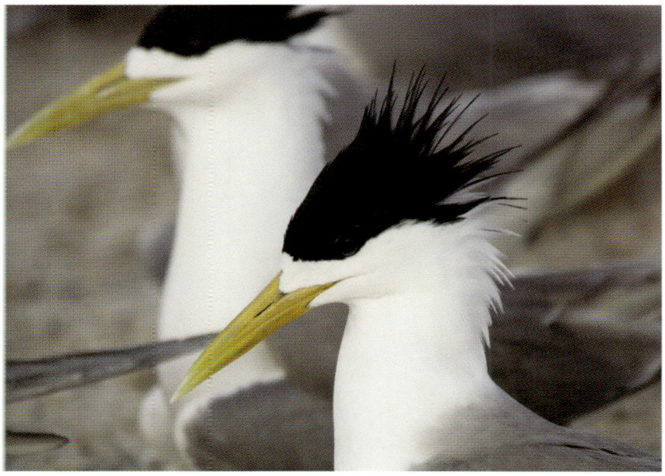

Below: *Great Crested Terns are commonly seen on the larger of Tubbataha's islets, particularly Bird Islet in the North Atoll.*

The Protected Forests of Negros

The surviving forests of Negros constitute one of the most critically important areas in the whole of the Philippines. This is because they represent one of the last remaining homes for a number of endangered species that are endemic not just to the Philippines but exclusively to the Western Visayas biogeographic region.

Consisting principally of Negros, Panay, Guimaras, Cebu and Masbate, deforestation and other human pressures have been more extreme in this region than in just about any other part of the Philippines, driving a number of regional endemic species to the very brink of extinction. Masbate, Cebu and Guimaras have almost no remaining natural forest areas, other than a few isolated patches. In Panay, some areas of old-growth forest hang on in the mountains that run down its western side, amounting to about 10 per cent of the island, though only a few areas are protected.

In Negros, natural forest cover is down from 90 per cent at the start of the twentieth century to roughly 4 per cent in the early years of the twenty-first. It is a staggering decline, brought on largely by the rapid

Below: *A view of dense rainforest from the top of Twin Falls, in Northern Negros Natural Park.*

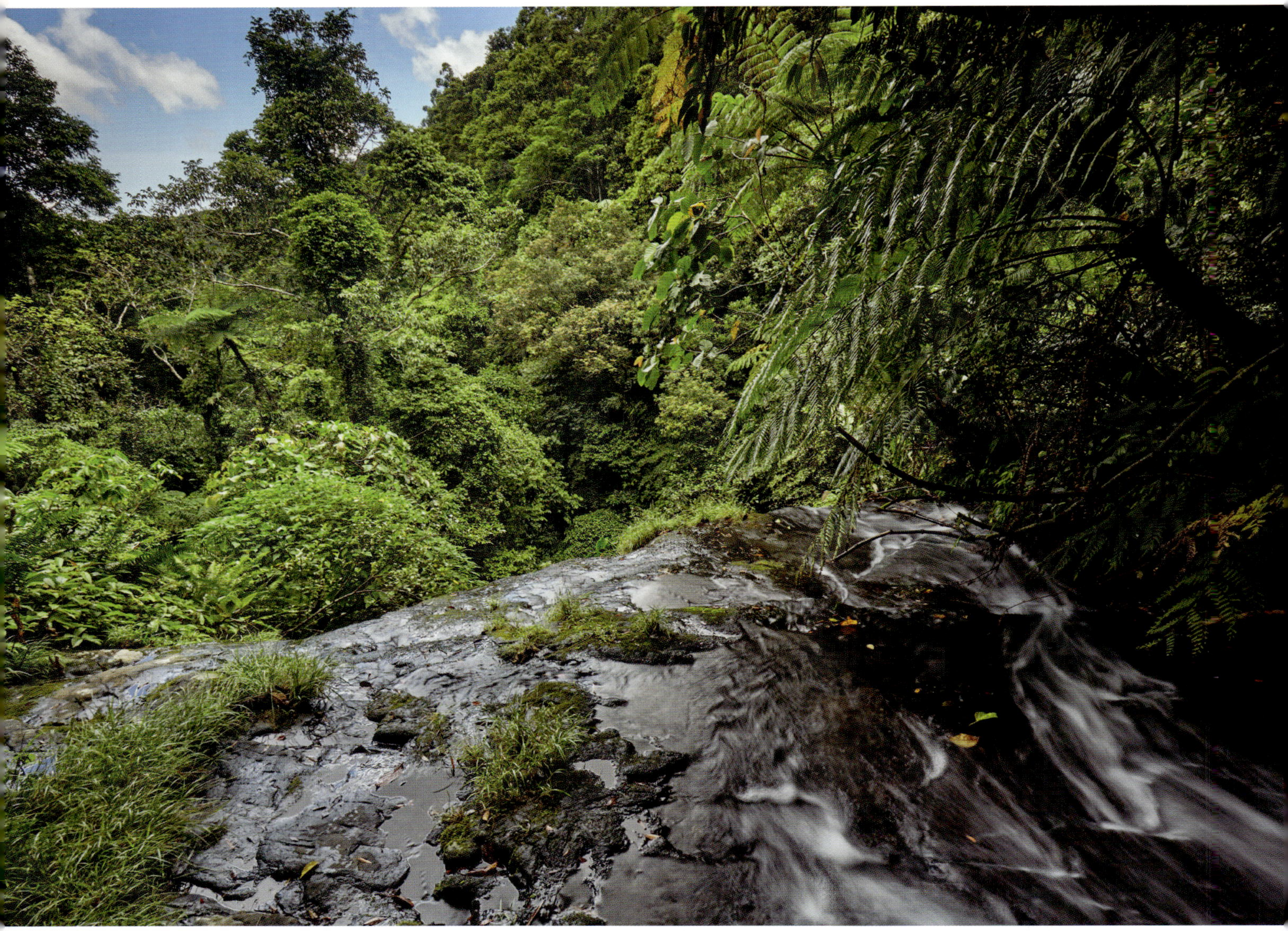

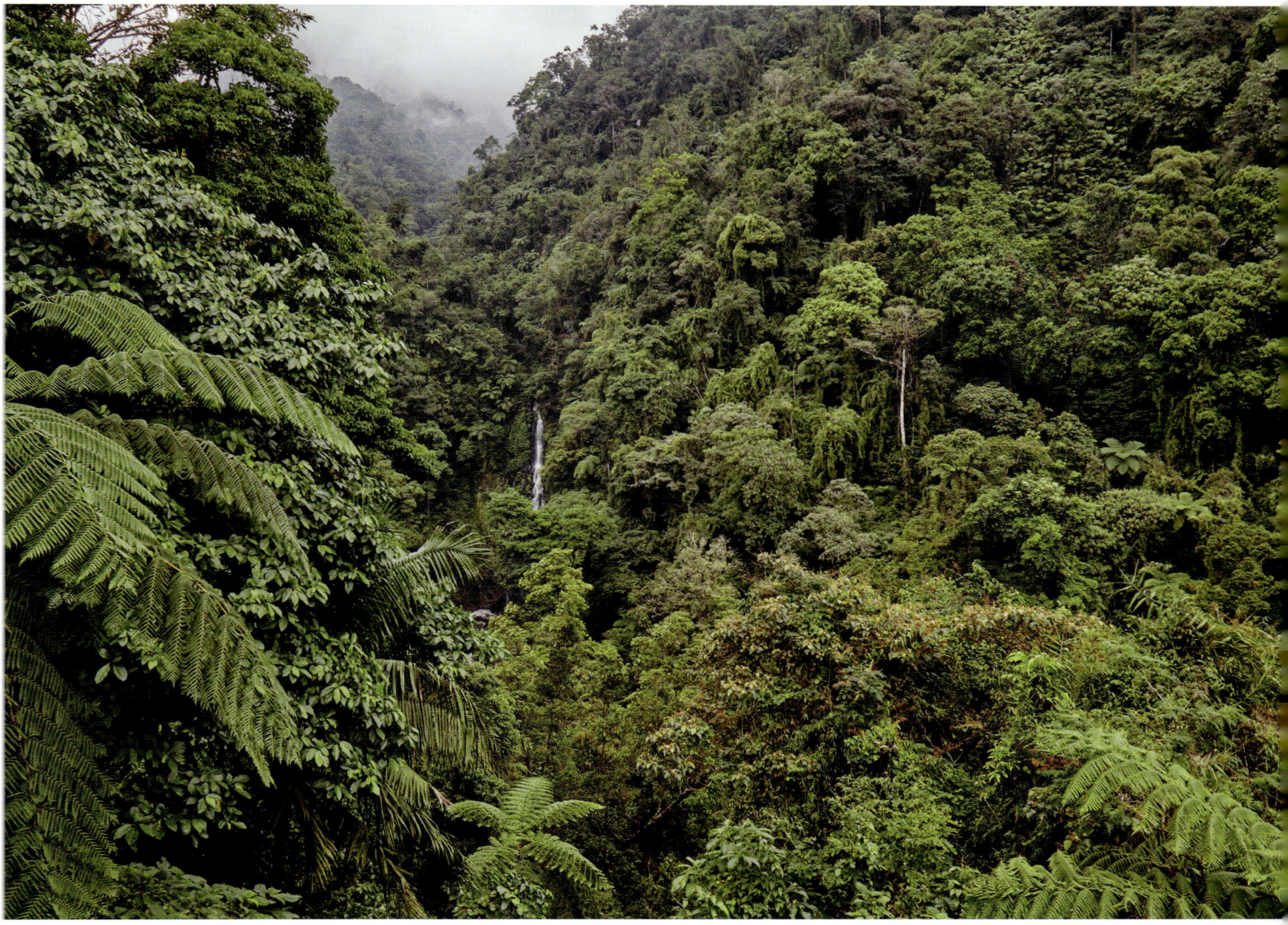

Above: The lower levels of montane rainforest at Guintubdan, in Mt Kanlaon Natural Park.

expansion of sugar plantations, commercial logging and conversion of land to farming. Today, the three largest remaining blocks of forest have all been protected as natural parks, though this has failed to save them from ongoing loss of forest habitat to farming. Along with hunting and plant gathering, these factors continue to exert huge pressure on Negros' plant and animal wildlife.

The three protected areas are Northern Negros, Mt Kanlaon and Balinsasayao Twin Lakes Natural Parks, all centred around volcanoes that form the core of the Negros Volcanic Belt, and which make up the island's rugged mountain spine.

Northern Negros Natural Park

Lying east of the town of Silay, Northern Negros Natural Park (NNNP) is by far the largest of the island's protected areas. Originally established as a forest reserve in 1935, it became a high-priority natural park in 2005. In the intervening years it spent long periods as simply a paper park, making rampant deforestation possible, huge areas being converted to sugar-cane plantations and general farmland. Today, of its 80,454 ha (198,721 acres) more than half is now occupied by farmland and villages. In 2001 it was estimated that its forest cover had dropped to about 22,300 ha (55,081 acres), though by 2011 this is believed to have risen again to about 30,180 ha (74,545 acres), thanks to a combination of a reforestation programme and secondary forest regeneration, a

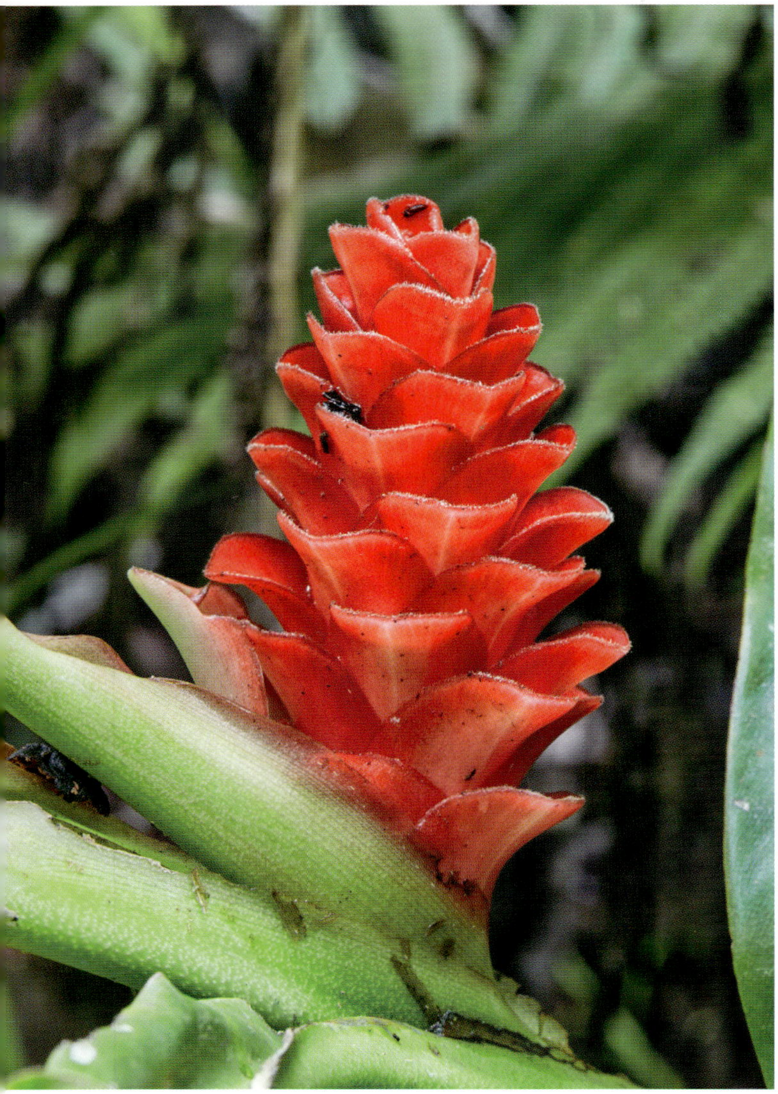

possible sign of hope for the future. Spread across two principal mountains, Mts Silay and Mandalagan, and six watersheds, the lower levels of the park contain much of Negros' remaining lowland forest, most of which is secondary growth. Much of the forest is montane in nature, the highest levels – up to the highest summit, 1,885-m (6,200-ft) Mt Mandalagan – becoming very stunted, gnarled and moss-draped, consisting of stone oak and laurel species, festooned with not only mosses but also ferns and orchids.

Mt Kanlaon Natural Park

At roughly 24,600 ha (60,762 acres) this natural park encompasses the forested slopes of Mt Kanlaon, lying just to the south of Northern Negros Natural Park. At 2,465 m (8,110 ft), this is the Visayas' highest mountain and the region's most active volcano, with over 30 recorded eruptions, the most recent in 2016.

Almost all of Kanlaon's forest is montane, there being very little forest remaining below an altitude of about 1,100 m (3,619 ft). That said, a couple of small pockets of lowland rainforest can be found, most especially around the hot-spring resort of Mambucal, at an altitude of about 750 m (2,467 ft) on Kanlaon's northern slopes. Kanlaon's mossy montane forest is particularly magnificent, consisting of dense and extensive stands of gnarled dwarf trees, enveloping a number of volcanic lakes to the north of the main summit. Forest gives way to grassland and bare rock on the highest slopes around the active crater and the highest summit.

Balinsasayao Twin Lakes Natural Park

This is the smallest of the protected areas, at just 8,015 ha (19,797 acres), lying in the far south of Negros, close to the city of Dumaguete. The park is on the northern side of Mt Talinis (also called the Horns of Negros), and encompasses two volcanic caldera lakes, Balinsasayao and Danao. The lakes are surrounded by

Left: *A vibrant wild ginger (Zingeraceae), a very common forest floor plant, seen in lowland forest in Northern Negros Natural Park.*

Below: *The Visayan Flowerpecker, endemic to the Western Visayas, in rainforest at Mambucal, on the edge of Mt Kanlaon Natural Park.*

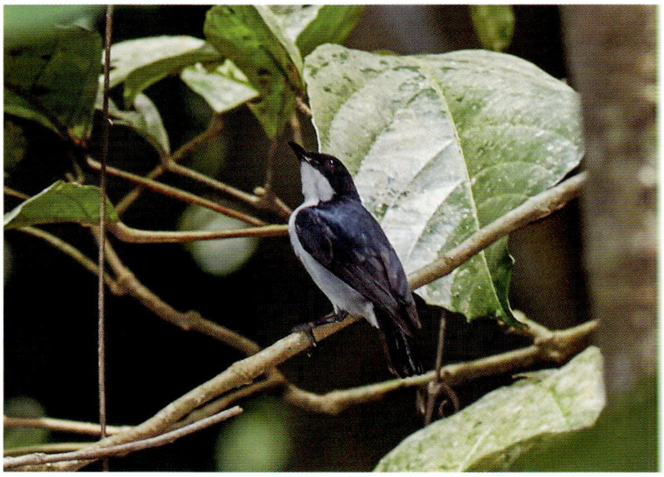

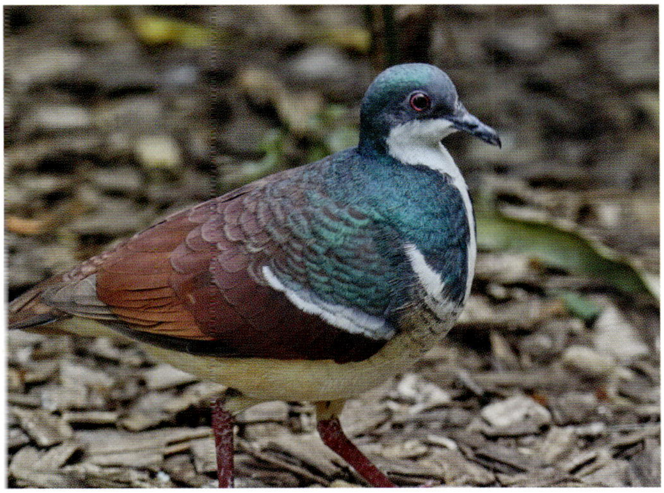

Above: The Critically Endangered Negros Bleeding-heart Pigeon, endemic to the Western Visayas.

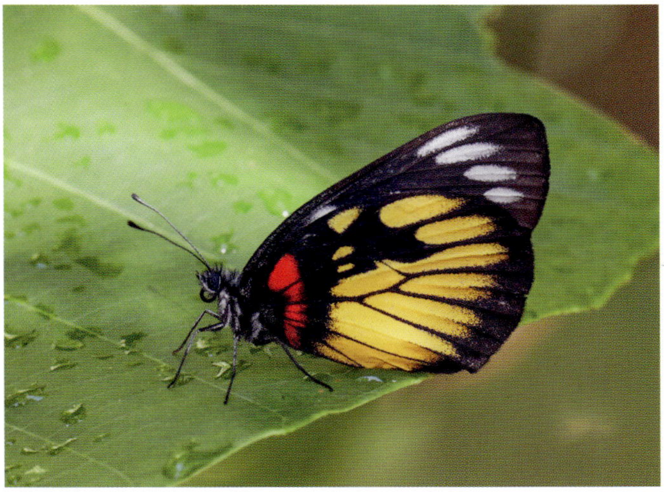

*Above: A Redbase Jezebel butterfly (*Delias henningia henningia*) in rainforest near Mambucal, in Mt Kanlaon Natural Park.*

healthy rainforest, most of it montane, though there are a few surviving areas of lowland dipterocarp forest, particularly around the park's lower edges.

To the south, Mt Talinis, at 1,870 m (6,152 ft), is southern Negros' highest peak. Many of its slopes are well forested, again mostly with montane forest, which is contiguous with the forests of the natural park. Although not protected as a NIPAS park, it is protected by the Energy Development Corporation (EDC), which considers it a vital watershed and possible geothermal energy source.

Negros forest wildlife

The trees of Negros' forests are fairly typical, the remaining areas of lowland forest rich in large tree types such as *Dipterocarpus*, *Shorea* and *Hopea* species, as well as *Pterocarpus indicus* and a variety of pandans and palms, including rattans. The montane forests are dominated by a number of stone oak, laurel and chestnut species. Beneath the tree canopy there is a huge diversity of smaller plants, including many orchids, ferns and pitcher plants, plus several species of *Hoya*, lipstick vines (*Agalmyla* species), *Begonia* and *Medinilla*, and at least one species of *Rafflesia* (*R. speciosa*).

A large proportion of the animal wildlife is endemic either to the Philippines as a whole or just to the Western Visayas. The species of the latter are almost universally seriously threatened due to both hunting and the loss of forest habitat, particularly those dependent on lowland rainforest.

Among the nearly 60 mammal species known to live on Negros, 33 per cent are endemic to the Philippines and many are seriously threatened. These include the Visayan Spotted Deer (*Rusa alfredi*), Visayan Warty Pig (*Sus cebifrons*), Visayan Leopard Cat (*Prionailurus bengalensis rabori*), Negros Shrew (*Crocidura negrina*) and several fruit bats, including the Philippine Tube-nosed Fruit Bat (*Nyctimene rabori*), Negros Naked-backed Fruit Bat (*Dobsonia chapmani*) and Golden-crowned Flying Fox.

Of the birds, more than 350 species have been recorded in Negros, over 80 of them endemic to the Philippines. A number are endemic to the Western Visayas region, almost all of which are threatened. They include the Negros Bleeding-heart Pigeon (*Gallicolumba keayi*), Negros Fruit-dove (*Ptilinopus arcanus*), Visayan Hornbill (*Penelopides panini*), Rufous-headed or Visayan Wrinkled Hornbill (formerly called Walden's Hornbill; *Rhabdotorrhinus waldeni*), Visayan (or Black-belted) Flowerpecker (*Dicaeum haematostictum*), Negros Striped Babbler (*Zosterornis nigrorum*) and Flamed-templed Babbler (*Dasycrotapha speciosa*), to name just a few.

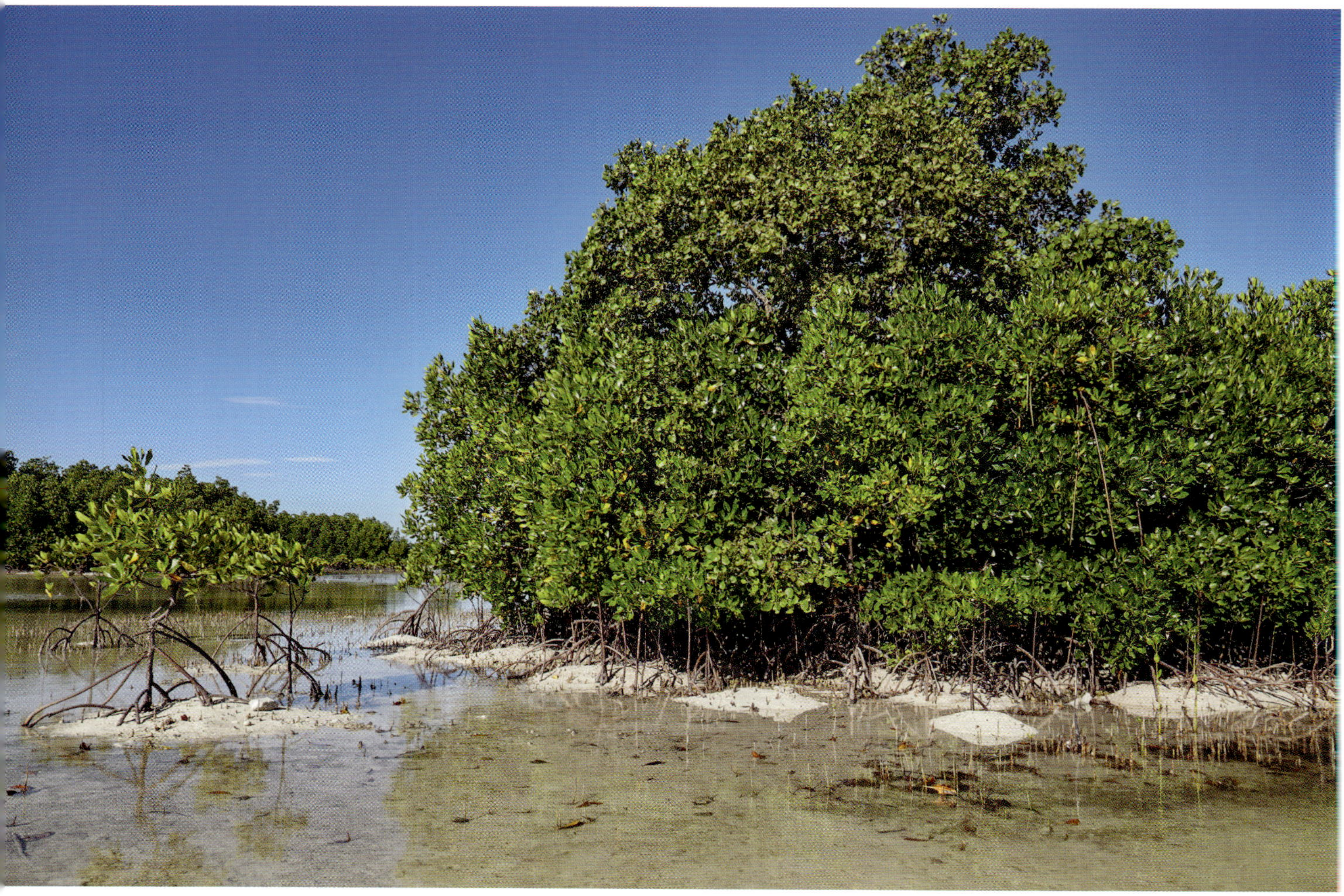

Olango Island Wildlife Sanctuary

Olango is a small island lying off the south-east coast of Mactan Island, close to Cebu City, in the western Visayas. Despite being so close to the Philippines' second largest metropolis, it is a world away; a highly rural place of rough, unpaved roads and small houses, many still roofed with Nipa Palm. Nevertheless, and despite being just 7.5 km (4.7 miles) long and 3.4 km (2.1 miles) wide at its widest point, it is home to 30,000 people.

The quiet southern end of the island faces on to a vast expanse of sand and mudflats, sprinkled with mangroves, that stretch for nearly 9 km (5.6 miles) to the south-west before reaching a fringing coral reef, then deep water. The Olango Reef Flat, as it is called, is a rich source of crustaceans, molluscs, echinoderms and worms, which live mostly buried in the sand and mud, offering a superb food source for both people and wildlife. A large part of these flats is exposed at low tide, and even at high tide the water covering it is mostly less than 1 m (3 ft) deep, making the pickings relatively easy. As a result, there is a huge amount of bird life here, both resident and migratory. To date, 97

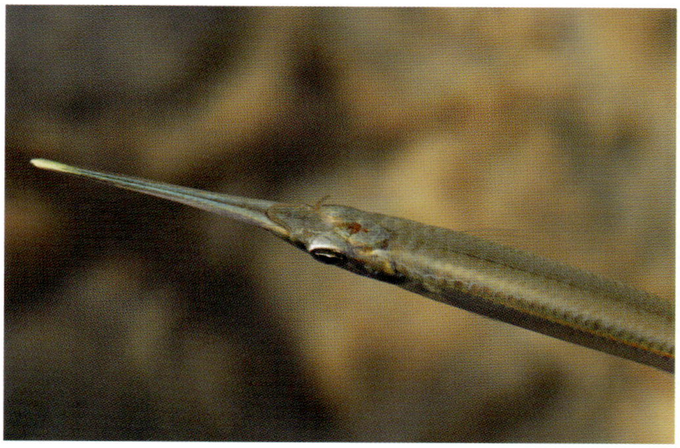

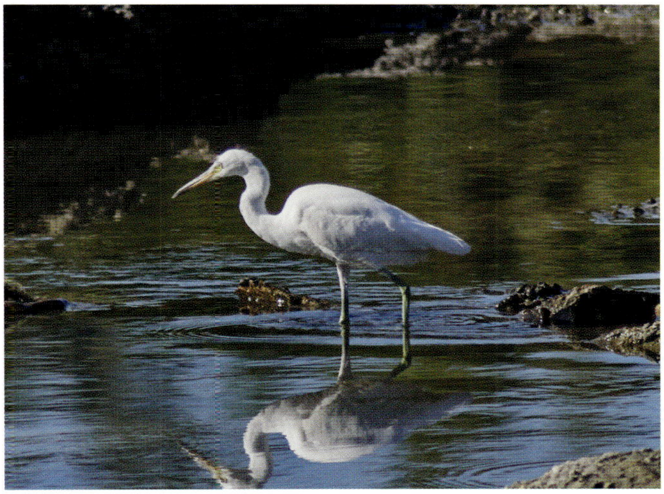

Above: *A Chinese Egret stalking fish in the shallows at low tide.*

Opposite: *Olango Island's mangroves at low tide, inside the wildlife sanctuary.*

Opposite below: *The Feathered River Garfish (*Zenarchopterus dispar*), known locally as Kansusuwit, is one of many fish seen in the mangroves' shallows.*

Right: *The Rufous Night-heron (*Nycticorax caledonicus*), though not a common bird, can be found on Olango Island.*

bird species have been identified, 48 of them migratory birds passing through en route between summer breeding grounds in Siberia, northern China and Japan, and overwintering lands in Australasia. The Olango Reef Flat is believed to be a vital part of the East Asian-Australasian Flyway, used by an estimated 77 species of migratory bird, over 60 per cent of which use Olango as a rest and feeding stopover site.

Protection of the Olango Reef Flat

The site was declared a wildlife sanctuary in 1992, covering 1,030 ha (2,544 acres) of the flats, and in 1994 it became the Philippines' first Ramsar Site, in recognition of its international importance to bird life. Olango is an increasingly popular birdwatching site. The best time of year is September–May, though peak months are November–March, a period when there are often up to 40,000 birds on the mudflats and in the mangroves. However, the site is so vast that at low tide the birds become very widely dispersed. For this reason, two hours either side of high tide is the best period for birdwatching.

Birds frequently seen on the reef flat include a variety of herons and egrets (including the rare Chinese Egret; *Egretta eulophotes*), several species of curlew (*Numenius* species) and plover (*Charadrius* species), as well as the Common Redshank (*Tringa totanus*), Common Greenshank (*T. nebularia*), Asian Dowitcher (*Limnodromus semipalmatus*), Ruddy Turnstone (*Arenaria interpres*), Common Sandpiper (*Actitis hypoleucos*) and Rufous-necked Stint (*Calidris ruficollis*). In the mangroves and some of the scrubby beach forest along the shore it is common to see several kingfisher species, and the Pacific Swallow (*Hirundo tahitica*), Yellow-vented Bulbul (*Pycnonotus goiavier*), Golden-bellied Flyeater (*Gerygone sulphurea*), Pied Fantail (*Rhipidura nigritorquis*) and Olive-backed Sunbird (*Nectarinia jugularis*).

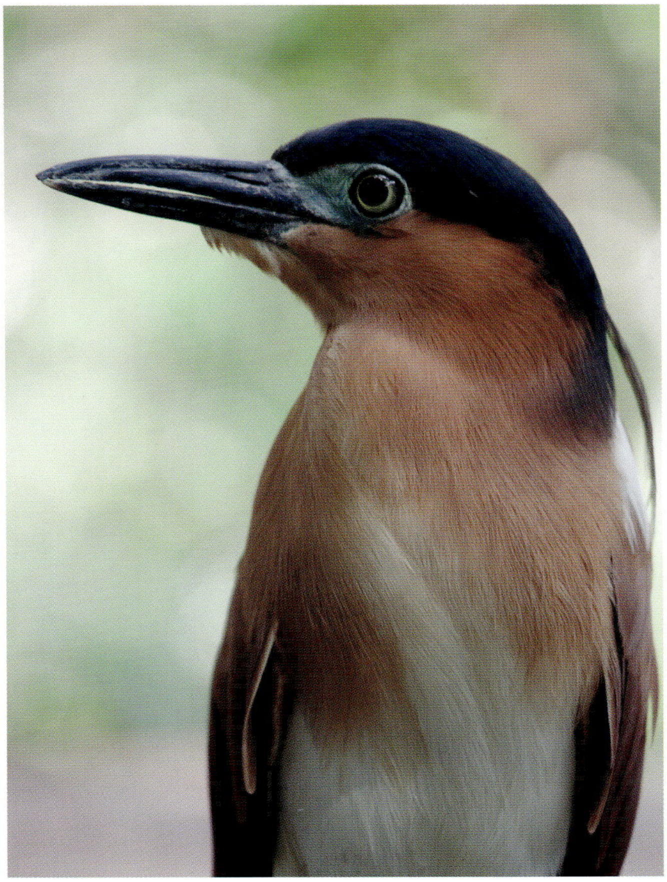

The Tañon Strait and the Coral Reefs of Moalboal

The islands of Negros and Cebu are separated by the Tañon Strait, a long, narrow stretch of water about 160 km (100 miles) from north to south. It is just 27 km (16.7 miles) wide at its widest point, and a mere 5 km (3 miles) wide at its narrowest, the southernmost end just before it joins the Bohol Strait to the south. It is also incredibly deep, reaching down to 500 m (1,645 ft) along parts of its length, making the Tañon Strait something of a marine wildlife highway.

Manta rays, several shark species, and shoals of tuna and barracuda, along with no less than 14 species of whale and dolphin, all either live and breed in the Strait or use it as a migration route. Overall, studies have shown the Strait to be home to 18,830 ha (46,510 acres) of coral reefs, along with about 70 species of fish, 20 of crustaceans and 26 of mangrove tree. Rare species found in the Strait include the Dugong, Chambered Nautilus (*Nautilus pompilius*) and Dalupapa (*Thysanoteuthis rhombus*), a giant squid that reaches up to 1 m (3.28 ft) in length.

Dolphins are a common sight in the Strait, particularly at its southern end, where dolphin-watching trips have become an important tourism industry. Early-morning spotting tours out from the Negros city of Bais regularly yield interactions with pods of Risso's Dolphins (*Grampus griseus*), Spotted Dolphins (*Stenella attenuata*) and Spinner Dolphins (*Stenella longirostris*), which take great delight in swimming along with the boats' bows. Less common, but seen from time to time, are Short-finned Pilot Whales (*Globicephala macrorhynchus*), while shier

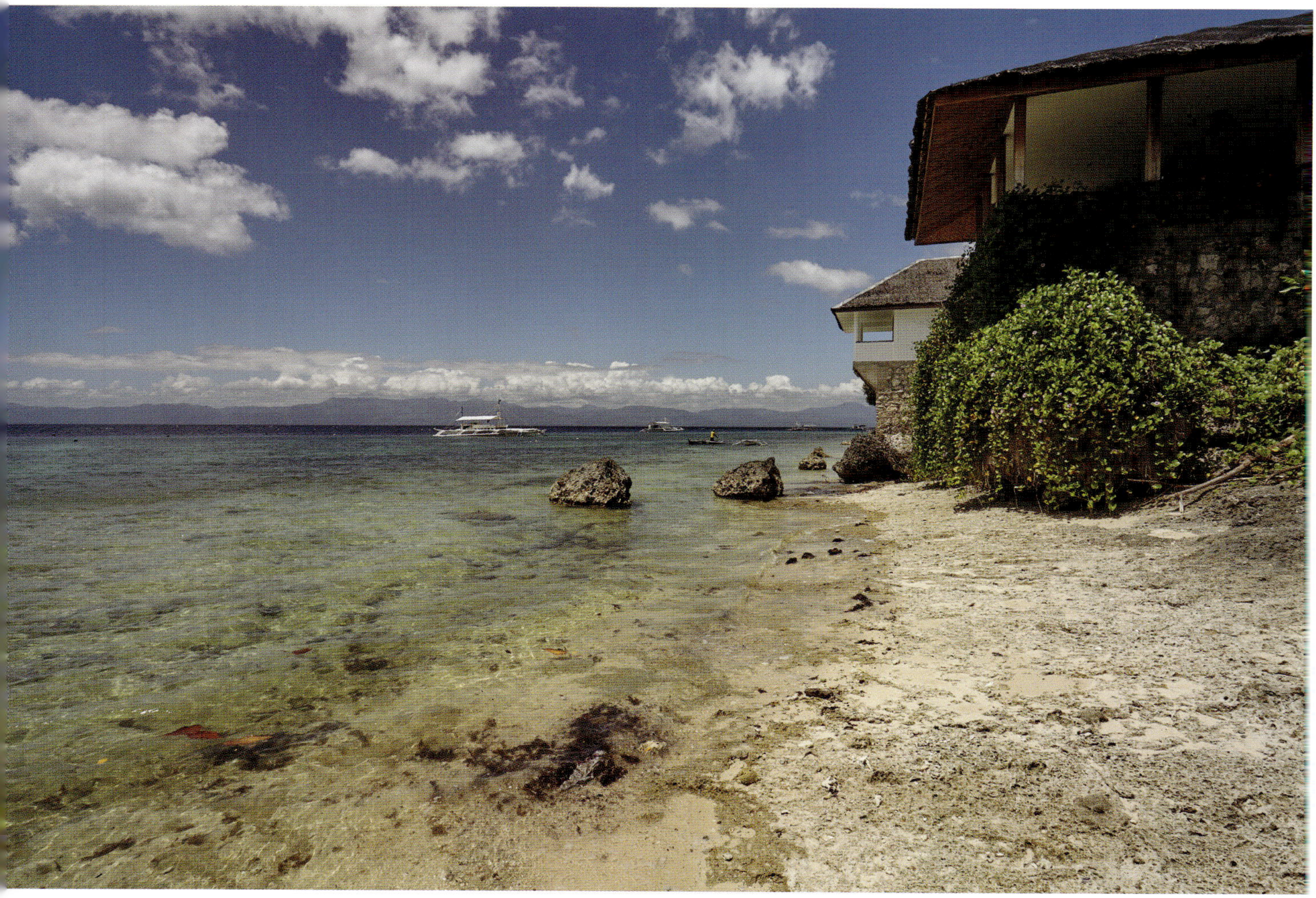

Opposite: *The shoreline at Panagsama Beach, Moalboal, on the Cebu shore of the Tañon Strait, is lined with dive resorts due to the superb coral reefs here.*

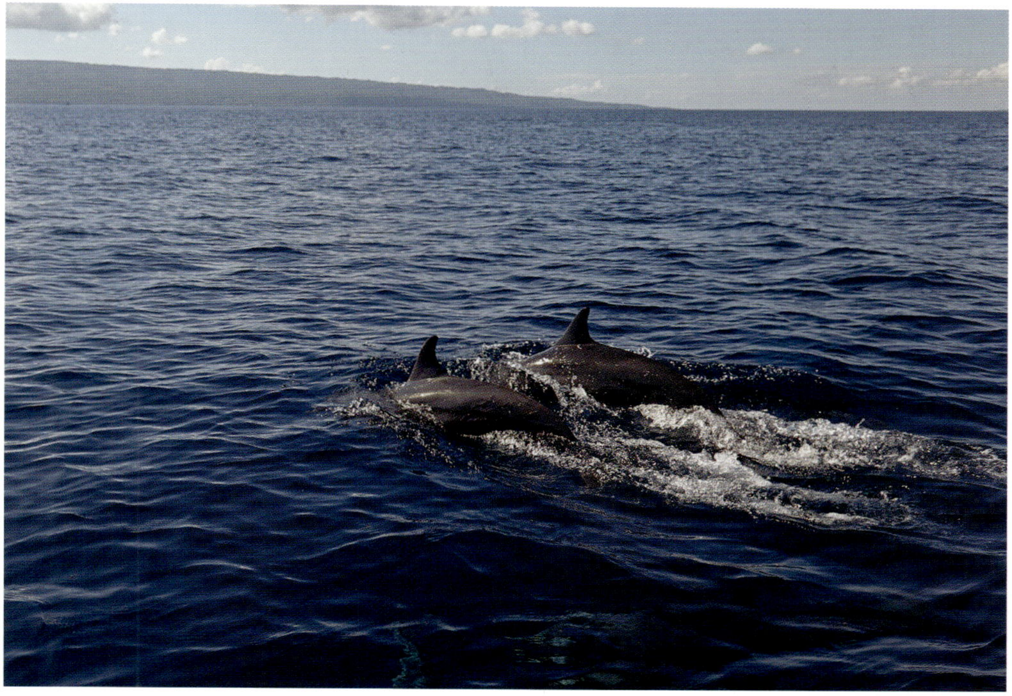

Right: *A pair of Spinner Dolphins enjoying the calm waters of the Tañon Strait. They are one of 14 species of cetacean that live here.*

cetaceans include Fraser's Dolphins (*Lagenodelphis hosei*), Dwarf Sperm Whales (*Kogia sima*) and Pygmy Killer Whales (actually a dolphin, *Feresa attenuata*).

When it comes to the Strait's coral reefs, the best known are along the south-west coast of Cebu, close to the town of Moalboal and around nearby Pescador Island. Over the past 25 years this area has become one of the Philippines' leading scuba-diving sites.

Moalboal's coral reefs

This relatively short stretch of Tañon Strait coastline has some of the country's most accessible and yet healthiest coral reefs. Recent surveys have found that in many places the reefs have more than 70 per cent coral cover, with a rich mix of species.

Many of the reefs popular with divers lie along the Copton peninsula, to the north of Moalboal town, as well as to the west at Badian Island. However, the star attraction is tiny Pescador Island, the tip of a sea mount that reaches barely 100 m (330 ft) in length and 6 m (20 ft) above the sea, lying just 2 km (1.25 miles) off the Cebu shore. At virtually all these locations the sea floor falls away steeply from the shore, quickly reaching depths of 50 m (165 ft), and just off Pescador dropping down to 165 m (543 ft).

As a result, most of the reefs are quite narrow, consisting of a short reef flat, a rich and robust reef crest, then either very steep slopes or sheer walls plunging immediately to a depth of 50 m (165 ft). This creates a very crowded scene, with solid 'blocks' of hard coral, such as *Acropora* species table and staghorn corals on the reef crests, and soft corals that include *Dendronephthya* species soft tree corals, as well as gorgonian sea fans, just a short distance away on the

Below: *A Golden Damselfish (*Amblyglyphidodon aureus*) on a reef just off Moalboal's Panagsama Beach.*

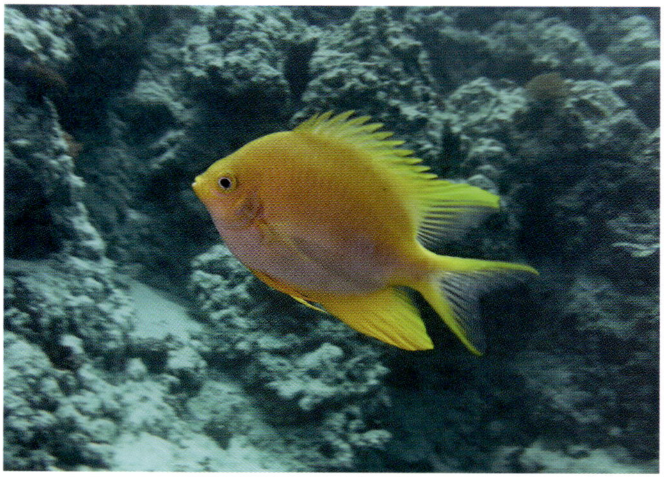

deeper walls and slopes. A vast array of fish and invertebrate reef animals crowds around the corals, while the surrounding deep waters often bring in shoals of large fish such as tuna, trevally and barracuda, while Reef Manta Rays and Whale Sharks (*Rhincodon typus*) are occasionally seen. Both Green and Hawksbill Turtles are common residents of the reefs. One of the most astounding sights is a vast shoal of sardines (*Sardinella* species), said to number over a million fish, that congregates every afternoon over the Saavedra reef, a short distance off Moalboal's Panagsama Beach.

Protecting the Tañon Strait's environment

Throughout the Strait there is a plethora of small, locally established and managed marine reserves. Anyone diving around Moalboal and Pescador Island will be aware of those just around there. These are local no-take zones designed to be safe havens for fish, which can then seed surrounding areas to help boost declining fish stocks. Even though many cover only a few hectares at most, there is growing evidence that they can be very effective.

On a far grander scale, central government has declared the entire Tañon Strait a protected area, the Tañon Strait Protected Seascape (TSPS). Covering a massive 521,018 ha (1.29 million acres), it is by far the Philippines' largest marine protected area, five times larger than Tubbataha Reefs Natural Park. Although the protected area was set up partly in recognition of its important marine biodiversity and the popularity of its dolphins and whales, a major reason is to promote conservation of the Strait's fish stocks, of critical importance to the 43,000 fishermen that live around its shores.

The Strait's waters have been coming under increasing pressure, with over-fishing resulting in rapidly falling fish catches, and rising pollution from coastal industries reducing water quality. With the advent of protection, commercial fishing (that is, with large offshore boats) has been banned in the Strait, in order to favour inshore fishermen working with small boats and traditional equipment. Moreover, the Philippines' Supreme Court has blocked moves to allow oil exploration in the Strait, something seen as a rare victory for environmental protection in the Philippines. Perhaps not surprisingly, the protected area's sheer size has hampered progress towards better protection. Although the Strait's Protected Seascape status was declared in 1998, it took another 17 years for a management plan to be agreed by the vast number of stakeholders involved. As a result, it is still early days for effective protection measures to be fully in place, and it remains to be seen what impact they will have.

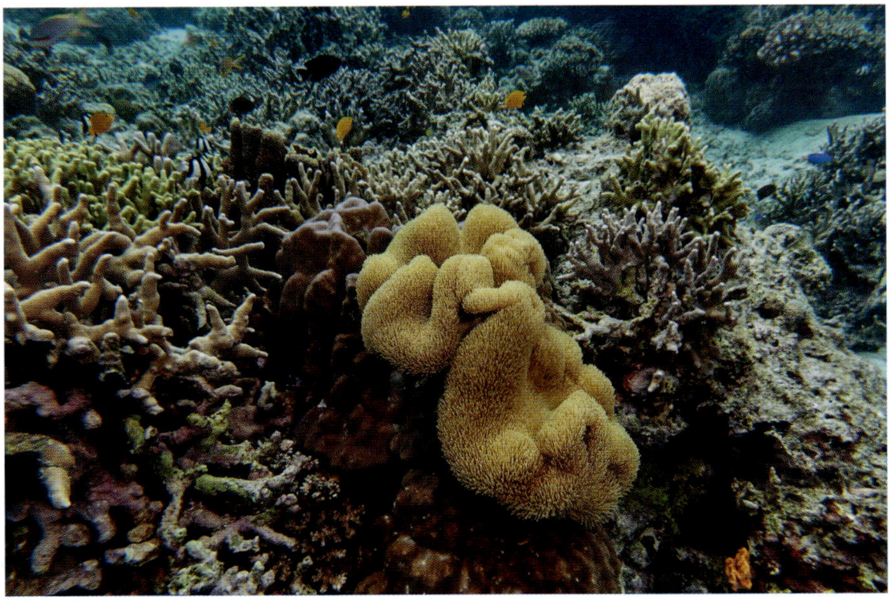

Left: *A mushroom leather coral (*Sarcophyton *species) growing among a crowd of* Acropora *species hard corals, close to the reef crest on Tuble reef, near Moalboal.*

Opposite: *A shoal of an estimated one million sardines congregates off Panagsama Beach every afternoon, creating a stunning spectacle for divers.*

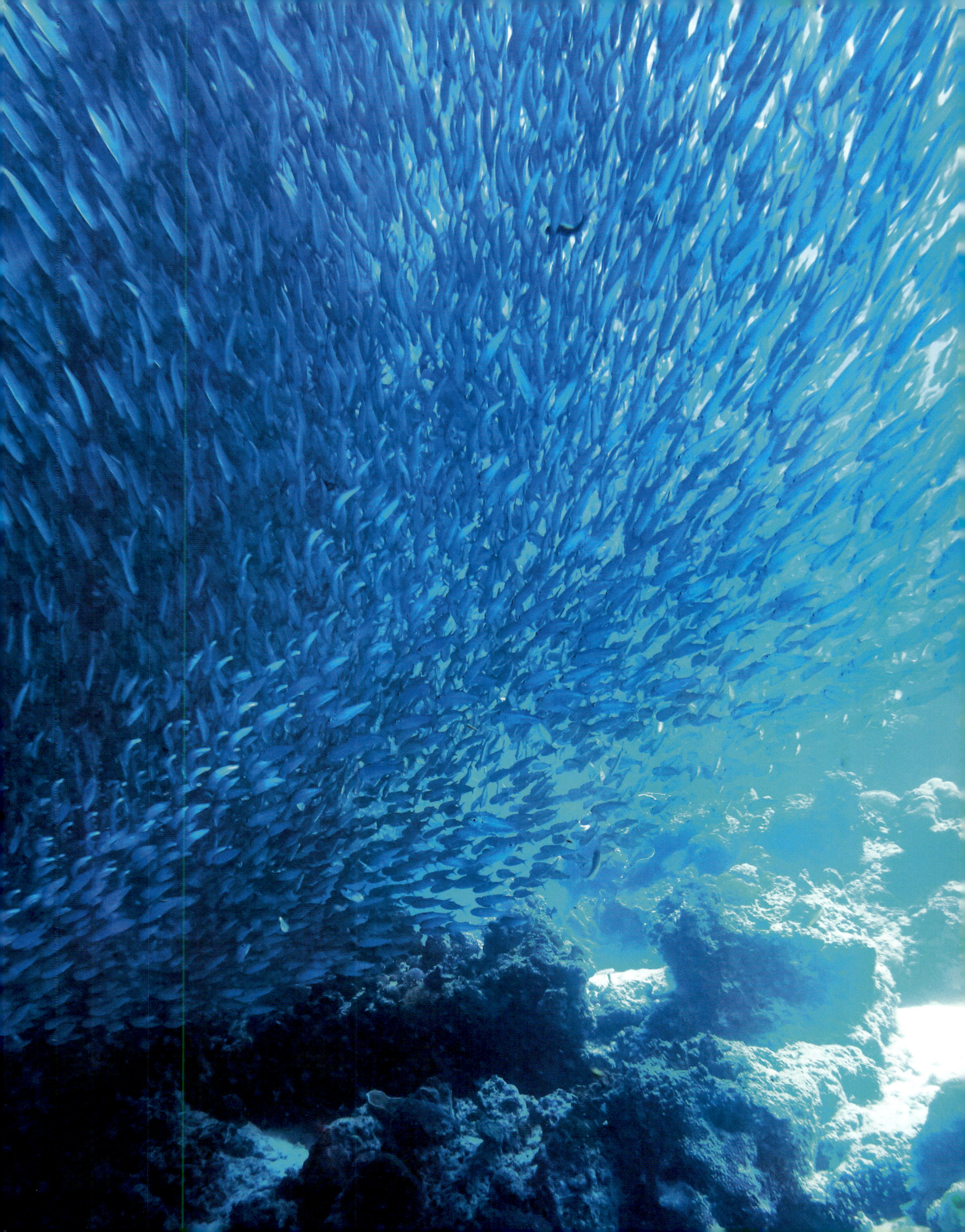

Samar Island Natural Park

This magnificent protected area encloses some of the Philippines' largest remaining tracts of lowland rainforest, covering 333,300 ha (823,251 acres), about a third of the whole of Samar Island, plus another 125,400 ha (309,738 acres) of buffer zone.

Samar Island itself is the country's third largest island, located in the Eastern Visayas but (along with nearby Leyte and Bohol) biologically part of the Greater Mindanao Faunal Region. It has always been a highly rural, relatively lightly populated region, something that has enabled its forests to survive so well. This does not mean, of course, that the forests are in pristine condition. A US military survey conducted in the 1950s found that more than 80 per cent of the island was forested at that time. However, by the 1970s that had shrunk to 33 per cent under an onslaught of logging and mining. By the 1980s barely a quarter was still forested, but when fatal floods were blamed squarely on the deforestation commercial logging was abruptly banned, though it still continues today on a small scale.

The remaining forest areas were declared a forest reserve in 1996, but in 2003 they were upgraded to a natural park. Four pre-existing protected areas have been incorporated into the natural park, the best known of which is the Sohoton Natural Bridge National Park, near the town of Basey.

Below: *Lowland rainforest close to the village of Tenani, site of the natural park's headquarters.*

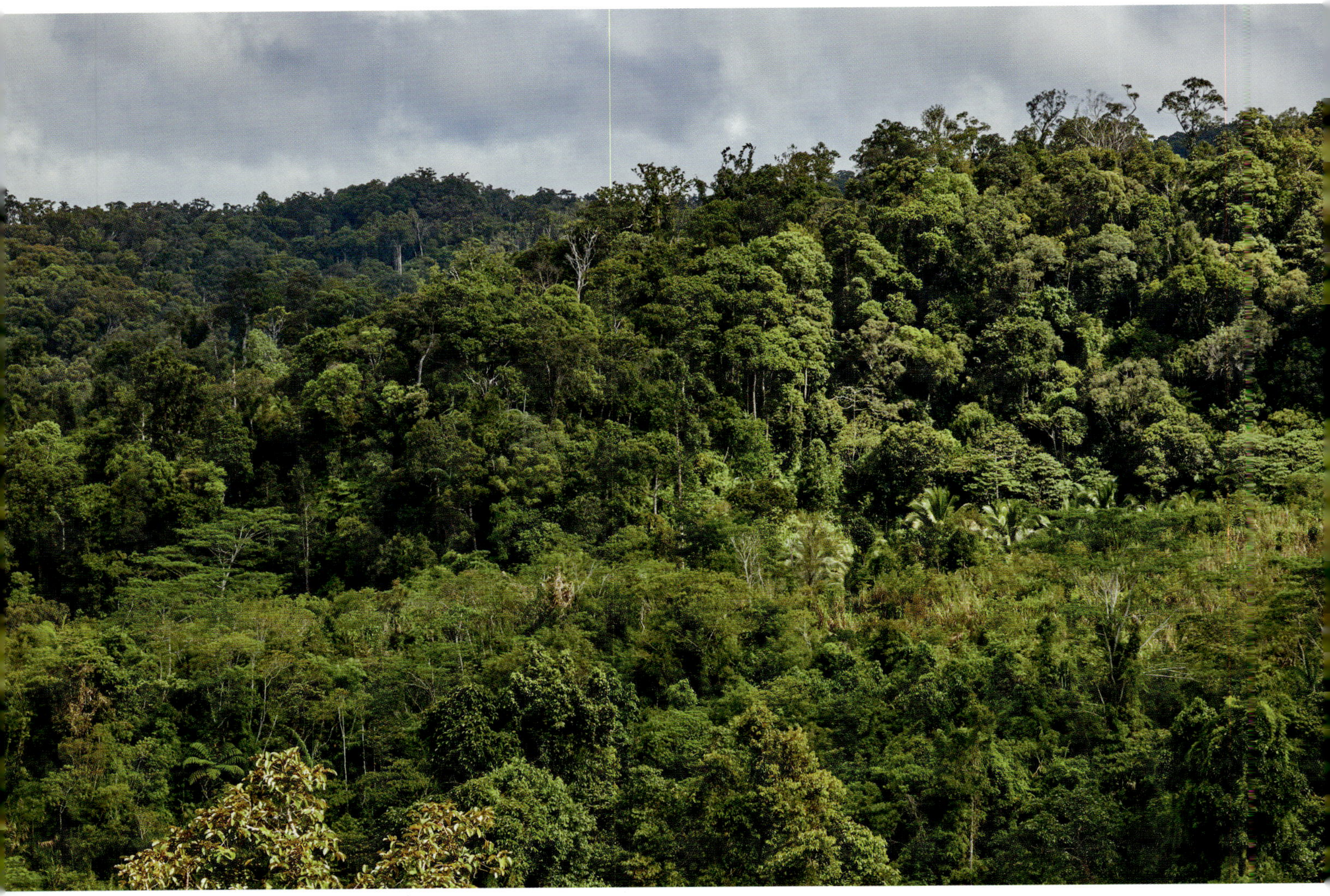

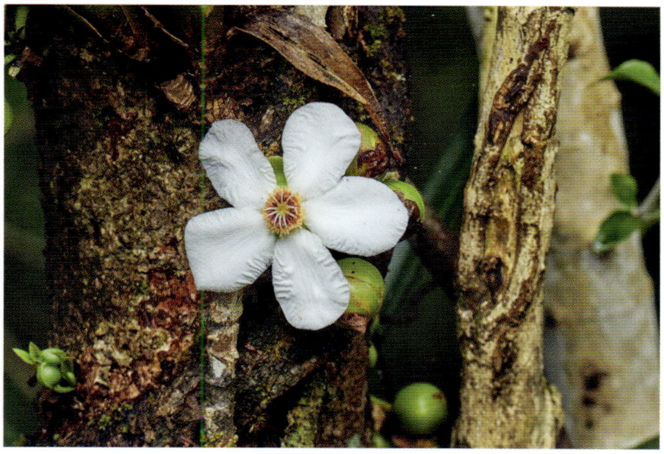

*Above: An example of cauliflory, flowers forming directly from a tree's trunk, here on a Katmon (*Dillenia cauliflora*) tree.*

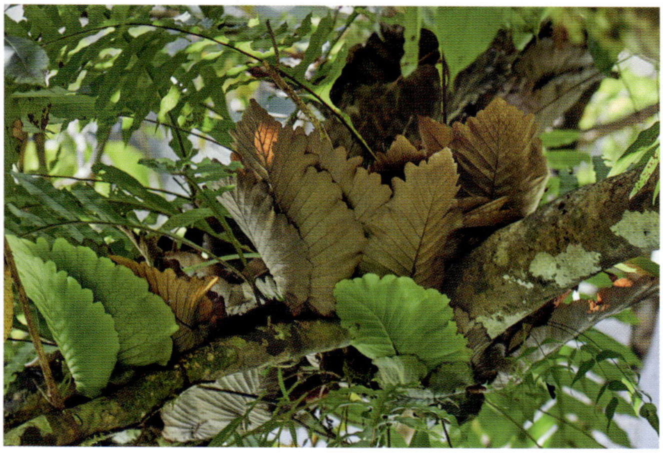

*Above: The native epiphytic Oak-leaf Fern (*Drynaria quercifolia*) growing in a tree in limestone rainforest, close to Sohoton Caves.*

The park's landscape

The park covers much of the rugged heart of Samar Island, penetrated by very few roads, and largely inaccessible to all but the most experienced forest walkers. The terrain consists mainly of rolling, mostly karst limestone hills, rising to a height of 853 m (2,806 ft), Mt Capoto-an, the park's highest point. The landscape is riddled with caves, most of which have been barely explored, though a few have become visitor attractions. The whole of Samar Island is exposed to both the north-east and south-west monsoons, so there is no distinct dry season, though the heaviest rains fall during November–January. The island is right in the Philippines' main typhoon zone, leading the park to be hit by an average of nine every year, usually between July and November.

Habitats and wildlife

Forest covers 63 per cent of the park's land, with most of the rest taken up by farmland, mainly coconut plantations mixed in with grassland and scrubby areas. Of this forest, about 45 per cent is estimated to be primary old growth that has never been logged, with the remainder consisting of secondary forest.

Several types of forest habitat occur, in both the primary and secondary forests. The main habitat is evergreen lowland rainforest, comprising largely dipertocarp trees and accounting for about two-thirds of the forest. Most of the remainder consists of limestone forest, areas where the soil is very thin and poor, and the vegetation is made up of plants adapted to the harsh environment and lack of water, such as *Dracaena* and *Sterculia* species, along with a variety of orchids and pitcher plants. Finally, there is a small amount of ultrabasic forest, typified by stunted trees adapted to survival over rocks that are rich in potentially toxic minerals.

The tree species that make up the park's forests constitute a significant proportion of the nearly 1,000 plant species so far identified here. More than 400 have been found to be endemic to the Philippines, with 40

*Below: A Chocolate Argus butterfly (*Junonia hedonia ida*), in forest in the Taft Philippine Eagle Sanctuary, near Taft.*

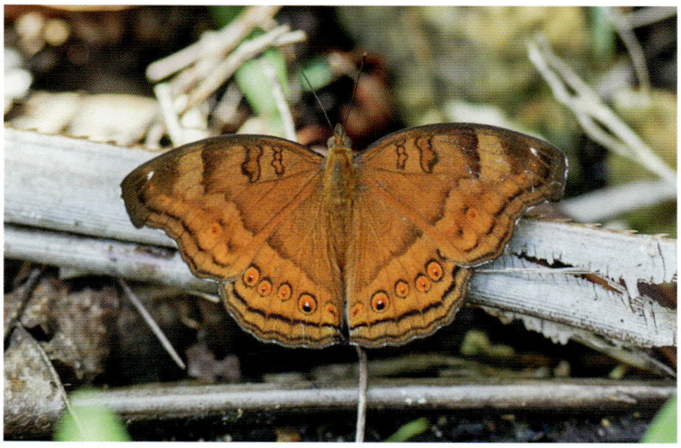

endemic to just Samar Island. Twelve of those 40 are trees that have become threatened, presumably due to excessive logging in the past.

It is a similar story with the park's animal wildlife. Almost 300 vertebrate species have been found, including more than 30 species of frog and toad, over half of them endemic to the Philippines. There are also 65 reptile species, more than half of them lizards and most of the remainder snakes.

Mammalian wildlife consists of about 50 species, over half of them bats, both insectivorous and fruit-eating. The non-flying mammals include a number that are endemic to the greater Mindanao region, such as the Philippine Flying Lemur (*Cyanocephalus volans*) and Philippine Tarsier (*Carlito syrichta*), as well as a number of forest rodents such as the Philippine Pygmy Squirrel (*Exilisciurus concinnus*). Over 70 per cent of the mammalian fauna is endemic to the Philippines.

The park's 150 bird species include the mighty Philippine Eagle, which is occasionally sighted here, and which was first identified in Samar in 1895. The area of that first sighting, near the town of Taft, close to the Pacific coast, is still classified as a Philippine Eagle sanctuary, though it is well within the boundaries of the park. The eagle is not, of course, the only endemic bird found here: there are about 60 including a number that are endemic just to greater Mindanao, and a handful restricted only to the eastern Visayan islands. The last of these includes the Samar Hornbill (*Penelopides samarensis*), which can be readily seen in forest along the Ulot River and close to the park headquarters at the village of Tenani.

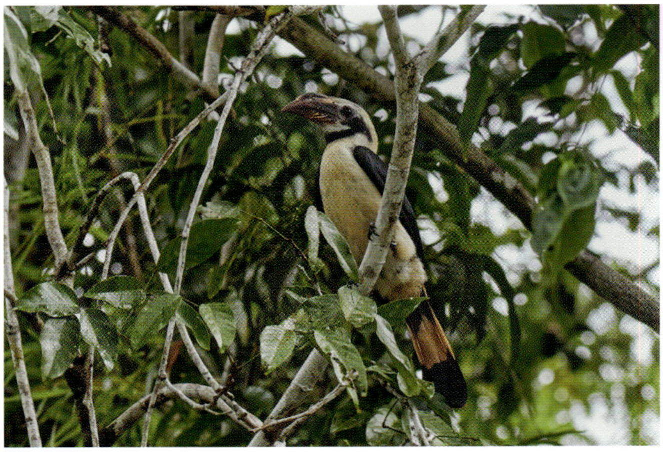

Above: *Male Samar Hornbill in forest along the Ulot River.*

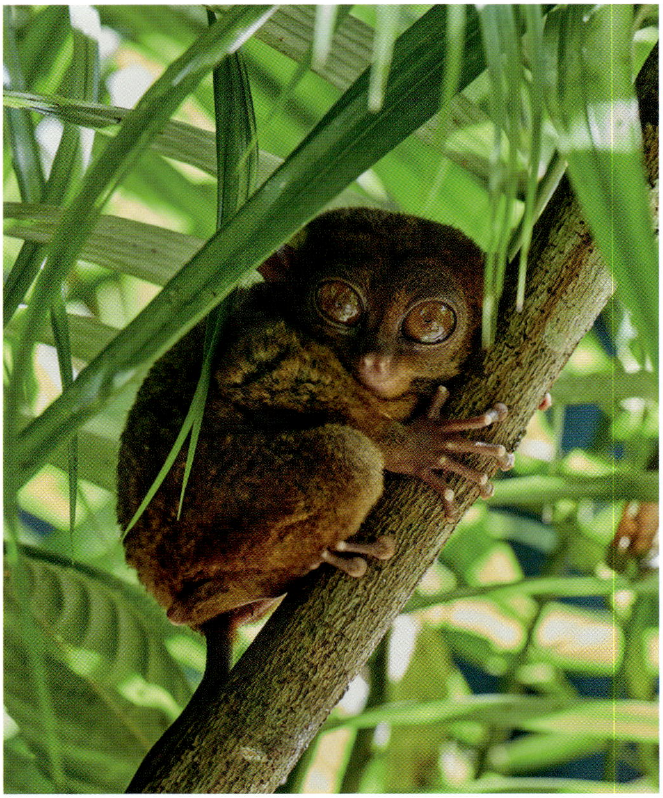

Below: *Being endemic to the Greater Mindanao Faunal Region, the Philippine Tarsier can be found in Samar Island Natural Park.*

Visiting the park

Visitor facilities around much of the park are quite limited, but the one area that has been seeing visitors for a number of years is the Sohoton Caves, near Basey, in the south of the island. Here it is possible to explore Panhulugan I Cave, filled with lovely stalactite and stalagmite formations, and see a limestone arch spanning across a gorge to link up opposite cliffs. The area is reached by boat along the Cadacan River from near Basey.

The forest around the village of Tenani, located on the cross-island road that cuts through the park from Paranas on the west coast to Taft in the east, is becoming a popular birdwatching area. A good range of birds can be seen from the headquarters itself, and

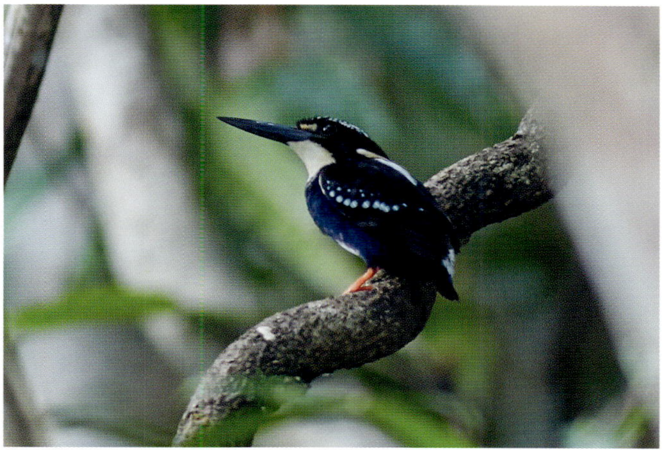

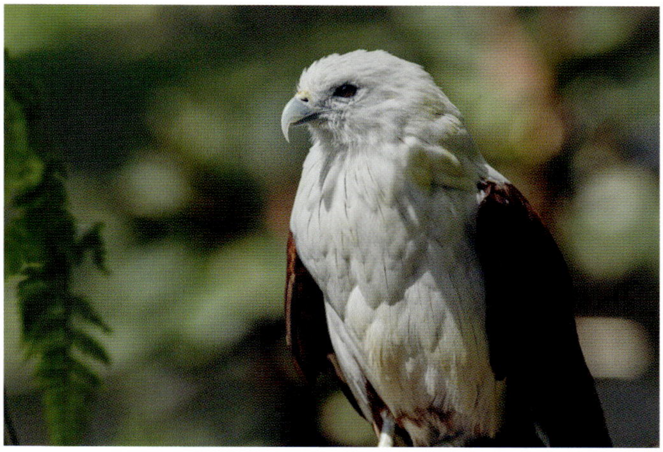

Above: *The beautiful and rare Silvery Kingfisher spotted perching on a tree along the Ulot River, near Tenani.*

Above: *Brahminy Kites (*Haliastur indus*) are a regular sight, soaring above the forest, particularly near open water.*

several trails make it possible to explore nearby forest. The headquarters has simple accommodation for birdwatchers.

Also in Tenani, it is possible to travel along the Ulot River into the heart of the forest, using fast motorized canoes that were once employed in carrying timber out of the forest. They give quite a thrilling ride, and provide one of the best ways to get into some good forested areas. Sightings of hornbills and various kingfishers, including the rare Silvery Kingfisher (*Alcedo argentatus*), a greater Mindanao endemic, are common along the Ulot River.

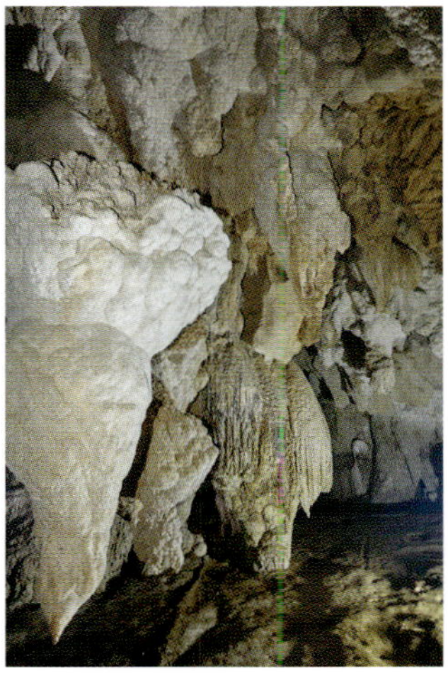

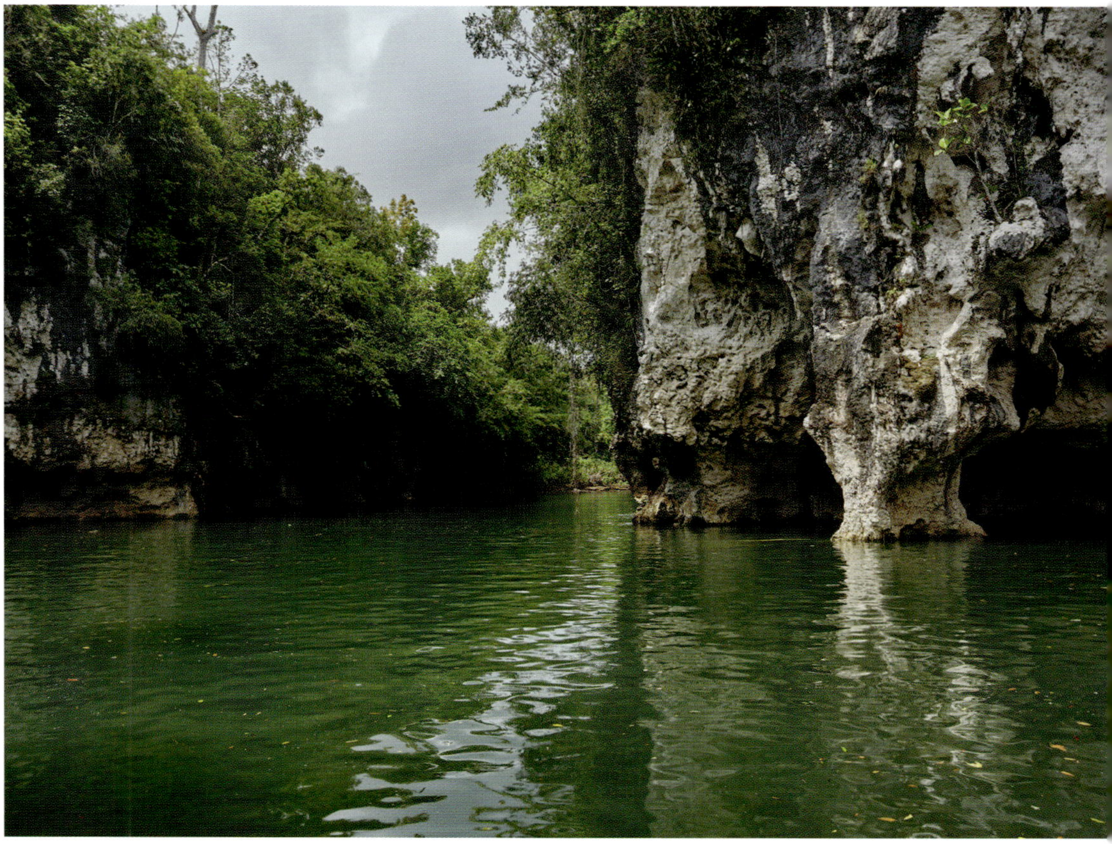

Above and right: *The limestone Sohoton Caves, in the natural park's far south, near the town of Basey, are reached along the Cadacan River. Inside is a magnificent show of white and coloured stalactites and stalagmites.*

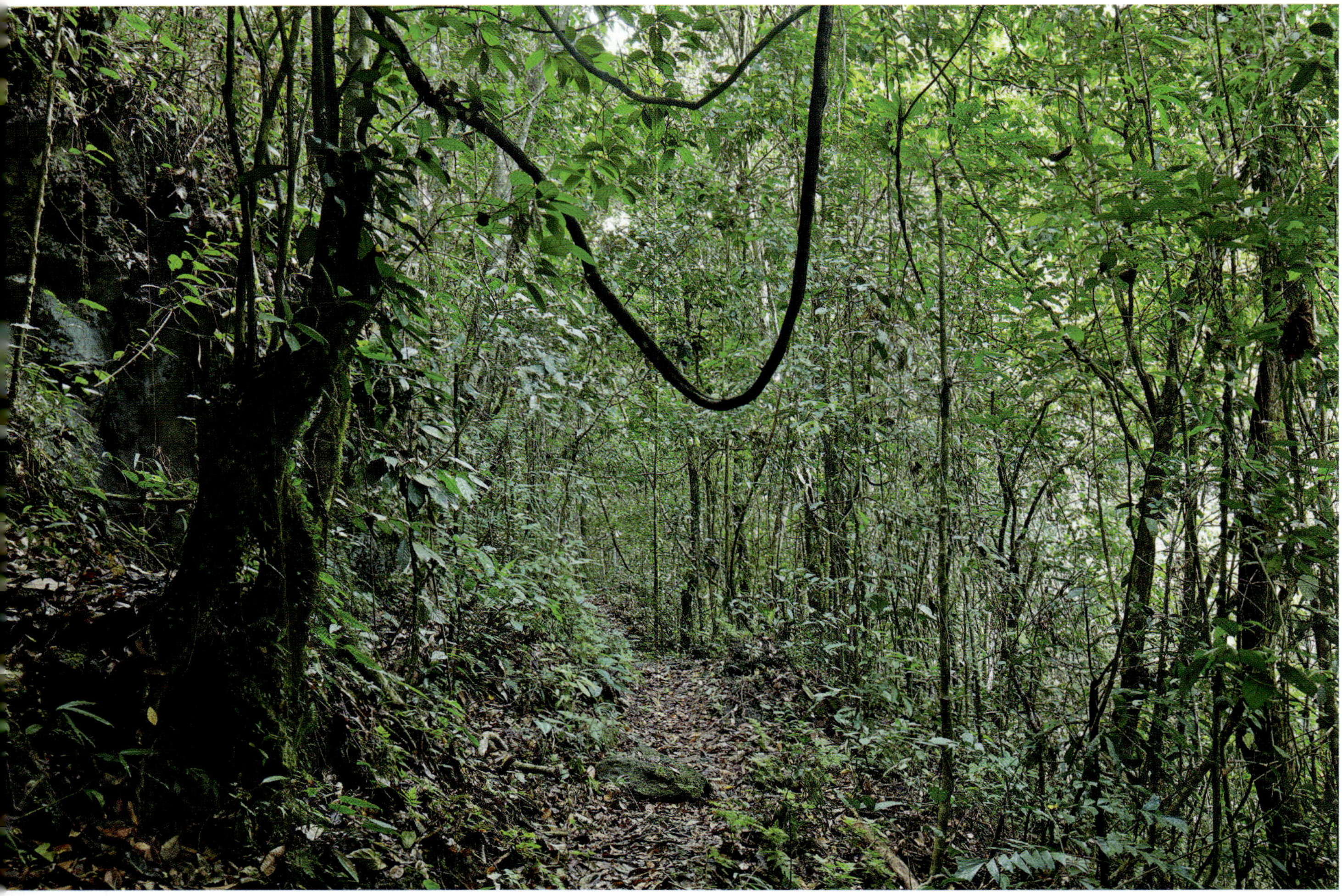

Rajah Sikatuna Protected Landscape

Encompassing an area of 10,452 ha (25,816 acres), Rajah Sikatuna Protected Landscape preserves the last extensive area of forest in the whole of Bohol, and is situated in south central Bohol, east of the town of Bilar. The region first came under protection in 1987 as a national park, at which time it covered a little over 9,000 ha (22,300 acres), but it was renamed and expanded to its current size in 2000.

The landscape consists largely of low, rolling karst limestone hills, with its highest point Bilar Peak at 510 m (1,678 ft). Much of the forest is secondary, although fragments of primary forest remain on the steepest, least accessible slopes. Both the primary and secondary forests are limestone in nature, and cover about 60 per cent of the park, with much of the remainder consisting of grasses and farmland. Since the 1960s there have been a number of reforestation programmes around the park's margins, but unfortunately these have used mainly mahogany trees – an alien species that is not very beneficial to local wildlife. Given the limestone base, there is very little surface water, but plenty underground, resulting in four natural springs around the park. The most accessible of these is Logarita, located close to the park's main entrance, near Bilar. The pool is popular with local people, particularly at weekends.

Rajah Sikatuna's wildlife

The park's flora has been found to consist of over 600 plant species, more than 400 of them trees, including 13 species of dipterocarp. Below the trees there is an extensive tangle of vegetation that includes more than 100 species of fern, along with a host of vines and

Opposite: *Classic limestone rainforest in Bohol's Rajah Sikatuna Protected Landscape.*

Right: *The Philippine Forest Dragon (*Gonocephalus cf. sophiae*), a Philippine endemic species, in the forest.*

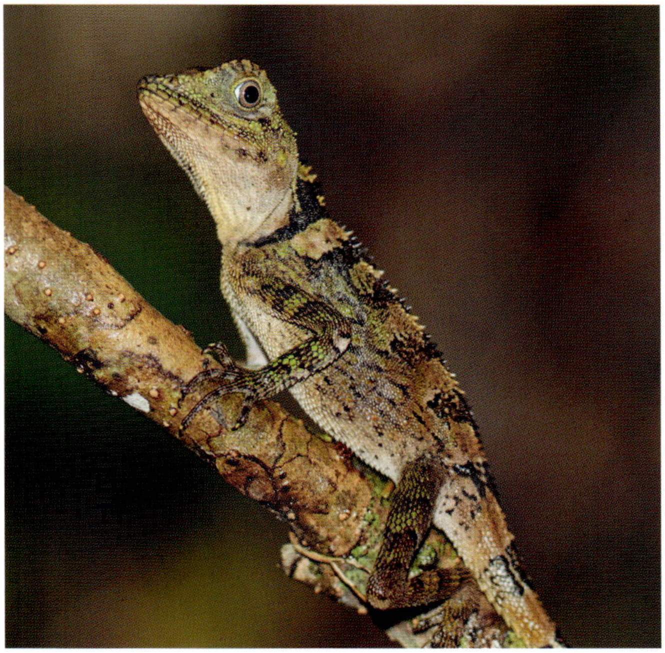

palms, including rattans. Although geographically part of the Visayas, Bohol's animal wildlife – and hence that of Rajah Sikatuna – reflects the fact that biologically it is part of the Greater Mindanao Faunal Region. For mammals this means the presence of the Philippine Tarsier and Philippine Flying Lemur. The latter can be seen quite regularly during the day, resting in some of the higher trees. The tarsiers, on the other hand, being tiny and nocturnal, are much harder to find, but can be readily seen at the nearby Corella breeding centre. Other, more widely distributed mammals are the Long-tailed Macaque, Philippine Warty Pig, and both the Malay and Common Palm Civet (*Viverra tangalunga* and *Paradoxus philippinensis*, respectively). There are also 25 species of bat, both fruit and insectivorous.

Bird life

Rajah Sikatuna Protected Landscape is well known for its range of rare birds, and is now one of the Philippines' most popular birdwatching sites. The forests here are home to 127 avian species, 52 of them Philippine endemics. The area is one of the few remaining refuges for the Mindanao Bleeding-heart Pigeon (*Gallicolumba crinigera*) and Azure-breasted Pitta (*Pitta steerii*), and is home to five subspecies that are endemic to Bohol. In addition, three species are found here that are endemic just to the eastern Visayan islands of Bohol, Samar and Leyte. These are the Samar Hornbill, Visayan Broadbill (*Sarcophanops samarensis*) and Yellow-breasted Tailorbird (*Orthotomus samarensis*).

Below: *Commonly seen in the forest is the Philippine Flying Lemur, a mammal endemic to Greater Mindanao.*

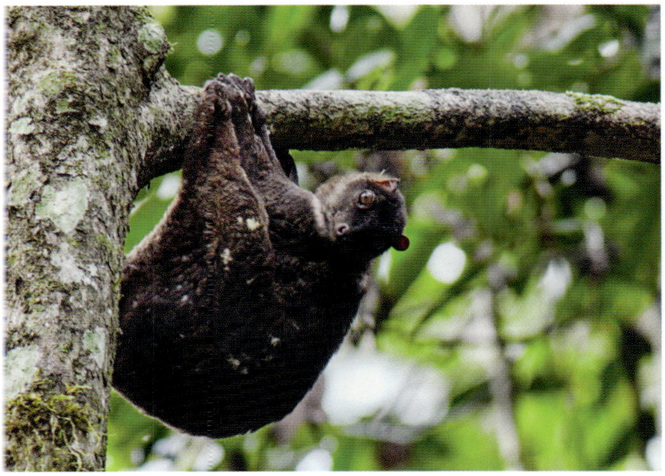

Below: *Rajah Sikatuna is believed to be one of the last safe homes for the Greater Mindanao endemic Mindanao Bleeding-heart Pigeon.*

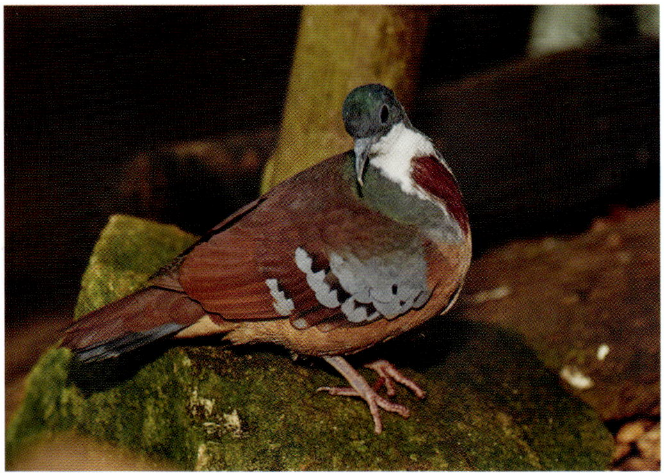

The coral reefs of Panglao and Balicasag Islands

Lying off the south-west coast of Bohol, Panglao Island is today one of the Philippines' leading beach destinations, while Panglao and Balicasag together constitute one of Southeast Asia's top scuba-diving areas. The quality of the coral reefs around these two islands is what initially drew the first visitors, with scuba divers arriving from the early 1980s onwards. From the beginning, they were based at attractive Alona Beach, on Panglao's southwestern shore, and from there this quiet dive resort has grown into a major beach destination, at which diving is just one attraction.

Panglao is quite a large island, lying close enough to the Bohol mainland for it to be connected by two causeways at its eastern end. Around here the water is shallow, with generally a sandy or muddy bottom and a few mangrove areas. The white-sand beaches, along with the fringing coral reefs, start further west, and extend along much of the southern shore, around Panglao's western promontories, and continue along the northwestern side of the island.

Balicasag Island lies 6 km (3.7 miles) south-west of Panglao, and by contrast is tiny; almost circular, just 600 m (1,970 ft) across and covering an area of only 30 ha (74 acres). It is also pancake flat, rising merely a couple of metres above sea level, is covered with some scrubby vegetation and is completely ringed by a

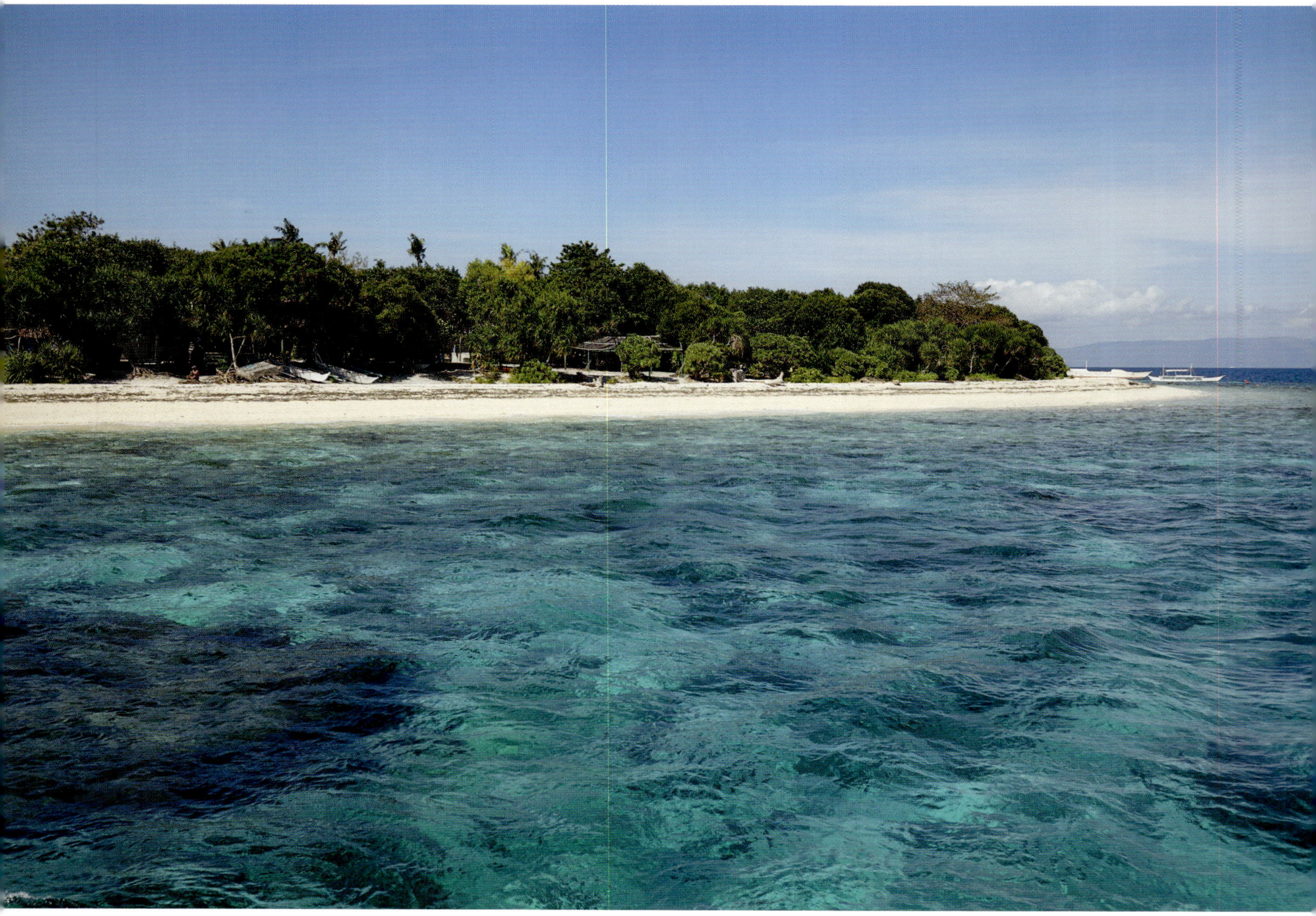

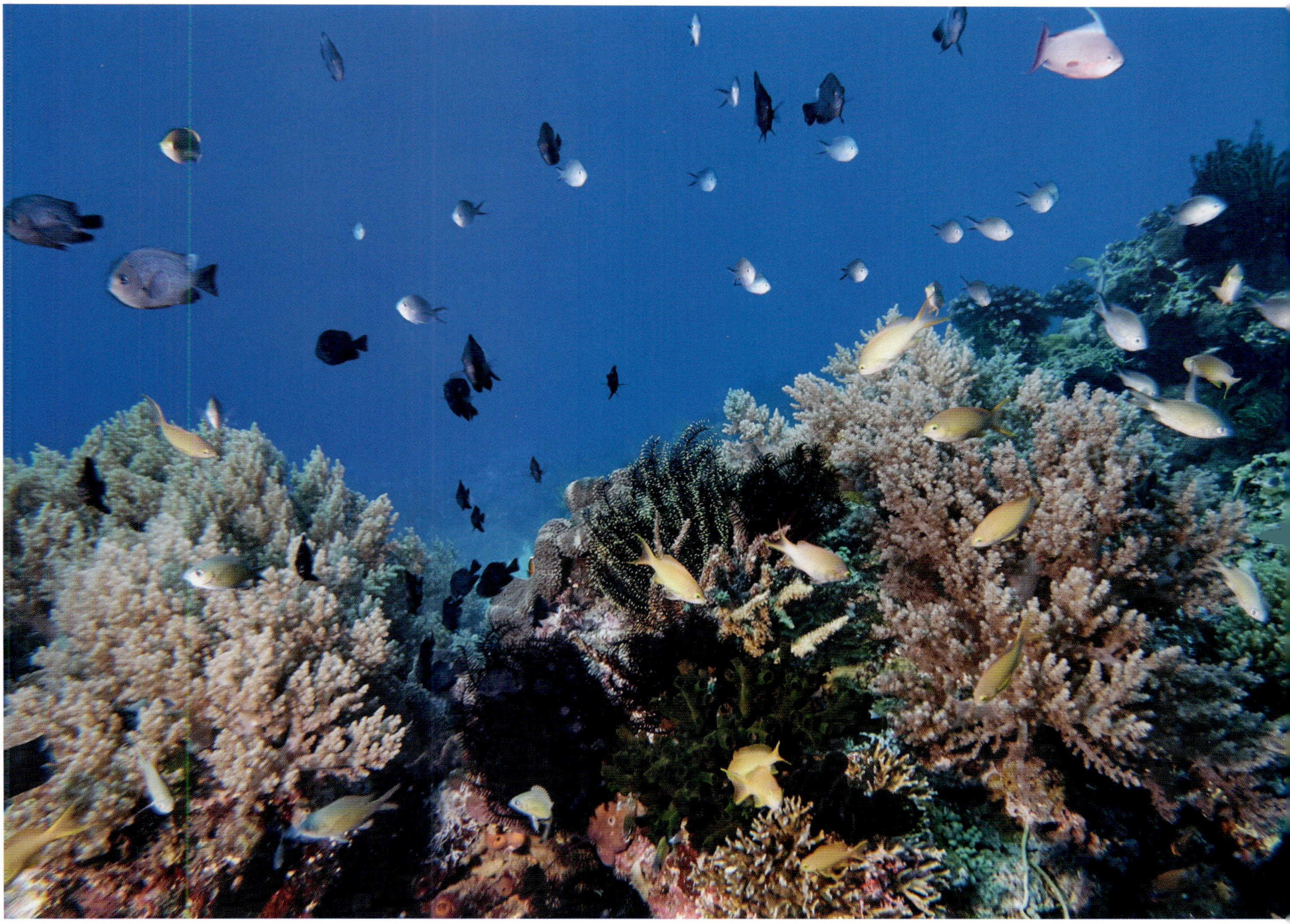

Above: A colourful array of life, with a mix of corals, feather stars and small reef fish, off the southern shore of Balicasag Island.

Opposite: Beautiful Balicasag Island, surrounded by crystal-clear turquoise waters, site of some of the Philippines' loveliest coral reefs.

blindingly white coral-sand beach. The island is simply the tip of a raised coralline sea mount, around which the shore drops steeply away into deep water.

The islands' coral reefs

Panglao's most beautiful reefs, and hence some of the most popular dive sites, occur at Tawala, which is almost directly offshore from Alona Beach, as well as around Pungtud Point, Pungtud Island and Doljo Point at Panglao's western tip, and then at Napaling on the north-west coast. In these areas the reefs consist mainly of a sandy, shallow reef flat close to the shore, followed by a reef crest built up from large hard corals, including a range of table, boulder, brain, plate and staghorn corals (among many others), followed by a steep drop-off down to a depth of about 20–25 m (66–82 ft). Here, intricate and delicate soft corals, such as tree corals, gorgonian sea fans and *Tubbastrea* species cup corals are common, along with an array of other non-coral invertebrates, such as sponges, anemones, nudibranchs and starfish.

The reefs all around Balicasag are typified by narrow reef flats in shallow water close to the shore, a strong reef crest in about 7–11 m (23–36 ft) of water, then very steep slopes or sheer walls dropping down to a sea floor at a depth of about 50 m (165 ft). As in the

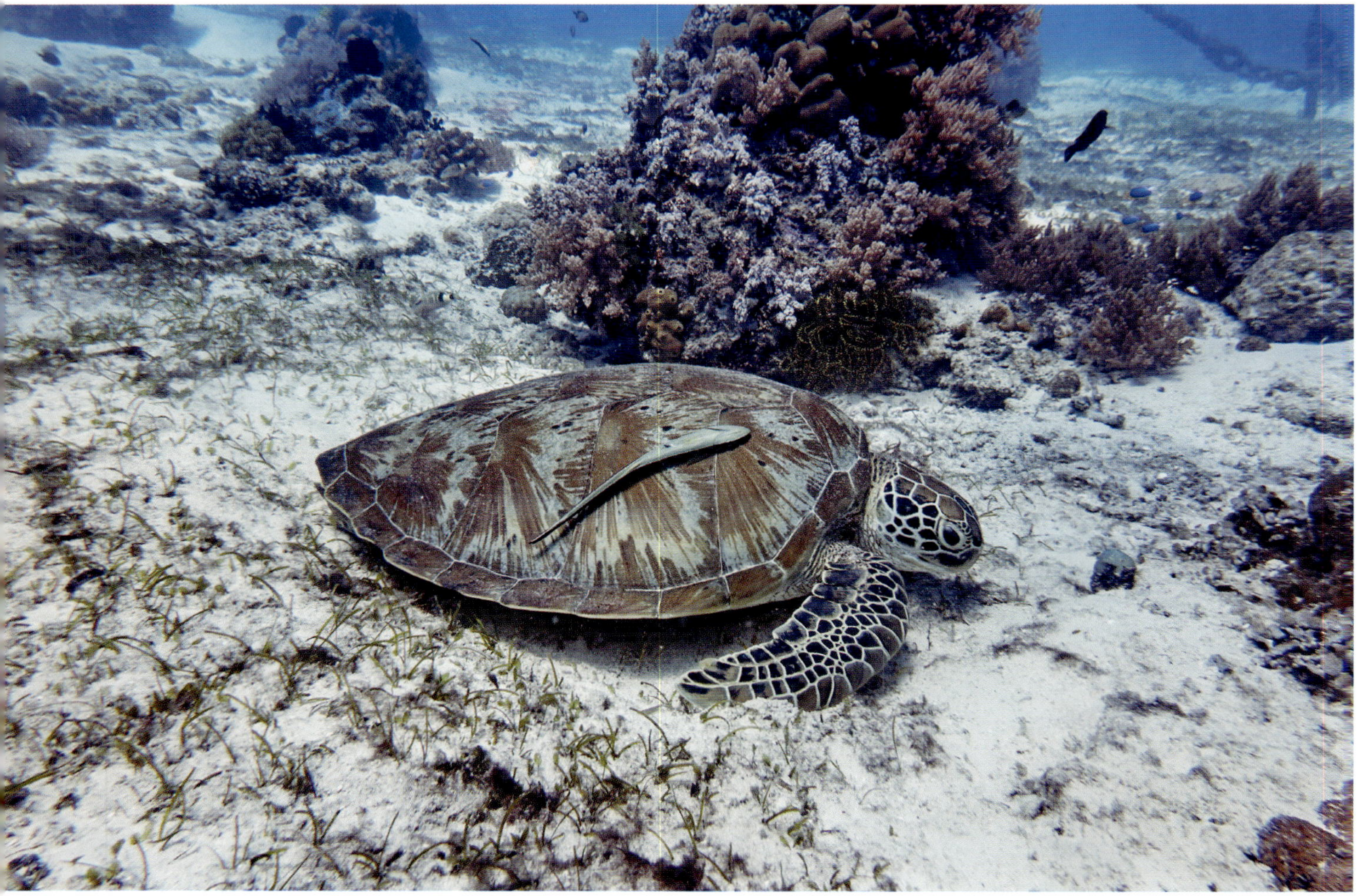

Panglao reefs, the reef crests are typified by large table, boulder, plate and staghorn coral structures, while the walls are festooned with a host of different coral types, including some large sea fans and (off the east shore especially) black corals.

In terms of fish life, both the Panglao and Balicasag reefs have a rich wealth of reef fish, ranging from tiny clownfish, up to large parrotfish and Humphead Wrasse, with everything in between. They can include a very colourful host of damselfish, angelfish, butterflyfish, fusiliers, Moorish Idols, triggerfish, batfish, sweetlips, pufferfish and wrasse, to name just the most obvious and most common ones. Moray eels, lobsters, crabs and shrimps can sometimes be found lurking in any of the many crevices and holes that litter the coral reef structure.

Balicasag's deep waters and strong currents create much better conditions for larger pelagic species than

Above: *A Green Turtle grazing on a bed of seagrass, in the shallows of the reef flat close to Balicasag Island.*

Below: *A Banded Sea Krait, a venomous sea snake, seen off the south-east coast of Panglao Island.*

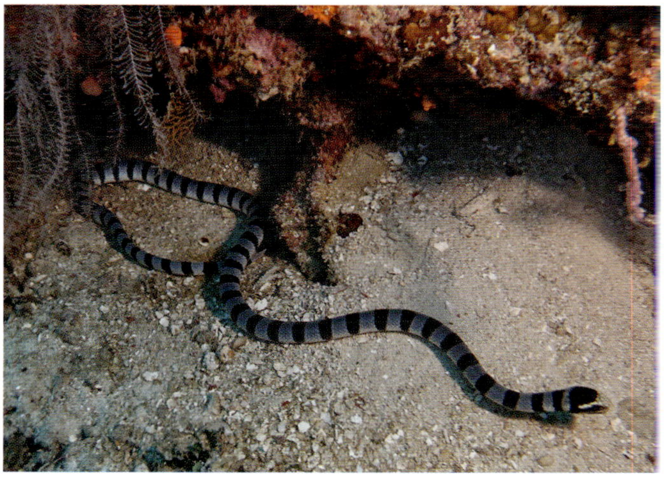

is generally the case around Panglao. A large shoal of Big-eye Trevally (*Caranx sexfasciatus*) is often seen off the southern shore, at a dive site known as Rico's Wall, while other shoaling fish frequently seen around the island include tuna and barracuda. Whitetip Reef Sharks and hammerhead sharks (*Sphyrna* species) are also sometimes seen, while Green Turtles are residents.

Pods of dolphins pass through the seas south of Alona Beach and south-east of Balicasag Island almost every morning en route to feeding grounds. In recent years, watching them has become a hugely popular tourist attraction.

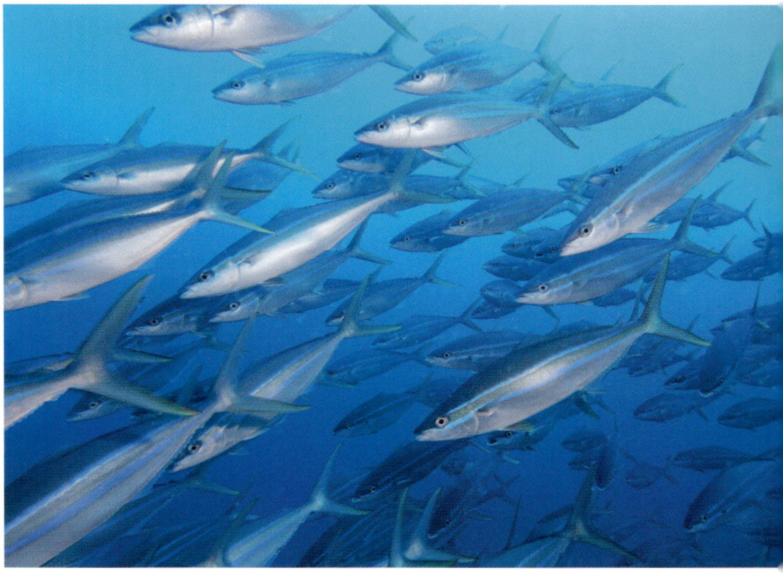

Above: A shoal of Rainbow Runners off the south-east shore of Panglao Island.

Protection of the reefs

Despite these reefs' rich biodiversity very little here has any formal protection. In 1978 it was proposed that Balicasag Island be declared a marine national park, though unfortunately this was never acted upon. Instead, in 1986 a small marine sanctuary, covering about 5.4 ha (13.3 acres) was created off its southwestern shore, a site that persists to this day as a well-managed no-fishing zone.

Since the creation of the Balicasag Marine Sanctuary, several more MPAs have been created along Panglao's north and south coasts, aimed largely at helping to maintain fish stocks for the local communities. In all these sanctuaries, fishing is strictly forbidden, at least in theory, and scuba diving and snorkelling are usually restricted. Although rather small in size (about 4–12 ha/10–30 acres), those that are well managed have been shown to result in improved fish populations within the sanctuaries, which have then helped maintain fish catches in nearby areas.

Below: A giant clam (Tridacna *species*), nearly 1 m (3.3 ft) long, on a reef off Panglao Island's south-east coast.

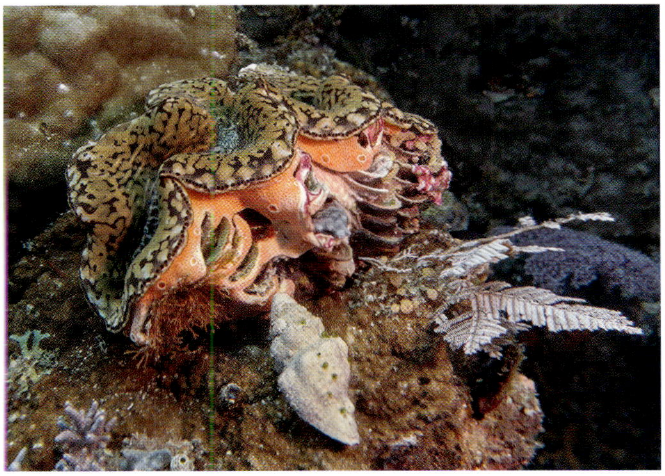

For scuba divers, the most obvious manifestations of these sanctuaries include the presence of mooring buoys due to a strict ban on the dropping of anchors (though this is also the case outside many of the sanctuaries), as well as environmental fees to help with sanctuary management. In recent years, Balicasag Island has become so popular with both divers and snorkellers that there is now a strict daily quota on the number of boats allowed to moor anywhere off the island.

With the huge increase in tourism that Panglao and Balicasag Islands have seen over the past 30 years, which is likely to continue to increase, the pressure on these reefs and their fish stocks is immense and growing. To help ensure their continued health, strong and effective protection measures need to be taken along much of the coast of both islands.

Agusan Marsh Wildlife Sanctuary

A vast freshwater swamp holding an estimated 15 per cent of the country's entire freshwater resources, Agusan Marsh was declared a protected area only in 1996, an event quickly followed in 1999 by its declaration as a Ramsar Site, recognizing Agusan Marsh as a wetland of international importance for migratory birds.

Lying in eastern Mindanao, Agusan Marsh occupies a flat, low-lying plain in an area where the Agusan River and a number of its tributaries merge. These rivers drain several mountain ranges, namely those of Surigao del Sur to the east, Bukidnon to the west and Compostela to the south, forming a vast sponge that soaks up the water pouring down off these mountains. This is particularly the case during the October–March wettest period (although rain falls in this part of Mindanao throughout the year), when the marsh's water levels rise by as much as 13 m (43 ft). Not only does the resulting vast wetland create a huge area for wildlife, it also protects cities downstream, especially Butuan, from floods that would otherwise occur.

Although the marsh as a whole covers about 90,000 ha (222,300 acres), the wildlife sanctuary originally protected only just under 19,200 ha (47,424 acres). However, in recent years its effective size has been more than doubled to 40,943 ha (101,129 acres), in order to take into protection some recently discovered habitats.

The marsh's habitats

The marsh is a stunningly beautiful, rather primeval area, a vast network of lakes, swamps, swamp forests and rivers, calm waters reflecting the huge skies, the only sounds those of birds and insects. Although surrounded by outside pressures, this remains one of the Philippines' great wilderness areas.

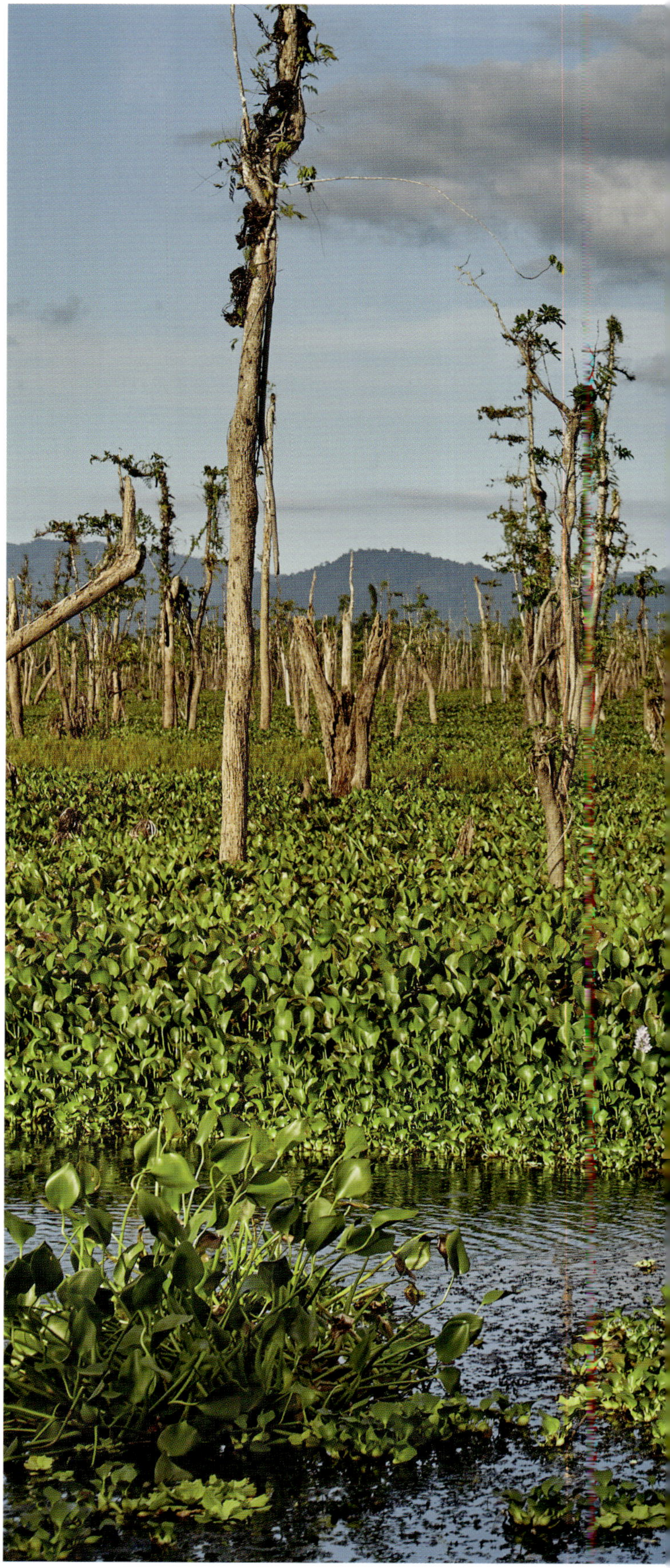

Right: *Typical scene of swamp forest consisting largely of* Terminalia copelandii *trees, growing straight out of the water, along the edge of Panlabuhan Lake, at the southern end of Agusan Marsh.*

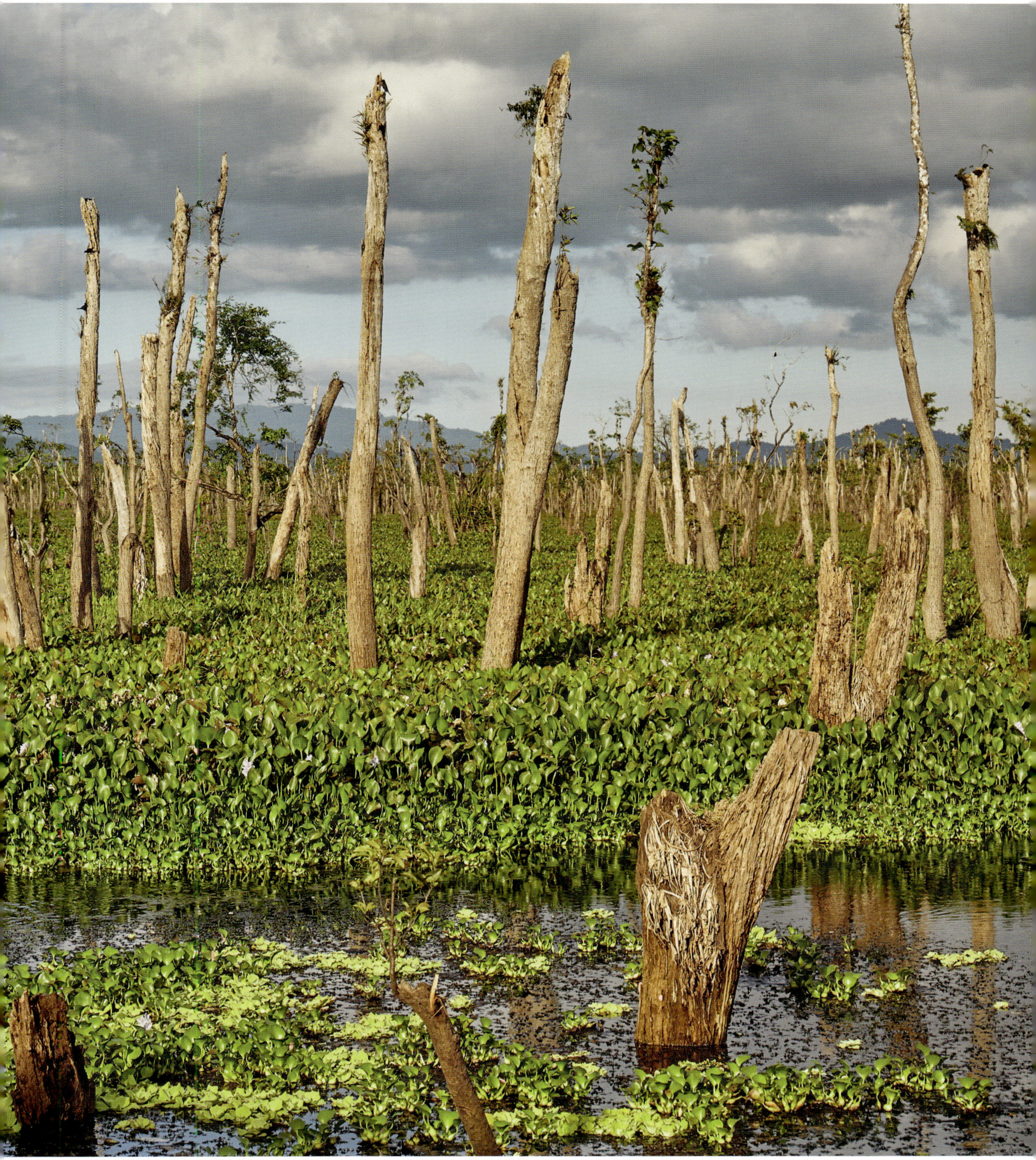

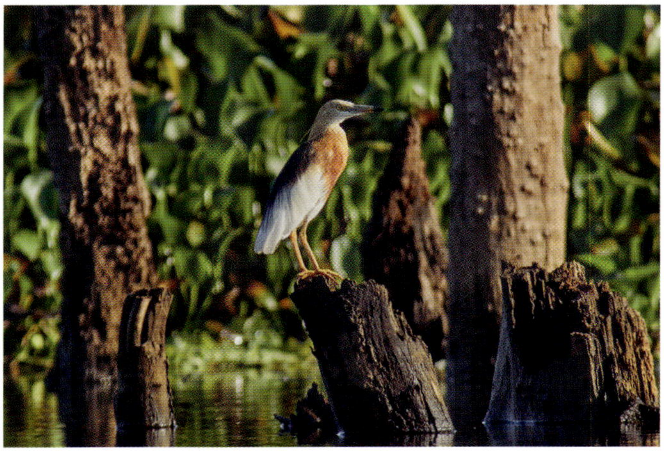

Above: A Chinese Pond Heron in full breeding plumage, in Terminalia *species swamp forest.*

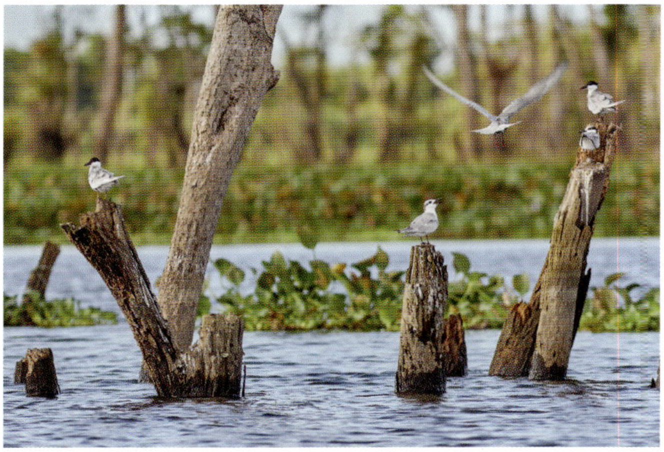

*Above: Gull-billed Terns (*Gelochelidon nilotica*), common in Agusan Marsh, perching on tree stumps in Panlabuhan Lake.*

There are an estimated 59 lakes scattered around the marsh, the best known and largest including Panlabuhan, Mihaba, Mambagongon and Kalobidan, clustered mainly in the south of the sanctuary. These are surrounded by a lattice of vegetative habitats that include herbaceous and scrub swamps, the former consisting of a mix of grasses, sedges and ferns, the latter mainly of small trees such as *Barringtonia* and *Nauclea* species. Unfortunately, the transitional zones between the lakes (and other waterways) and the vegetation are increasingly choked by the Water Hyacinth (*Eichhornia crassipes*), a fast-growing aquatic plant that is a foreign invader originating in the Amazon. Its nuisance and potentially damaging effects are only mildly mitigated by its stunning violet flowers that do at least add some beauty.

The most striking of the vegetative habitats are the swamp forests, of which there are two types in Agusan Marsh. The first is a form that has been well known for many years and is normally seen around the lakes, dominated by *Terminalia copelandii* trees and Sago Palms (*Metroxylon sagu*), trees whose roots grow right out of the flooded swamp floor. The second habitat has only been discovered in recent years, and consists of peat forests, not dissimilar to the peatlands of Borneo. Three such forests have been discovered in Agusan Marsh to date, one small area near the town of Bunawan in the south of the marsh, and two much larger areas in the north, called Caimpugan and Talacogon Peatlands. The former is by far the larger, covering 5,325 ha (13,153 acres) and consisting of species that develop stilted roots, including a mix of *Calophyllum sclerophyllum* and *Tristaniopsis* species trees, as well as *Pandanus* species, together creating a relatively open canopy about 25–30 m (82–98 ft) high. A few areas are characterized by stunted forest, which may indicate the presence of a peat dome, an area of very deep, nutrient-poor peat. If this turns out to be the case, this will be the first peat dome found in the Philippines.

It was largely to bring the two new Philippine habitats into protection that the wildlife sanctuary was expanded to its present size. Since their discovery, a handful of similar peat forests have been found in other parts of the country.

Animal wildlife

Not surprisingly, Agusan Marsh is an important centre for wetland birds, in particular waders, and from January to March it is a home, or stopover site, for migratory birds escaping the winters of Siberia, China and Japan. In total, 127 bird species have been identified in the marsh, including the Purple Heron (*Ardea purpurea*), Chinese Pond Heron (*Ardeola bacchus*), Black Bittern (*Ixobrychus flavicollis*), Little Egret (*Egretta garzetta*), Cattle Egret (*Bubulcus ibis*) and

Purple Swamphen (*Porphyrio porphyrio*), to name just the most commonly seen. The endangered Philippine Duck and Oriental Darter (*Anhinga melanogaster*) also live in the marsh and can occasionally be spotted. Thirty-one of the marsh's bird species are Philippine endemics, and 10 are theatened.

Reptiles and amphibians are, of course, quite common in the marsh, with a currently estimated 60 species, 38 of which are endemic. The Saltwater Crocodile (*Crocodylus porosus*) is known to live in these waters, and it is claimed that the Critically Endangered freshwater Philippine Crocodile does too. However, according to the Mabuwaya Foundation, an NGO in the vanguard of the fight to save this species, there have been no surveys and no recorded sightings in Agusan Marsh in recent years. In terms of fish, 16 species are currently known to live in the marsh's waters.

Fourteen species of mammal live here, seven of them bats, and five of these fruit bats, including the endangered Golden-crowned Flyng Fox.

Below: A Purple Swamphen skips across Water Hyacinths choking one of the waterways in the south of the marsh.

Threats to Agusan Marsh

Within and immediately around the marsh the most serious threats include over-hunting of animal wildlife and over-collection of plants, by both settlers living around the marsh's edges and the native Manobo people who live in the marsh itself. As usual across the Philippines, conversion of the swamps to agriculture is also a major issue, many areas having already been turned into rice fields.

Threats originating outside the marsh include siltation due to deforestation of the surrounding mountains that supply the marsh's water, as well as chemical pollution from mines in those same mountains. Moreover, many of the *Terminalia* species trees appear to be under stress, having lost branches and leaves. This is believed to be due to rising summer temperatures, one of the many effects of global warming.

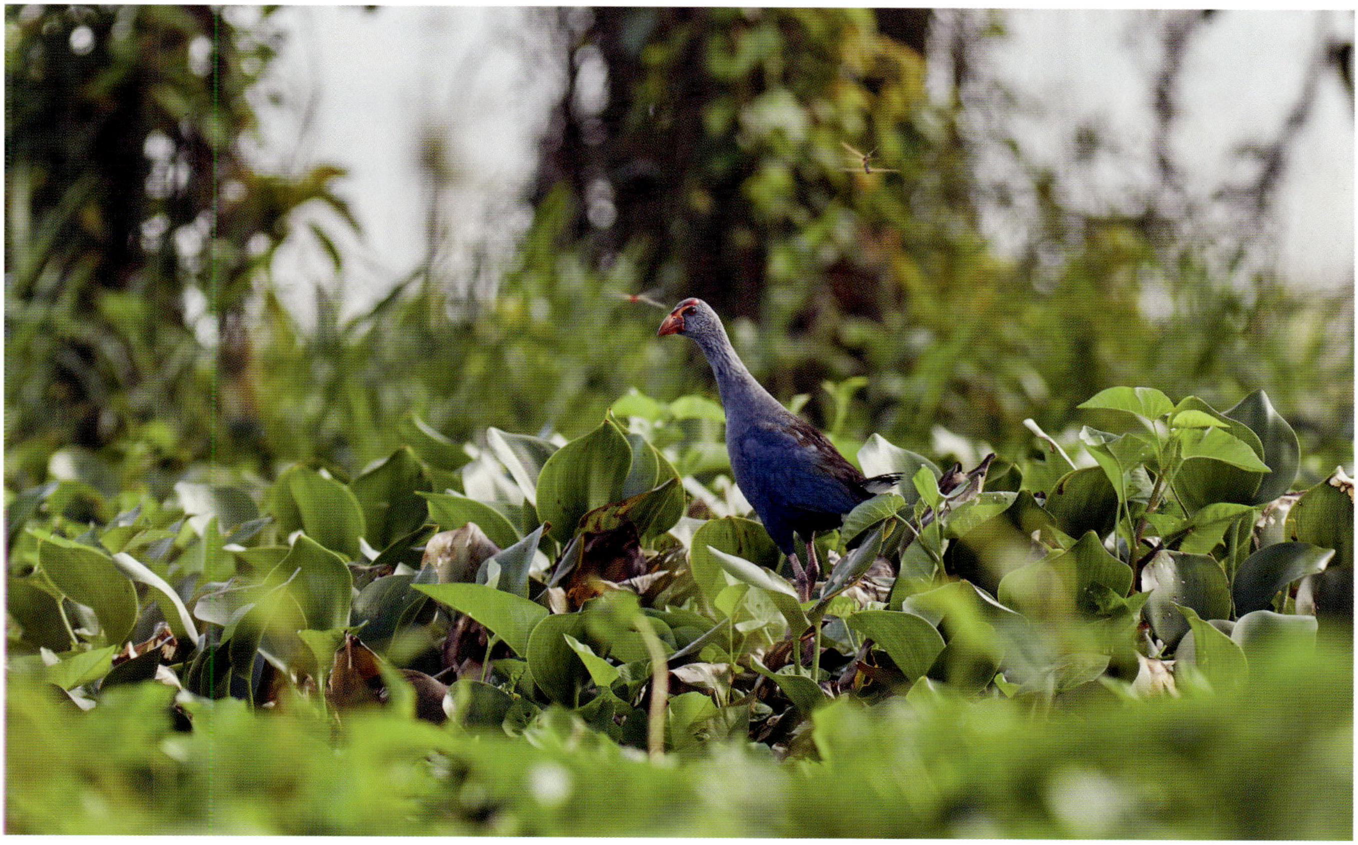

Mt Apo Natural Park

Lying south-west of Davao, Mindanao's largest city, Mt Apo is the Philippines' highest mountain, at 2,954 m (9,689 ft). This quiescent volcano is at the centre of Mt Apo Natural Park, one of the country's most important protected areas, its forest a major centre of Mindanao's biodiversity. The mountain range consists of several secondary peaks, the highest of which is Mt Talomo, at 2,674 m (8,770 ft), lying north-east of Mt Apo.

The mountain range was declared a national park as far back as 1936, with an initial area of 76,900 ha (189,943 acres), though over the years this was reduced to 72,112 ha (178,117 acres). In 1996 Apo was renamed a natural park, though this was not fully formalized until 2004. At that time the park's size was reduced to just under 54,975 ha (135,788 acres). In 1984 Mt Apo was recognized as an ASEAN Heritage Park, using the 72,112 ha (178,117 acres) area, a coverage that continues today. This is also the area recognized by Birdlife International as an IBA.

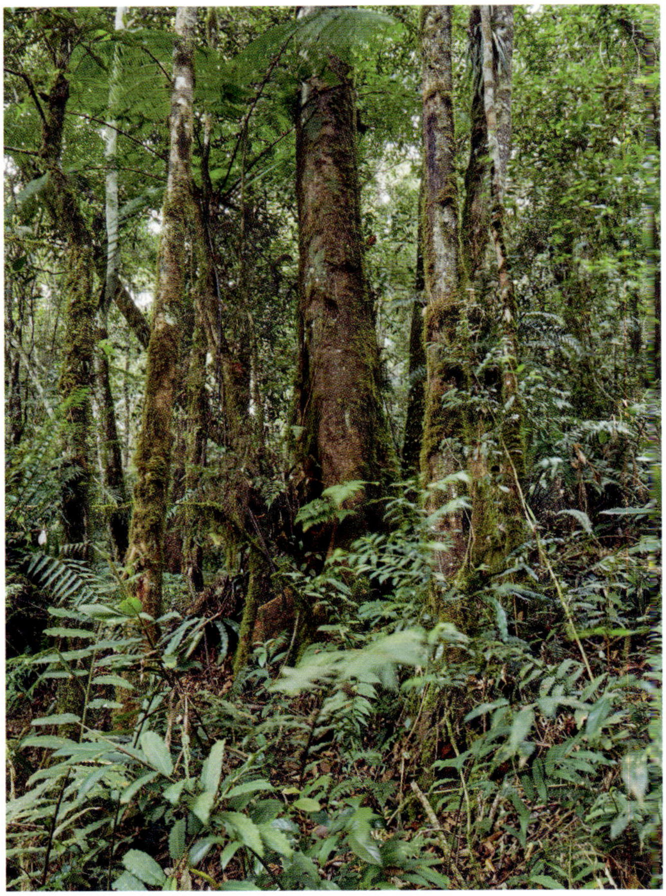

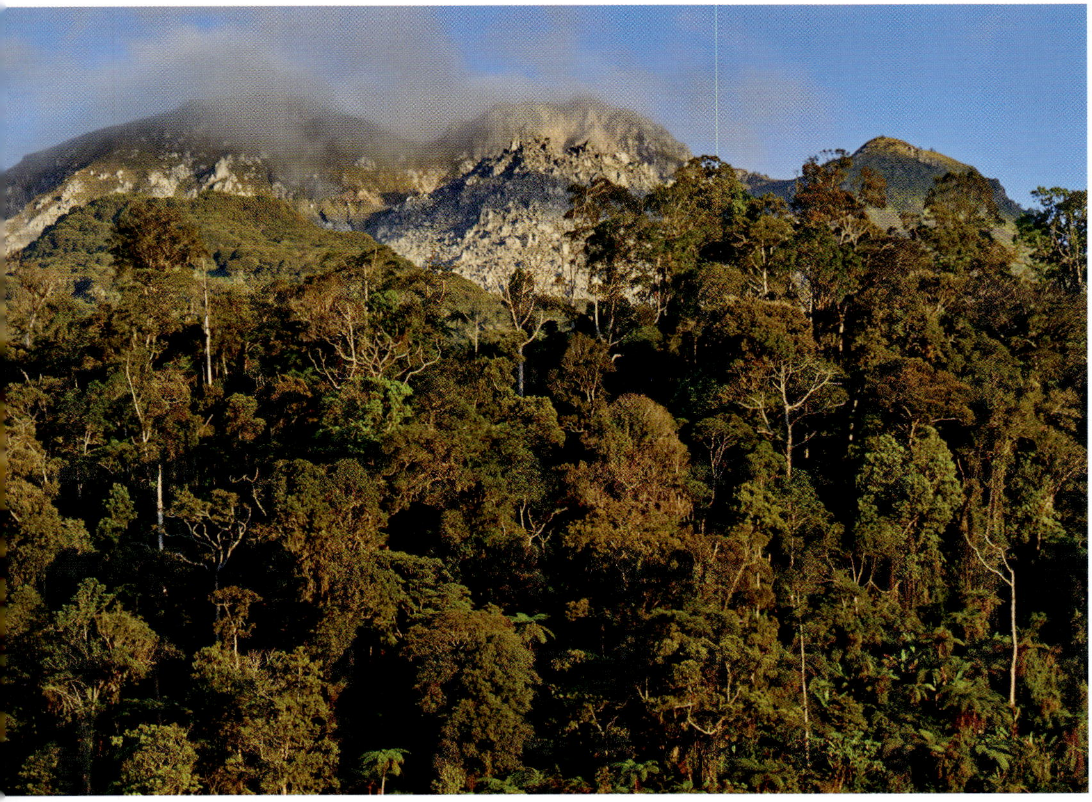

Above: *Typical montane rainforest with a jumble of tree ferns and moss-covered trees, at 1,500-2,100 m (4,940-6,900 ft), along the Sibulan/Sta Cruz trail in the park.*

Left: *An early morning view towards Mt Apo's summit, looking across montane rainforest from the hamlet of Tumpis, one of the highest settlements in the natural park, on Mt Apo's south-east side.*

Opposite, near left: *The flowers of a lipstick vine (*Aeschynanthus *species), a rare group of rainforest plants, in upper montane forest at about 2,100 m (6,900 ft).*

Habitats and wildlife

With rainfall distributed evenly throughout the year, the park's natural habitats should range from lowland evergreen rainforest on its lower slopes, through lower and upper montane forests higher up, and finally to grassland around the summits. However, almost all the forest has been cleared up to an altitude of about 1,000 m (3,300 ft), and replaced with farmland and villages. As a result, very little lowland forest – which would have been dominated by dipterocarp trees – survives. However, there are areas of upland dipterocarp forest, reaching up into the montane forest areas, up to about 1,500 m (4,920 ft). Above this height, montane forest takes over, with generally much smaller trees, although the transition zone contains large Almaciga (*Agathis philippinensis*) trees. Typically, the montane forest contains such small, often gnarled trees as stone oaks (*Lithocarpus* species), *Podocarpus* species, and the Red Tea Tree, known locally as Tinikaran (*Leptospermum flavescens*), tree ferns, pandans, vines (including occasional lipstick vines) and epiphytic ferns, and orchids. The rare and Mindanao-endemic orchid Waling-waling (*Vanda sanderiana*) survives in this forest. Many of the trees are festooned with large amounts of moss. In all, the park is believed to contain more than 800 species of higher plants.

Above the forests, on the southern slopes there is a sulphatara area, mostly restricted to a narrow gorge at

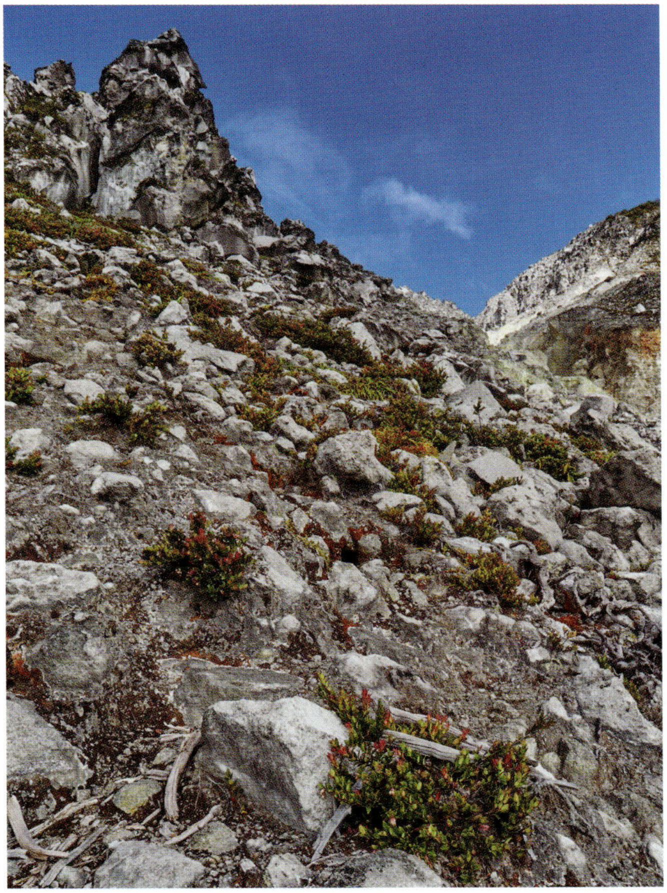

Above: *The steep, rugged Boulders section of Mt Apo, at about 2,400-2,600 m (7,900-8,550 ft), with the sulphatara area in the gorge off to the right.*

Below: Vaccinium microphyllum, *a relative of the temperate climate heathers, growing across rocks in the Boulders region of Mt Apo.*

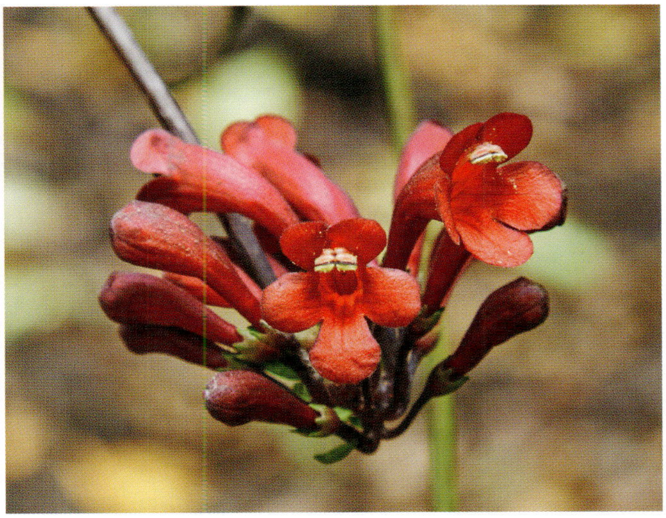

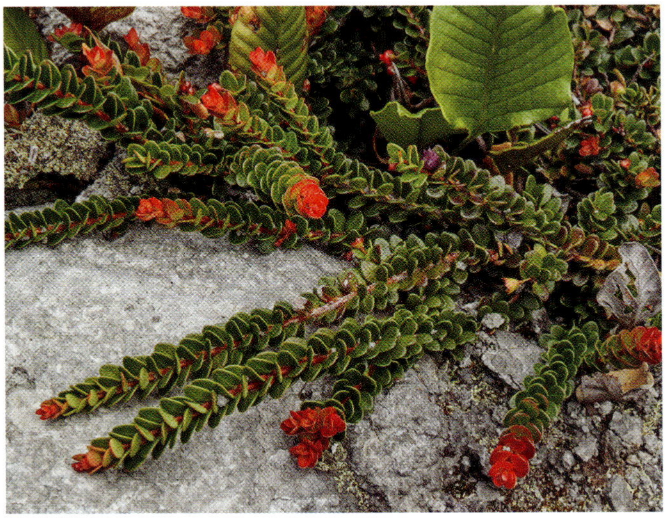

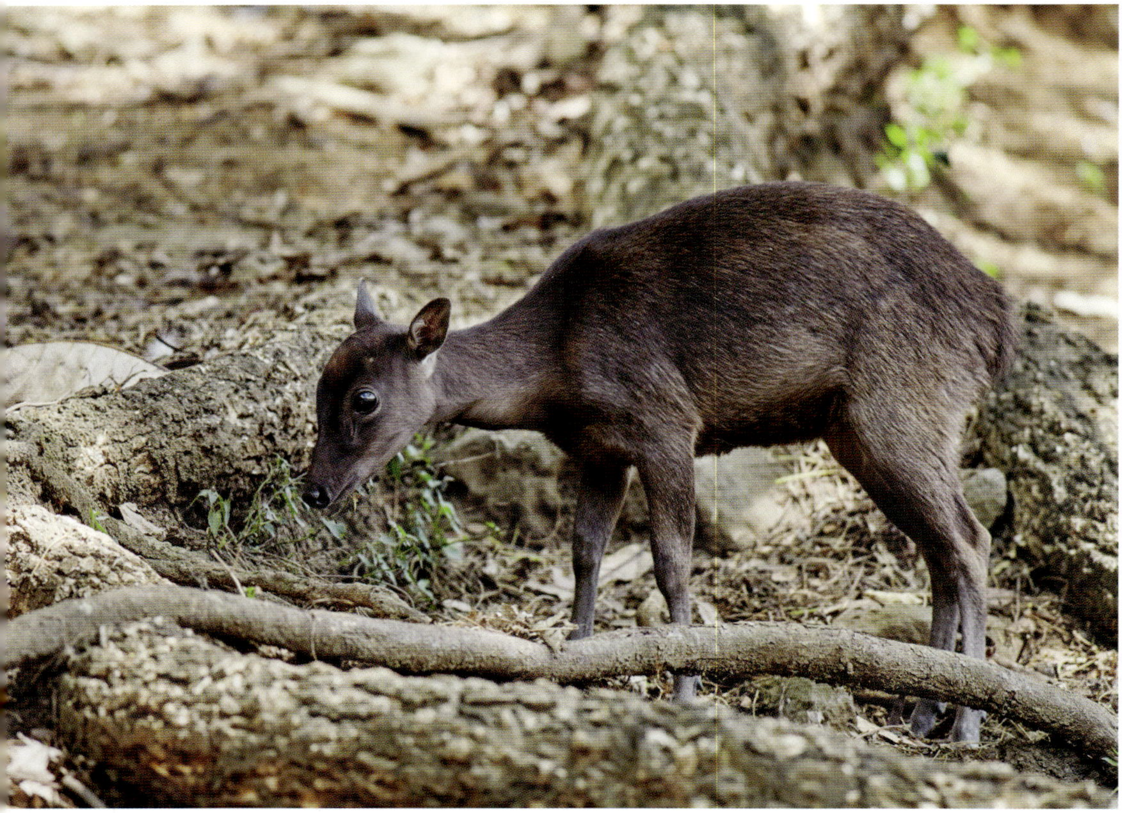

Left: *A young Philippine Brown Deer. The dense forests of Mt Apo are a valuable and increasingly rare sanctuary for this species.*

Below: *A temperate weather specialist, the Chinese Yew, growing at high altitude in the Boulders region of Mt Apo.*

about 2,400–2,500 m (7,872–8,200 ft), although also associated with the adjacent Boulders area. This is a very steep section of the mountain covered with little vegetation, other than a few very low-growing shrubs, including some rhododendrons and Chinese Yew (*Taxus sumatrana*). The surface is mainly bare rock and loose boulders, many of which are quite hot with geothermal heat. The sulphatara vents constantly pour out steam and gases, and the surrounding rock is coated with yellow sulphur crystals, active proof that Mt Apo the volcano is far from extinct.

There are also a number of lakes. One of the best known is Lake Agco, a hot geothermal pool at about 1,250 m (4,100 ft), on the north-west slopes, and close to the village of Illomavis. Higher up is Lake Venado, at about 1,984 m (6,508 ft), partially surrounded by forest, but also with an adjacent grassy plain.

In terms of animals, the most famous resident is the Philippine Eagle, which is believed to still survive in a few of the park's remoter forests. Altogether, the park has been found to contain 272 bird species, 40 per cent of them endemic, including forest species such as the Apo Myna (*Basilornis mirandus*), Apo Sunbird (*Aethopyga boltoni*), Slaty-backed Jungle-flycatcher (*Vauriella goodfellowi*) and Red-eared Parrotfinch (*Erythrura coloria*).

Apo's mammal population is believed to include many of the species endemic to the Greater Mindanao Faunal Region, including the Philippine Tarsier and Philippine Flying Lemur. There are also several rodent species, most of which are Mindanao endemics, including the Mindanao Wood Shrew (*Podogymura*

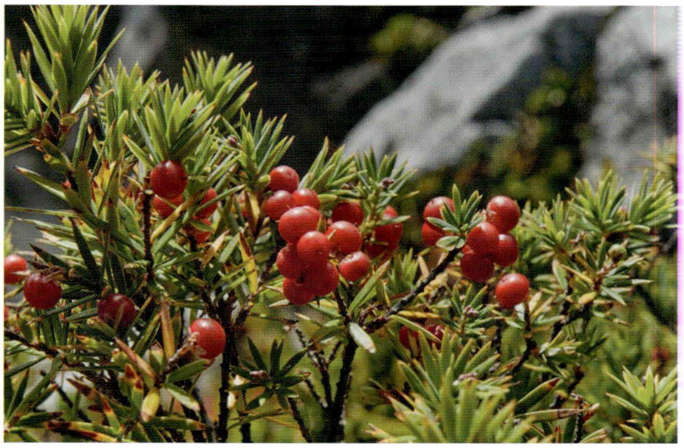

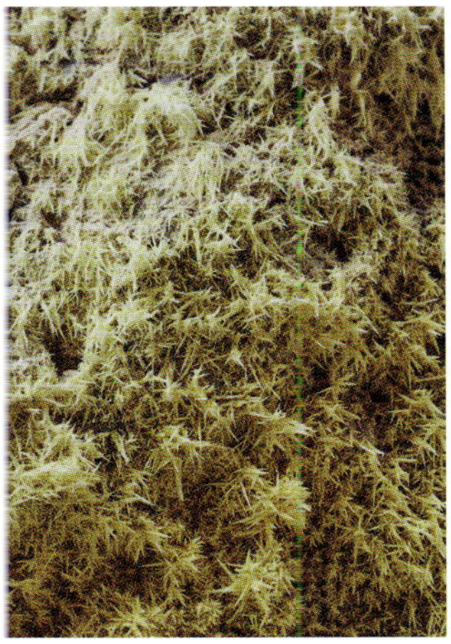

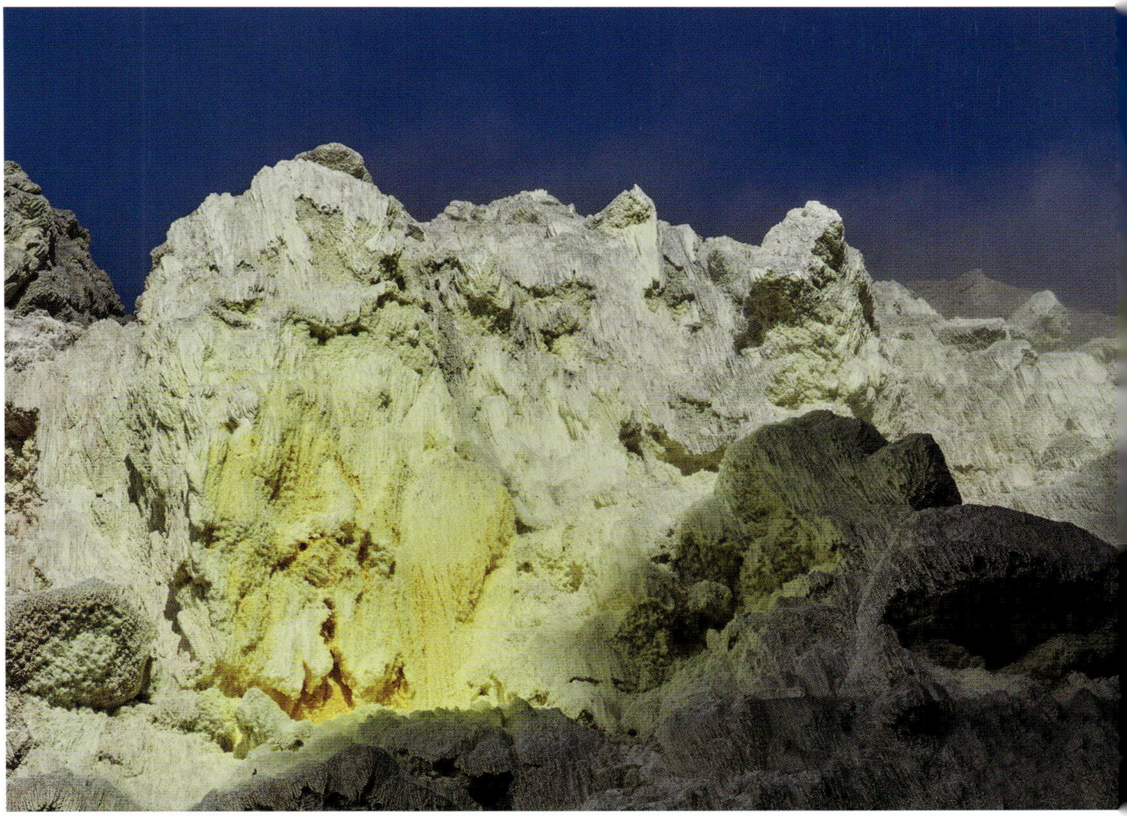

Above and right: Mt Apo's sulphatara area, at about 2,400 m (7,900 ft), is a narrow gorge filled with smoke and sulphurous fumes. The cliff walls are covered with yellow sulphur, whose intricate crystal patterns can be seen close-up.

truei) and the Greater Mindanao Shrew (*Crocidura grandis*). Other, more widely spread mammals, include the Philippine Brown Deer and Mindanao subspecies of the Philippine Warty Pig. Bats include the nationally endemic Golden-crowned Flying Fox.

Threats to the park

Over the years Mt Apo has suffered from large-scale logging, conversion to farmland and fires, to the extent that now more than 50 per cent of the park's land is either agricultural or brushland, the latter dominated by Cogon Grass (*Imperata cylindrica*). Furthermore, in the 1990s, 700 ha (1,729 acres) of forest in the north-west, close to Lake Agco, were taken over for the construction of a geothermal power plant.

In recent years there have been growing efforts to reforest some of the brushland areas using native tree species. Both local government units and community groups have launched tree nursery and planting projects, while one of the largest projects has been run by the EDC, the company responsible for the Lake Agco geothermal plant. It has successfully reforested an area around Lake Maag with Tinikaran/Red Tea Trees, has been working to reforest a 700-ha (1,729-acre) area of brushland near Lake Agco, and has recently agreed to reforest an area near Lake Venado that was destroyed by fire in 2016. It is still early days for many of these reforestation projects, and it remains to be seen how successful they will be, but they definitely point a way forwards to future possible improvements in the park's protection.

The park is very popular with hikers, the climb to the summit being one of the Philippines' leading challenges. There are several routes to the top, one starting from Illomavis in the north-west, and another – currently the most popular route – leading from the village of Sibulan in the south-east. Rapidly growing numbers of climbers have led to concerns about damage to trees and the problem of litter. Fire is also a constant risk. The 2016 fire that destroyed forest around Lake Venado is believed to have been caused by campers in the area. Controls on the number of hikers allowed at any one time are coming into force and hopefully will help to improve the situation.

A Wildlife Tour

Ecotourism is growing rapidly in the Philippines, as more people come to understand the uniqueness and immense variety that is on offer in the country, both on land and beneath the waves. Although terrestrial wildlife can be very hard to spot in the dense forests, there are a number of locations where it is perfectly possible to see it, especially when in the company of a skilled local spotter. It would be unusual, however, to be able to find mammalian wildlife, other than fruit bats and macaques, and for the most part birds are the most readily traced form of animal wildlife.

Beneath the sea it is a different story, with wildlife in abundance. One day you can be diving over a spectacular coral reef, following turtles as they feed across the reef crest or a sandflat, the next day snorkelling with Whale Sharks, or following dolphins as they ride along on your boat's bow wave. It can be a magnificently colourful and wildlife-filled world.

This quick guide provides pointers to some of the best wildlife-watching places in the Philippines, both on land and in the sea.

Below: A boardwalk on the coast of Siquijor makes exploration of the mangrove forest significantly easier.

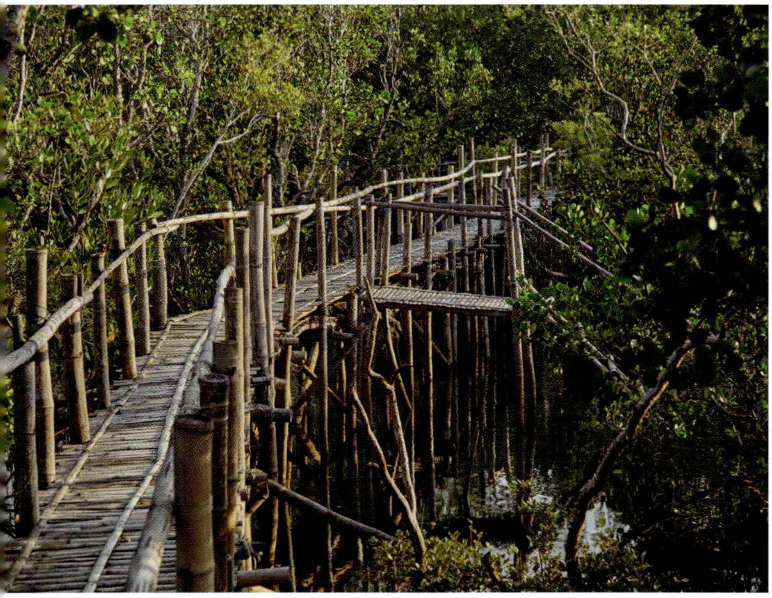

Manila: the starting point

Surprisingly, wildlife watching can kick off right in Manila, firstly with an exploration of the mangroves of the Las Piñas-Parañaque Critical Habitat, in the south-west of the city. Though under huge pressure from pollution and surrounding developments, it is still a good place to spot typical mangrove birds such as various egrets, herons and kingfishers.

At the other end of the city, in the north-east, is La Mesa Eco Park, a partially forested parkland where it is possible to spot such birds as the Emerald Dove, the Black-naped Oriole and the usually rather shy Ashy Ground-thrush.

Not too far away the DENR maintains the Ninoy Aquino Parks and Wildlife Center, a zoo-cum-wildlife rescue centre that is home to an array of species, not all of them Philippine. Many of the animals are not housed in particularly attractive settings, but there are some quite large and naturalistic enclosures for Philippine Eagles, Philippine Serpent-eagles and White-bellied Sea-eagles, providing a rare opportunity to get close to such birds.

Further out, in Rizal province to the city's north-east, Avilon Zoo is a good place to see some captive Philippine wildlife, including both species of crocodile, numerous hornbills, the Luzon Bleeding-heart Pigeon and a number of the endemic mountain rodents.

Northern Luzon

Travelling north from Manila, the first place possibly of interest is Candaba Swamp, an area of marshland that is a good place to see an assortment of egrets. Further to the north-west is Subic Bay and its lowland rainforests, one of the country's best birdwatching sites. The old military roads that cut through the forest here give excellent clearings and vantage points from which to spot a wide variety of birds, including the Luzon Hornbill, various pigeons, Dollarbird, bee-eaters, Philippine Frogmouth, and many more. Subic Bay also offers one of the best opportunities in the Philippines to see a large colony of fruit bats – a mix of Giant Fruit Bats and Golden Crowned Flying Foxes. Macaques are commonly seen along the roadsides. As

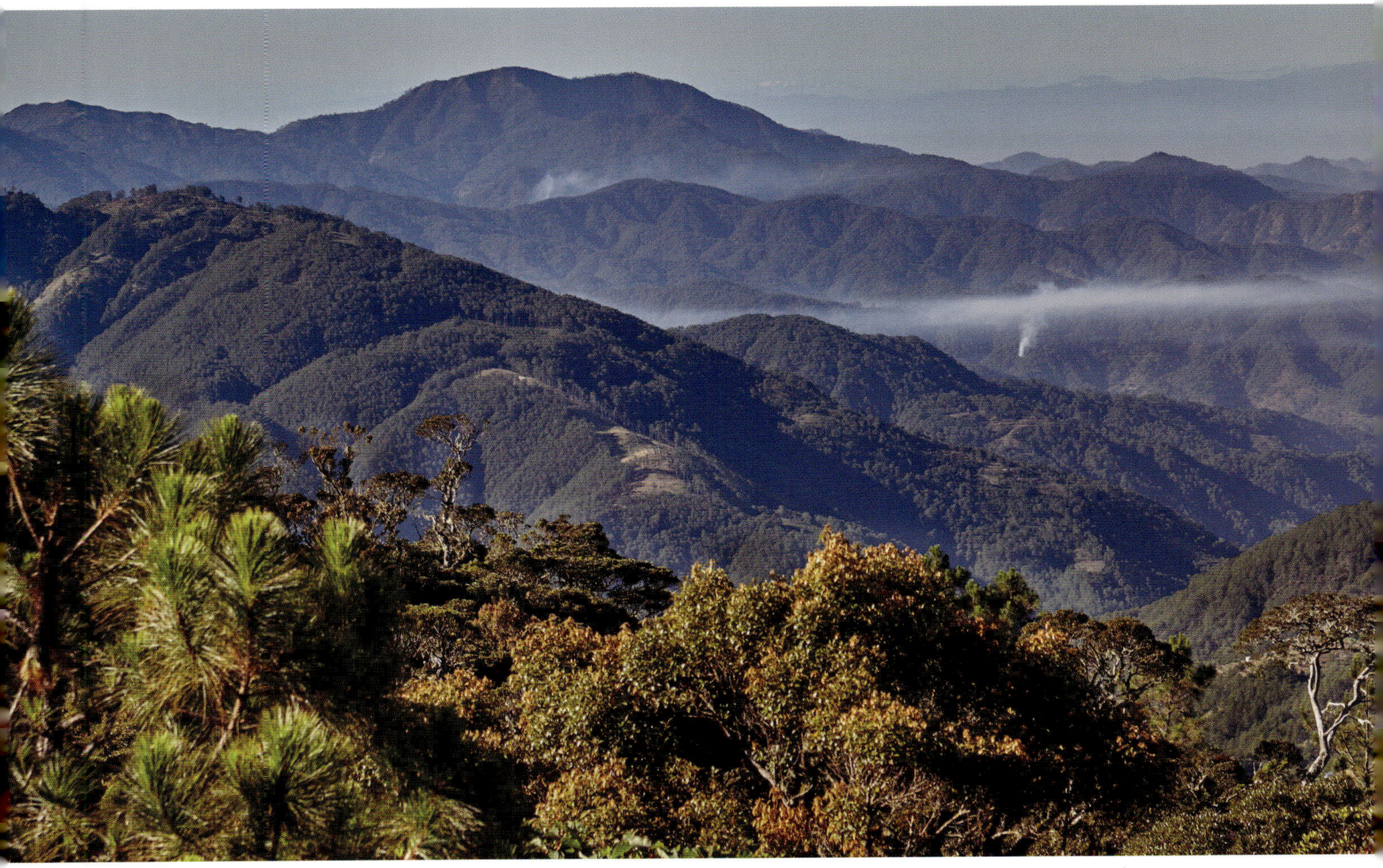

one of the country's most accessible lowland forests, this location also provides a good opportunity to have a close look at some truly large dipterocarp forest trees. In the Cordillera Central mountains the main attraction is Mt Pulag, where both bird and plant spotting is possible along the main trail from the Babadak ranger station to the mountain's summit. More remote, the Tabeyo trail on Pulag's northern slopes gives an excellent view of undisturbed upper montane rainforest.

Southern Luzon

The first point of interest south of Manila is Mt Makiling, like Subic Bay one of the Philippines' best birdwatching sites. The mountain is forested from the campus of the University of the Philippines at the northeastern foot all the way to the summit, starting with some excellent lowland rainforest and finishing

Above: *A superb view of the Cordillera Central mountains of northern Luzon from a clearing in montane rainforest on the upper slopes of Mt Pulag.*

with montane rainforest. Birdwatching on the mountain is superb, and includes regular sightings of Luzon Hornbills. Lizards and forest frogs are also regularly seen, and for plants this is a good site for the lovely Jade Vine. The university campus itself is a good place to start, followed by the adjoining Mt Makiling Botanic Garden. A trail leads up the mountain from near the Botanic Garden, the hike to the summit and back manageable in a day.

In the far south of Luzon, Whale Shark-watching at Donsol is almost a must. Snorkelling alongside one of these behemoths is quite an experience. For scuba diving, the main top-quality area is around Anilao, near Batangas, a couple of hours south of Manila.

Mindoro

Across the Verde Island Passage, a narrow stretch of water separating Luzon from Mindoro, Puerto Galera is one of the Philippines' leading scuba-diving locations, with some superb reefs filled with corals and alive with a wide assortment of reef fish. Inland, Mts Malasimbo and Talipanan are still partially forested, offering some good hiking and possibly birdwatching opportunities. To the south, Mt Halcon, though unprotected, still has some very wild and pristine montane forest, a haven for orchids, pitcher plants and some of Mindoro's montane birds.

Palawan

Palawan's most important protected area is Puerto Princesa Subterranean River National Park. Although most visitors head straight for a tour of the underground river, there is also some good birdwatching in the adjacent forest and in the mangroves along the Poyuy-Poyuy River. Close to the ranger station, Long-tailed Macaques and Palawan Water Monitor Lizards are very commonly seen. The Tabon Scrubfowl and Palawan Peacock Pheasant can also occasionally be seen.

In the far north of Palawan, El Nido and the islands of Bacuit Bay are among the Philippines' most beautiful places. Good scuba diving is available around several of the islands, where both Green and Hawksbill Turtles are commonly seen. On the islands, monitor lizards, Palawan Hornbills, Tabon Scrubfowl and Reef Egrets are regularly seen, along with several kingfishers.

Some of the world's best scuba diving is available on the coral reefs of Tubbataha Reefs Natural Park. Rather remote and accessible only via live-aboard dive safari vessels, many divers dream of coming here. The atolls' islets are home to birds like boobies, noddies and terns, but landing on the islets is rarely allowed.

Below: A huge shoal of sardines at Moalboal, on the south-west coast of Cebu.

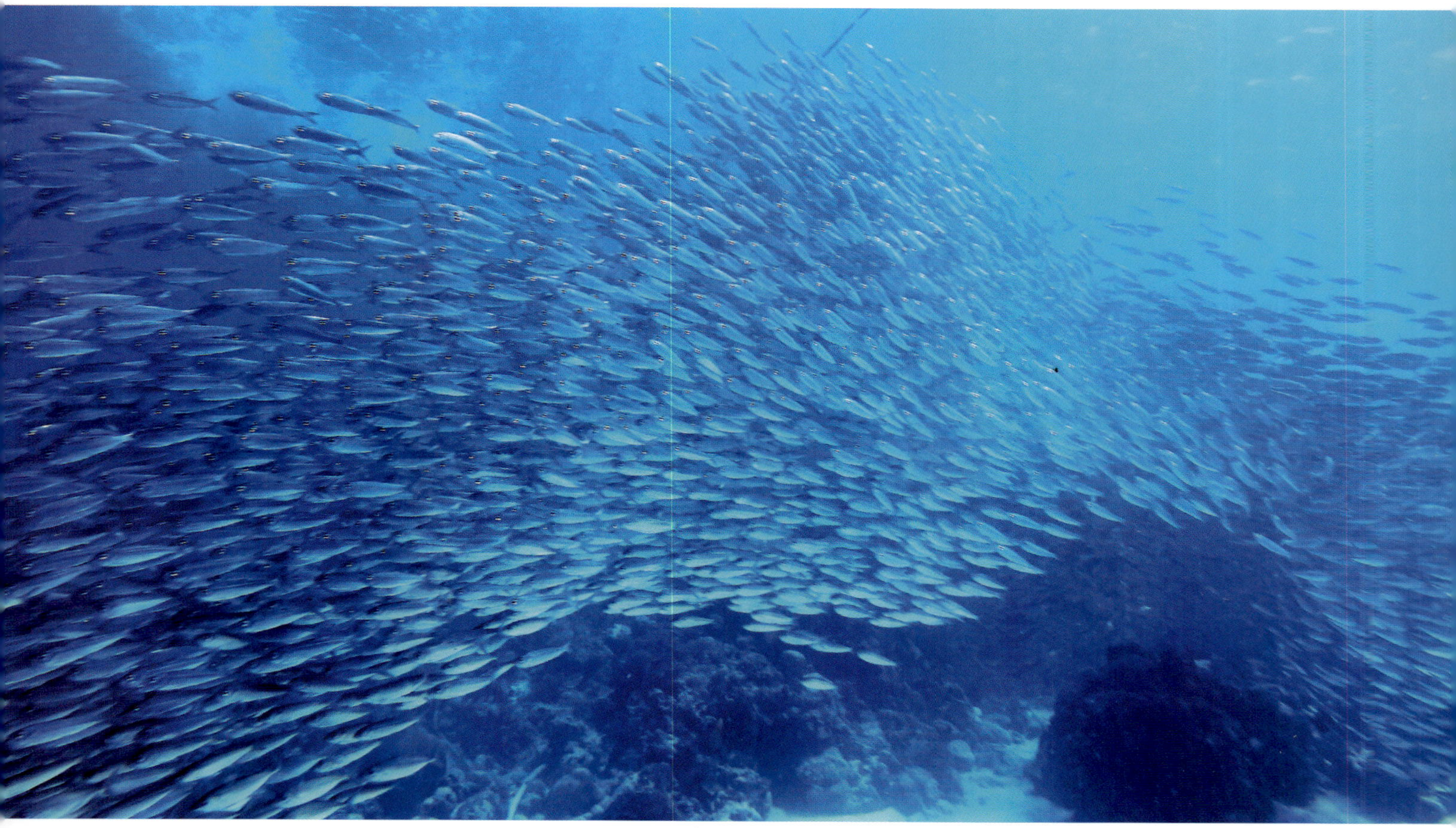

The Visayas

The main entry point into the Visayas is via Cebu City, and close to here the mangroves and sandflats of Olango Island are a wonderful place to explore and watch for a diversity of waders, particularly during the November–March migration season.

Further south, on Cebu Island, Oslob has become a very popular place to see Whale Sharks. However, here the local fishermen feed the Whale Sharks, and the whole daily event appears to have been turned into something of a tourist circus that many conservationists find unethical and distasteful.

Some of the Philippines' best scuba diving is on Cebu's south-west coast at Moalboal, where almost pristine reefs, fronting on to very deep water, create perfect conditions for shoals of pelagic fish, as well as turtles and a host of reef fish.

Nearby, on the south-east coast of Negros, early-morning boat trips to watch dolphins are almost always successful, heading out from the town of Bais. In the same area, close to the city of Dumaguete, the Balinsasayao Twin Lakes Natural Park is worth a visit for possible birdwatching. Mt Talinis, usually accessed via Valencia, just south-west of Dumaguete, has some good hiking through dense montane rainforest.

At the northern end of Negros, the city of Bacolod is home to the Talarak Foundation's zoo-cum-breeding centre, offering an opportunity to get close to the Visayan Spotted Deer, Visayan Warty Pig, Negros Bleeding-heart Pigeon, Blue-naped Parrot, Colasisi, Philippine Eagle-owl, and other endangered animals.

To the south-east, the forests of Mt Kanlaon can usually be explored. There are several trails to the summit, passing through dense and pristine montane rainforest, though they are often closed due to the danger of volcanic eruption. Instead, the mountain resort of Mambucal is worth visiting as home to one of Kanlaon's few remaining fragments of dipterocarp rainforest. Scattered colonies of fruit bats are easily seen here, and birdwatching is excellent, particularly if you are with one of the very skilled local guides.

In the eastern Visayas, Bohol has some of the Philippines' best scuba diving on the reefs around Panglao and Balicasag Islands. Inland, Rajah Sikatuna Protected Landscape includes some fine limestone rainforest, where it is very common to see Flying Lemurs, Long-tailed Macaques and Samar Hornbills. Nearby, in the village of Bilar, Habitat Bohol is an excellent place in which to see a large number of Philippine butterflies in a small area, including the world's largest moth, the Atlas Moth. In Bohol's south-west, close to Corella, the Philippine Tarsier Sanctuary offers the best opportunity to see the tiny Philippine Tarsier in some dense forest.

Further east, the island of Samar has by far the Visayas' largest remaining tract of lowland rainforest, much of it still very wild and inaccessible, protected within the Samar Island Natural Park. There is some very good birdwatching along forest paths near the park's headquarters in the village of Tenani. Nearby, boat rides along the Ulot River enable regular sightings of Samar Hornbills and the rare Silvery Kingfisher.

Mindanao

In the south of Mindanao, south-west of the city of Davao, stands Mt Apo, the Philippines' highest mountain, offering the country's greatest hiking challenge. The trails pass through some excellent montane forest (all the lowland forest has been cleared). Although birdwatching here is possible, more popular are the slopes of the Mt Kitanglad Range Natural Park, where many Mindanao species can be found. This park also offers probably the best chance of seeing a Philippine Eagle in the wild (though still only a slim chance).

North of the city of Davao, the headquarters of the Philippine Eagle Foundation is worth a visit, to see not only some of its Philippine Eagles but also various other animals housed there, including White-bellied Sea-eagles and Philippine Deer.

Finally, in eastern Mindanao, Agusan Marsh Wildlife Sanctuary is a stunningly beautiful place. Its lakes and swamp forests are home to terns and an array of waders, including Purple Herons, Chinese Pond Herons and Black Bitterns, plus Philippine Ducks.

Resources

It is beyond the scope of a book such as this to provide a full list of all the references used in putting together the text. Listed below are the most important publications and ones that may be the most useful, along with contact details for a number of organizations involved in Philippine conservation.

Books

Fisher, Tim & Hicks, Nigel. *A Photographic Guide to Birds of the Philippines*. New Holland Publishers, 2000.

Gosliner, Terrence, Behrens, David & Williams, Gary. *Coral Reef Animals of the Indo-Pacific*. Sea Challengers, 1996.

Heaney, Lawrence & Regalado, Jacinto, Jr. *Vanishing Treasures of the Philippine Rain Forest*. The Field Museum, Chicago, 1998.

Heaney, Lawrence, Balete, Danilo & Rickart, Eric. *The Mammals of Luzon Island*. John Hopkins University Press, 2016.

Hicks, Nigel. *The National Parks and Other Wild Places of the Philippines*. New Holland Publishers, 2000.

Kassem, Ken & Madeja, Eric. *The Coral Triangle*. John Beaufoy Publishing, 2014.

Tañedo, Maia, Hutchinson, Robert, & Constantino, Adrian & Trinket. *A Naturalist's Guide to Birds of the Philippines*. John Beaufoy Publishing, 2017.

Online documents (PDFs)

Coral Triangle Initiative National Action Plan 2009. Biodiversity Management Bureau of the Philippine Department of Environment and Natural Resources. www.bmb.gov.ph/elibrary/publications/action-plans.

Habitats of Philippine Dipterocarps. Edwino Fernando and Nemrod Delotino. Soil and Water Conservation Foundation, 2009. www.rainforestation.ph/.../Fernando_2009_Habitats_of_Philippine_Dipterocarps.pdf.

Red List of Threatened Species. International Union for the Conservation of Nature (IUCN). www.iucnredlist.org.

The National Wetlands Action Plan for the Philippines 2011–2016. Biodiversity Management Bureau of the Philippine Department of Environment and Natural Resources. www.bmb.gov.ph/elibrary/publications/action-plans.

Philippine Aquatic Wildlife Rescue and Response Manual Series: Marine Turtles. Marine Wildlife Watch of the Philippines, 2014. http://mwwphilippines.org/2015/04/14/philippine-aquatic-wildlife-rescue-and-response-manual-to-marine-turtle-incidents.

Philippine Biodiversity Strategy and Action Plan 2015–2028. Biodiversity Management Bureau of the Philippine Department of Environment and Natural Resources. www.bmb.gov.ph/elibrary/publications/action-plans.

Philippine Crocodile *Crocodylus mindorensis*. Merlijn van Weerd. Extract from 'Crocodiles: Status Survey and Conservation Action Plan'. Third Edition, 2010. By S. Manolis and C. Stevenson, Crocodile Specialist Group, Darwin. www.iucncsg.org/365_docs/attachments/protarea/13_C-511712ed.pdf.

Philippine Plant Conservation Strategy and Action Plan. Biodiversity Management Bureau of the Philippine Department of Environment and Natural Resources. www.bmb.gov.ph/elibrary/publications/action-plans.

The State of Philippine Birds. Haribon Foundation and Birdlife International, 2014. http://datazone.birdlife.org/userfiles/file/sowb/countries/StateofPhilippineBirds.pdf.

Conservation organizations

Biodiversity Management Bureau (BMB) Government arm of the DENR responsible for conservation of plant and animal wildlife in the Philippines. www.bmb.gov.ph.

Birdlife International UK-based, internationally working organization specializing in bird conservation. Very active in the Philippines. www.birdlife.org.

Conservation International (Philippines) Philippine branch of a famous US-based, internationally working conservation organization. www.conservation.org/global/philippines/Pages/Nature-Is-Speaking.aspx.

Department of Environment and Natural Resources (DENR) Philippine government department responsible for all aspects of protecting the environment. www.denr.gov.ph.
Haribon Foundation Long-established and much-respected Philippine non-government conservation organization, working throughout the Philippines. www.haribon.org.ph.
International Union for the Conservation of Nature (IUCN) The global authority on conservation and threats to wildlife around the world. Publisher of the Red Lists, listing theatened species around the world. www.iucn.org.
Katala Foundation Palawan-based non-government conservation organization, established primarily to work on conservation of the Philippine Cockatoo, but which is now reaching into broader Palawan conservation issues. www.philippinecockatoo.org.
Large Marine Vertebrates Research Institute, Philippines (Lamave) Non-government conservation research organization that studies Whale Sharks, sharks, rays, turtles and marine mammals in Philippine waters. www.lamave.org.
Mabuwaya Foundation Non-government conservation organization based in north-east Luzon, working primarily in the Northern Sierra Madre region, focusing on saving the Philippine Crocodile from extinction. www.mabuwaya.org.
Palawan Council for Sustainable Development Branch of the Palawan provincial government responsible for all things to do with the natural environment in Palawan. https://pcsd.gov.ph/igov.
Philippine Eagle Foundation (PEF) One of the Philippines' best-known conservation organizations, working to save the Philippine Eagle, based near Davao, on Mindanao. www.philippineeaglefoundation.org.
Philippine Tarsier Foundation (PTFI) Conservation organization working to protect the Philippine Tarsier, concentrating mainly on Bohol. www.tarsierfoundation.org.
Talarak Foundation Negros- and Panay-based conservation organization that has taken over the Negros Forests and Ecological Foundation. Works to protect and conserve the many endangered species of the western Visayas. www.talarak.org.
World Wide Fund for Nature (Philippines) Philippine branch working throughout the Philippines. https://wwf.org.ph.

Protected areas

Most of the Philippines' protected areas do not have their own websites, but here are a few that do.

Mt Kitanglad Range Natural Park
www.mkrnp.org.
Olango Island Wildlife Sanctuary
www.olangowildlifesanctuary.org.
Tubbataha Reefs Natural Park
www.tubbatahareef.org.

Dive operators

El Nido One of the most professionally run dive centres in El Nido, coupled with undoubtedly among the area's most luxurious accommodation, is that at Lagen Island Resort, run by the El Nido Resorts company. www.elnidoresorts.com/lagen-island.
Moalboal Quo Vadis Dive Resort, at the heart of Moalboal's Panagsama Beach, runs one of the area's best and most active dive centres, and comes with excellent accommodation. https://quovadisresort.com.
Panglao and Balicasag Islands Situated on one of Panglao Island's most beautiful beaches, and providing both excellent accommodation and a superb diving centre, is the Bohol Beach Club. www.boholbeachclub.com.ph.
Puerto Galera Asia Divers runs one of Puerto Galera's best dive operations, along with its resort, El Galleon Beach Resort. http://asiadivers.com.
Tubbataha Reefs Natural Park The only way to scuba dive on the Tubbataha Reefs is to visit via a live-aboard dive boat, and then only in the March–June calm season. Possible companies to use include: Discovery Fleet, based in Metro Manila, www.discoveryfleet.com, and Seadoors, also based in Metro Manila, www.seadoors-liveaboard.com.

Index

Page numbers in **bold** denote illustrations.

First published in the United Kingdom in 2019 by
John Beaufoy Publishing,
11 Blenheim Court, 316 Woodstock Road,
Oxford OX2 7NS, U.K.
www.johnbeaufoy.com

10 9 8 7 6 5 4 3 2 1

Photo credits:
All photography by Nigel Hicks, except the following:
Bob Natural/MBCFI (pp98 and 99); Florante Cruz (p136); Errol Gillang (p137 top); Prof Juan Carlos Gonzalez (p169 top and bottom).

ISBN 978-1-912081-78-3

Editor: Krystyna Mayer
Designer: Nigel Partridge
Cartographer: William Smuts
Project management: Rosemary Wilkinson

Printed and bound in Malaysia by Times Offset (M) Sdn. Bhd.